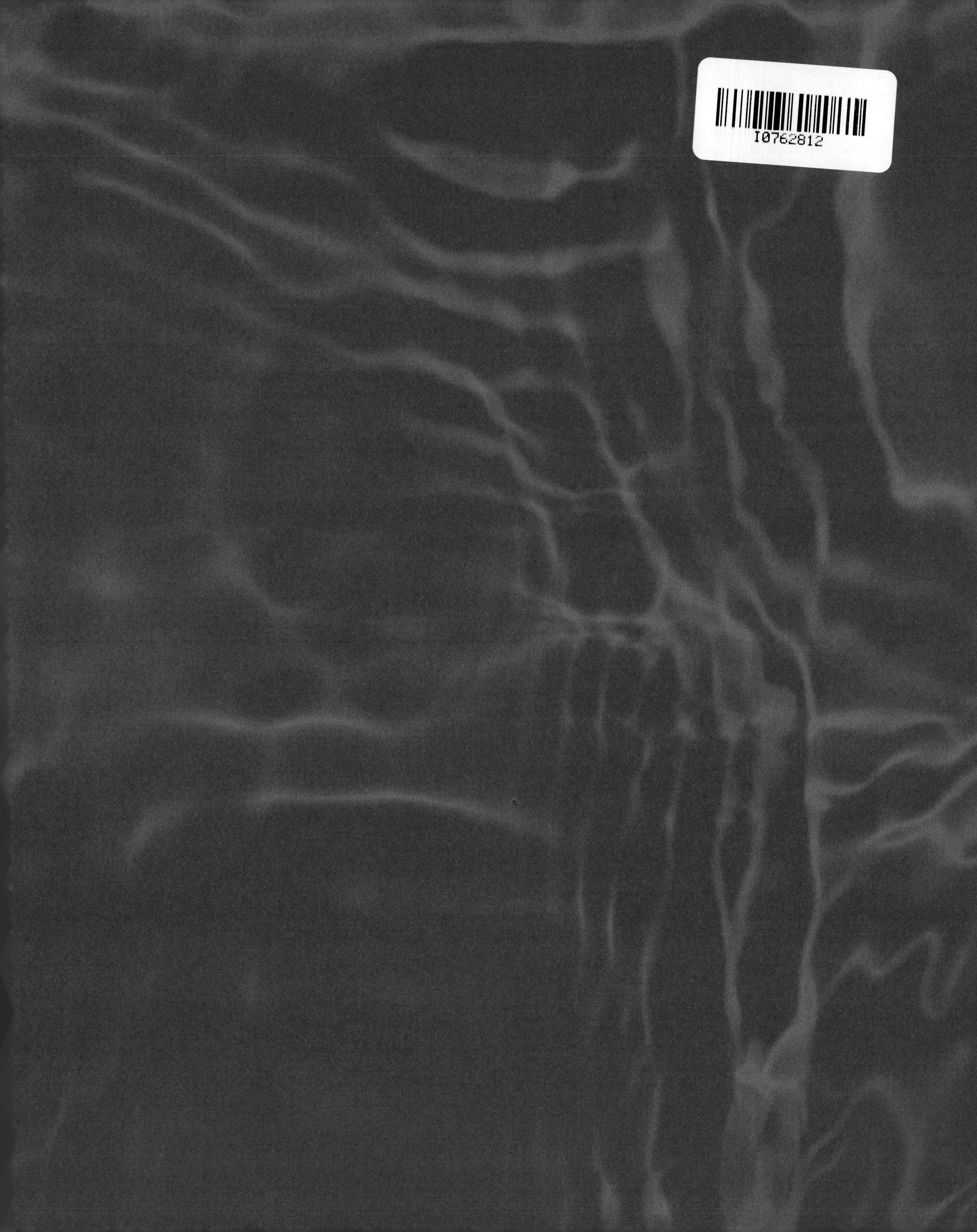
I0762812

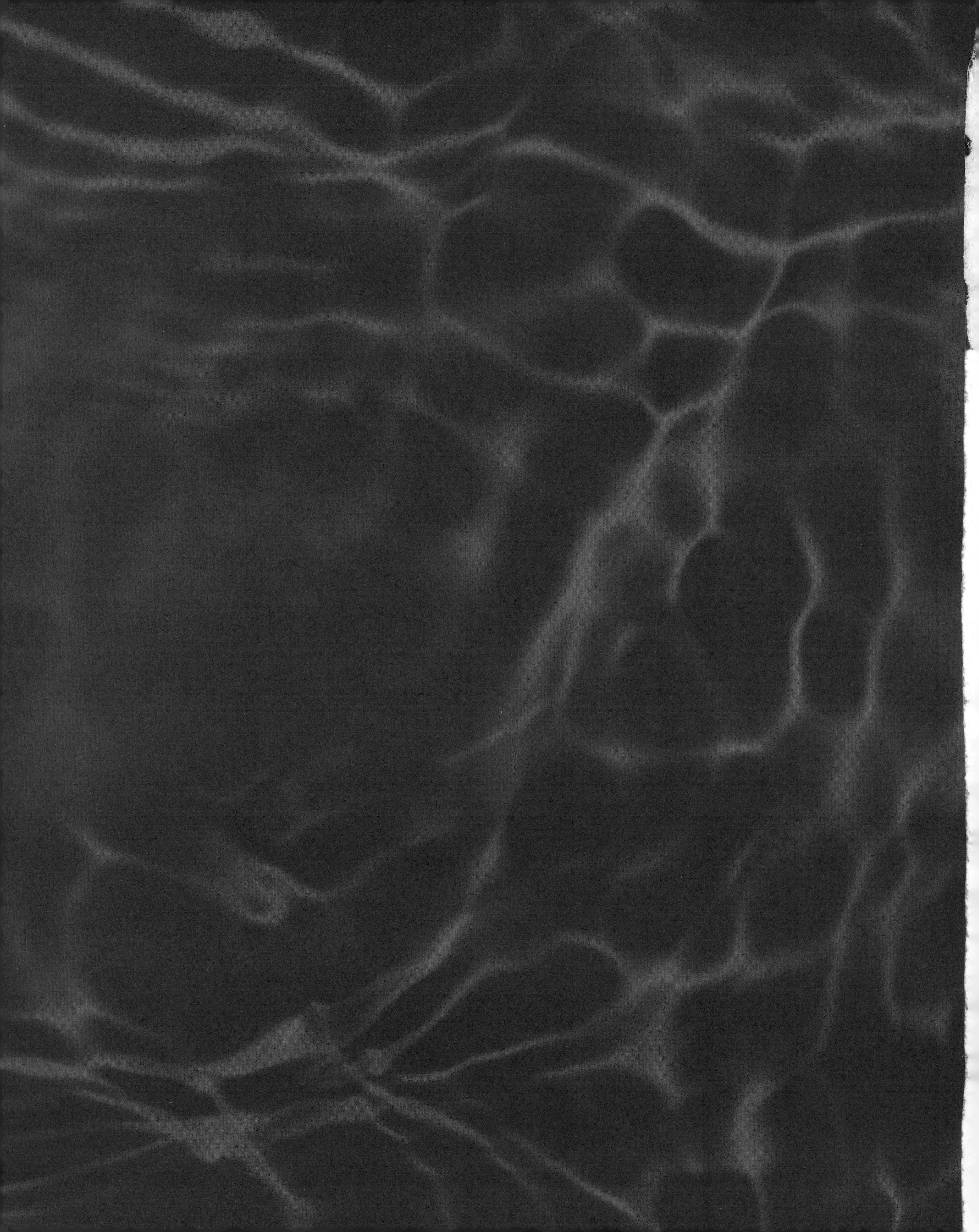

THE RADIANT SEA

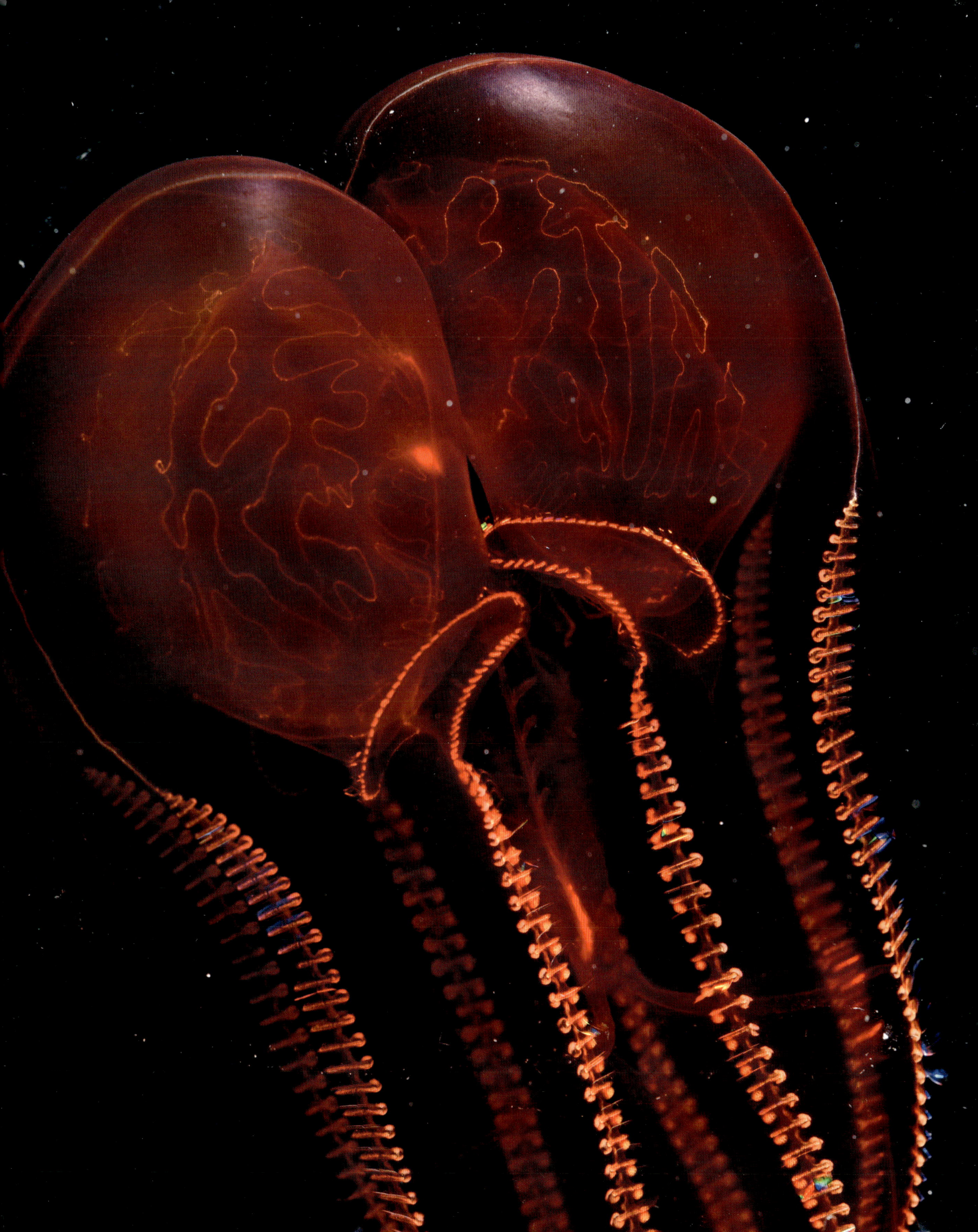

THE RADIANT SEA

Color and Light in the Underwater World

Steven Haddock
and Sönke Johnsen

Foreword by Helen Scales

ABRAMS, NEW YORK

Published in 2025 by Abrams Books,
an imprint of ABRAMS.

Library of Congress Control Number 2022945351

ISBN 978-1-4197-7237-5
eISBN 979-8-88707-226-5

Printed and bound in China
10 9 8 7 6 5 4 3 2 1

This book was conceived, designed, and produced by
UniPress Books Limited.

Publisher: Jenny Manstead
Managing Editor: Slav Todorov
Project Manager: Ruth Patrick
Editor: Beth Dymond
Designer: Alexandre Coco
Illustrator: Robert Brandt
Picture Researcher: Steve Behan
Proofreader: Dawn Bates

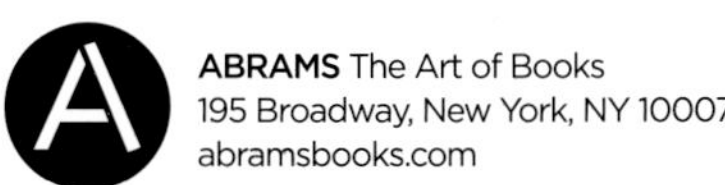

ABRAMS The Art of Books
195 Broadway, New York, NY 10007
abramsbooks.com

FOREWORD

My love of the ocean began, as it does for many of us, at the shallow edges of the sea. I was one of those kids at the beach who spends hours poking around in seaweed and gazing into rock pools to find seashells and crabs, and maybe catch a glimpse of a gleaming fish. Then, as a teenager, I learned to scuba dive and plunged for the first time beneath Britain's cold, wild waves. I became completely addicted to the ocean and felt a fierce calling to help protect it from the human troubles I was starting to witness.

A while later, after traveling the world, studying endangered fish and mapping coral reefs, I felt drawn to deeper parts of the ocean. By then I had discovered the power of storytelling and was writing books to share my passion for the seas. It dawned on me that many of the most exciting ocean stories come from the deep. I became mesmerized by animals such as the furry-looking yeti crabs that thrive among scorching hydrothermal vents, and the gossamer worms that pirouette through deep open waters and produce puffs of golden sparkles. This was how I came to know the work of many deep-sea biologists, including Steve and Sönke.

Whenever I hear about a new study of life in the deep, I get a giddy sense of the ocean divulging yet another big secret. It's also reassuring to know there are experts, like Steve and Sönke, and so many of their colleagues, who are figuring out how and why deep-sea species often look and behave so differently to anything we see in shallower seas.

If you're picking up this book for the first time, I recommend you start by doing what I did: turn through the pages and bask in this collection of exquisite animals. As you do, don't forget these are all real species that live right here on Earth. There's no need to label them as aliens or wonder if Artificial Intelligence was involved.

When you've seen the ruby red shrimp swirling blue goo into the water, the intricate tendrils of *Umbelulla* sea pens, the shadowy vampire squid, and so many glittering, gelatinous creatures, then go back and dive into the text. Steve and Sönke are ideal guides to the deep sea. Reading the picture captions alone feels like having them looking over your shoulder and telling stories about the animals they've encountered and studied. For each species and phenomenon on display, they deftly explain what's going on, and offer their perspectives on what scientists currently think, as well as what mysteries remain.

Most of us won't get to visit the deep sea, which is why books like this are so important. The deep is vast and remote but it's not beyond the reach of human impacts. Mining corporations are eager to scrape profit from the deep seabed and trawlers are sinking their nets ever deeper. Plastic pollution and the climate crisis are there too. Now is a critical time for as many people as possible to know and care about the living wonders of the deep sea, and to join the growing global effort to keep this part of our planet intact and healthy, for the benefit of everyone alive today and for generations to come.

HELEN SCALES

Marine biologist, writer, and broadcaster

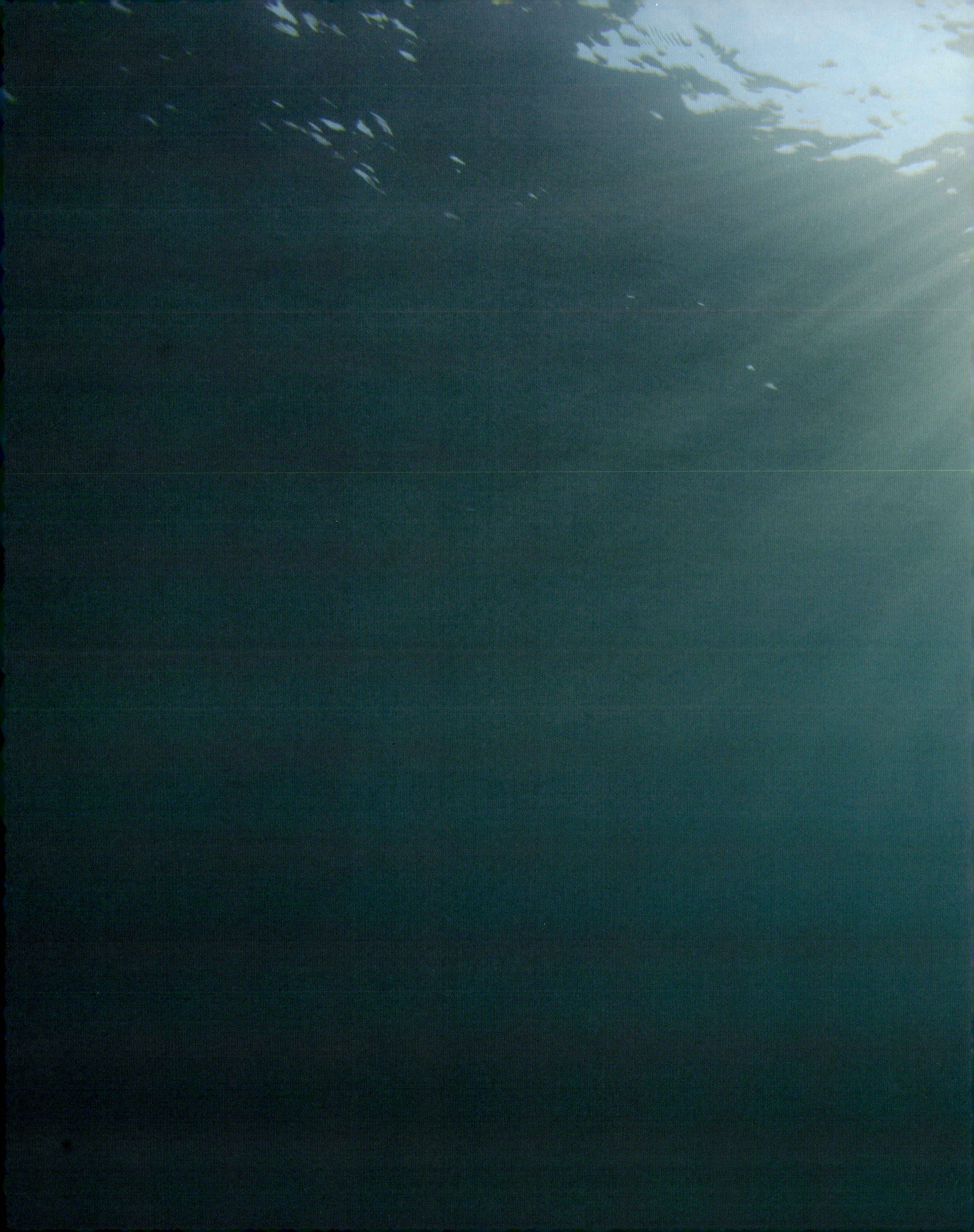

Living in the ocean presents challenges for those trying to hide. It also presents opportunities for animals seeking a meal or a mate.

INTRODUCTION: LIVING UNDERWATER

Of all the gifts we as humans have received, perhaps the most wonderful is the beauty of our natural world. The ocean is no exception, being the largest and most mysterious of them all. Even the beach is full of the sights of shells, the sound of the surf, and the satisfying feel of sand between your toes. Look up, and just beyond the horizon is a larger and much deeper ocean that contains forms of life that few of us ever see. If you take a short boat ride from Lisbon, New York, Buenos Aires, or Tokyo, you will be floating above an enormous wilderness teeming with beautifully strange organisms.

Many of the challenges that these animals encounter involve visual interactions. Imagine you are a fish, darting about a coral reef. Is that distant shadow something to eat, or is it going to eat you? Where can you hide? How can a lonesome fish find a friend around here? The underwater optical environment has several distinct properties very different from what we experience on land. In the pages that follow, we will explore five ways that organisms interact with light: transparency, pigmentation, iridescence, bioluminescence, and fluorescence. Each of these provides stunning tricks and tools that would help your fishy self live to see another day.

Like most people, we might only experience the underwater ocean world when we accidentally open our eyes while swimming in the surf. Before the salt sting forces us to quickly shut them again, we may see a blue or greenish blur. If you're lucky enough to snorkel or scuba dive, you will see a riot of colorful fish, crabs, sponges, and coral, surrounded by what can be a very blue background. But the vast majority of divers swim near to shore, with the seafloor only 10–20 meters below. The ocean is, of course, much larger than this. It is also much deeper, with an average depth of about 4,000 meters. This huge, deep realm is foreign to us air-breathing land dwellers, but it has been explored by diving near the surface or via human-occupied or remotely operated submersibles. Even this limited exploration has shown

The visual rainbow is just one special section of the total electromagnetic energy emitted by the sun. The higher- and lower-energy portions (shorter and longer wavelengths) cannot pass effectively through water. Within the visual range, blue-green light penetrates best below the surface. The result is that as you go deeper in the water, violet and red photons penetrate less, and the ambient light narrows to a monochromatic blue-green, depending on where you are in the world.

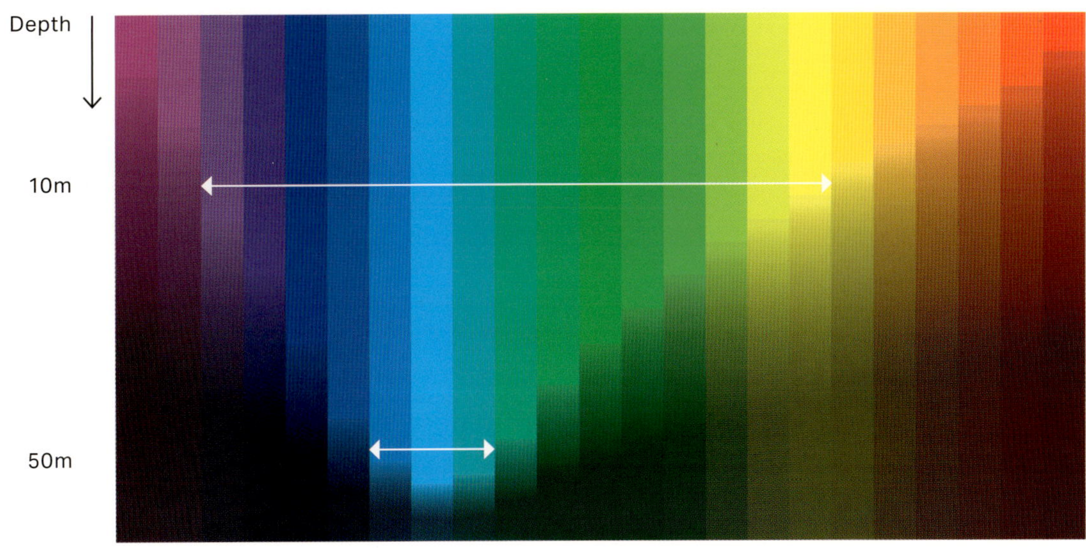

us that oceanic and deep-sea animals interact with and even produce light in diverse and beautiful ways. We hope to share some of the wonder of this radiant sea with you, but first we need to discuss the habitat itself.

One of the many miraculous aspects of water is that it allows only a small part of the electromagnetic spectrum to pass through it. Visible light is just a fraction of a larger range of electromagnetic energy. Radio waves, cell phone signals, microwaves, and most infrared wavelengths are all quickly absorbed when passing through water, as are most ultraviolet rays, X-rays, and gamma rays. In between all of these is a small window of what we call "visible light." Without this relatively transparent band, even the water in our eyeballs would absorb everything and render us blind. Water, however, does not absorb all visible wavelengths equally, which affects not only what the underwater world looks like but also how animals see and how they appear.

It is not just a coincidence that nearly all organisms can see somewhere within the same rainbow of light. Some can see into the ultraviolet, or sense beyond red into the infrared, but most of the action (from vision to bioluminescence to photosynthesis) seems to occur within this narrow band of colors. That these wavelengths pass furthest through water (and through aqueous tissues and even the atmosphere) is only part of the explanation. In addition, shorter wavelengths are more energetic and can damage biological molecules, and longer wavelengths have less energy to fire off a chemical reaction or trigger a cellular cascade of events.

Imagine you are diving in the tropics. At the surface, you observe a blue sky and a yellow-white sun, which together provide the full range of colors you can see, in addition to all the electromagnetic radiation you can't see, such as the heat on your face from the sun. Immediately upon dropping below the surface, things change rapidly. Your face cools to the temperature of the water and all the other unseen parts of the electromagnetic spectrum are gone; for example, if you forgot to remove your cell phone from your pocket (it can happen), it would no longer receive calls. However, near the surface, the visible world still looks about the same: the background water

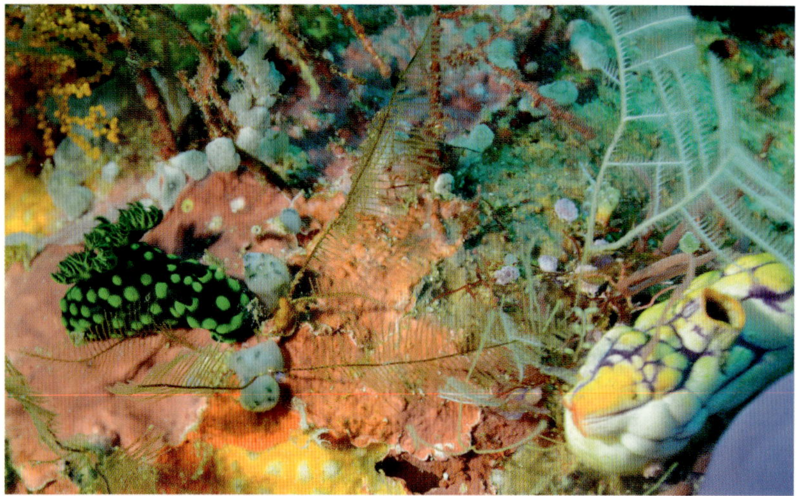

How humans see the ocean is not the same as how animals will see it. Many have eyes with one type of cone, meaning those animals see the world in shades of gray.

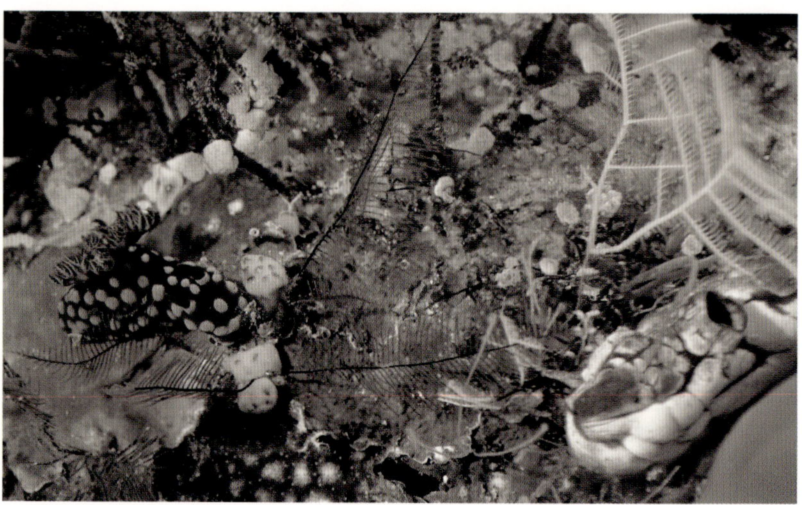

Note that you can get a monochromatic view either by having one type of color sensor or one color of illumination—both of which happen in the deep sea.

is blue, and your red gloves still look red. As you drop further—even down 5 meters—you notice that your gloves now look dark brown. By 10 meters, your gloves are black. At the same time, the violet end of the rainbow is being absorbed and is approaching black. By 20 meters, any orange colors would also be gone, along with more of the violet. In short, the rainbow of colors we see at the surface is being nibbled away at both ends, and by 30 meters, the world around you would be almost entirely composed of different shades of blue. If you were to look up, the spectrum of light you see would be broader, but still quite blue. Looking down, the watery background below would be blue, with a small amount of purple added. Below 30 meters, diving starts to become dangerous, but in a submarine, by 200 meters, what light remains is entirely blue in all directions. In clear tropical water at this depth, the light level is about one hundred to one thousand times dimmer than at the surface. On a clear day, this would be about the same as room lighting.

In the tropical ocean, almost all of this is due to the fact that water passes blue light more readily than it does other colors. It passes green and blue-violet less well, yellow and violet less well than that, and there is a steep wall at orange where it and red are absorbed almost immediately. But even the blue light is absorbed to some extent, reducing its illumination by tenfold every 70–100 meters in depth. This means that by about 500 meters—even in the clearest water on the brightest day—the ocean world is a dim indigo, much like late twilight. Deeper than this, and the remaining sunlight—while still blue—is seen by our eyes as a very dim gray, since we have now adapted to black-and-white night vision. By 800 meters, the last little bit of light we see when looking up has disappeared and the only light seen is generated by the animals themselves.

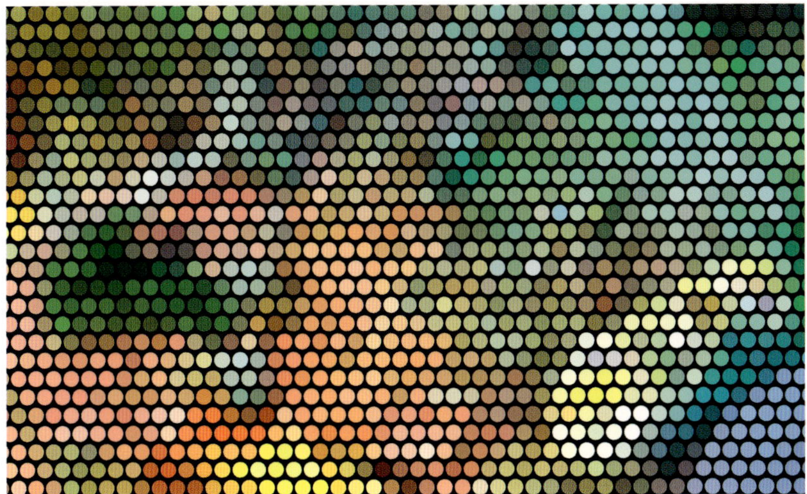

Most animal eyes have fewer photoreceptors (think of a Light-Brite toy or an LED sign in a storefront) so they see a low-resolution view that lacks detail.

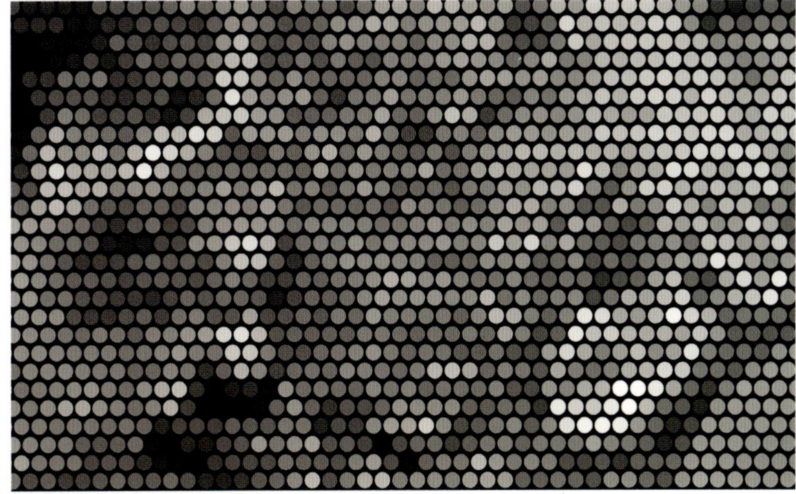

When we think about visual interactions in the ocean, it's important to see things from the animal's perspective: for most, it's a grainy one-color world. (At the other extreme, mantis shrimp can have fourteen different color receptors and perceive both ultraviolet and polarization!)

Closer to the coast, the sea is often full of algae, and probably some sediment as well. Because both absorb light more strongly than water, and pass green and even yellow light better than blue, it doesn't take much to make the water green or brown. For the same reason, it also gets dark much faster. In murky areas near the shore, at 20 meters it can be as dark as the tropical deep sea. This dim habitat is a challenge to vision but can benefit those who want to hide and those who make their own light.

HIDE AND SEEK IN AN EMPTY WORLD

The optics of water and the algae and sediment within it create a very different world from that experienced on land. While a butterfly deep within the heavy canopy of a tropical rainforest may notice that the light is greener and darker, the effects in the ocean are much stronger. In addition to this is the fact that—away from some parts of the bottom—the ocean is, on average, empty. This has a few important consequences. First, it's difficult to hide. While animals on land and on coral reefs can try to blend in with their complicated surroundings, animals in the open need to essentially look like water. As we've seen while experimenting (or playing) with underwater laser pointers, anything that stands out even a little may be investigated as potential food. Second, the emptiness allows for certain things to be seen from a long distance, simply because nothing gets in the way. A butterfly within a forest or dense patch of meadow may not be seen at any great distance, but a flash from a small bioluminescent animal underwater may be seen from over 50 meters away, even accounting for the fact that the water is absorbing its light. This bioluminescence (or possibly fluorescence under the right conditions) could

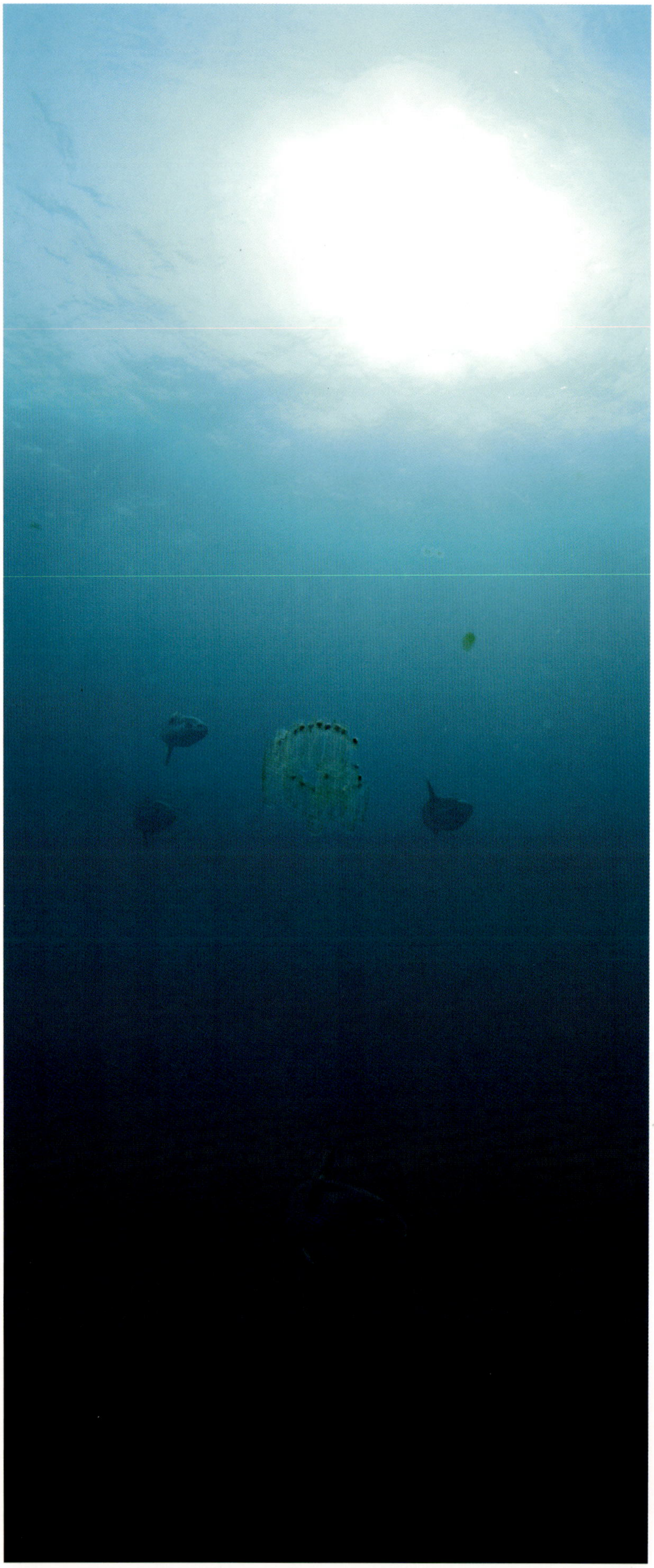

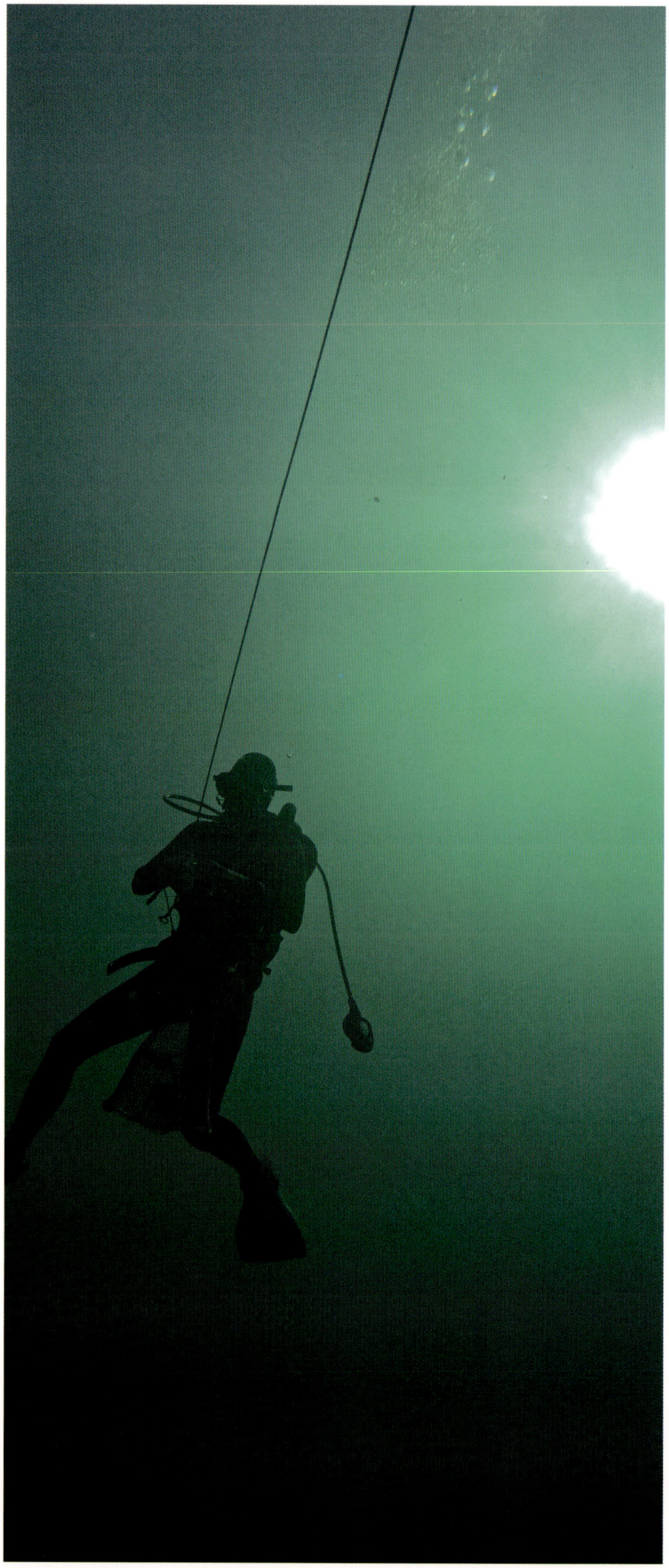

be used for luring, communication, or certain types of defensive purposes—sometimes all three in one animal.

Marine animals make up for their challenging optical arena, limited visual resolution, and one- or two-color vision in several ways. As with land animals, they may have high sensitivity, contrast enhancement, and edge detection hardwired into their visual neurons, and respond rapidly to moving objects, like a fishing lure trailed behind a boat. However, they also have larger eyes with more sensitive retinas and the ability to sacrifice detail for sensitivity. As you look at the examples to follow, try to see them from the perspective of a potential prey, predator, or mate

TRICKS OF LIGHT IN THE SEA

Animals in the open ocean and in the deep sea need to solve the same problems of life as animals on land: they need to find food, avoid being found as food, defend themselves, and reproduce (which may involve finding mates). Many land animals use their appearance to solve these problems by luring prey, hiding themselves, communicating, and signaling to mates, and scientists have spent many decades studying how they do this. The optics of the water and the vast emptiness of the ocean, however, require a different set of solutions. Differing challenges and opportunities arise, new to us as humans but likely much older than those found on land. Here in this book we focus on the exciting uses of light in the ocean, many of which are so rare on land as to be almost nonexistent. First we discuss whole-body transparency, which appears to be a common form of camouflage. Then we look at the ways in which oceanic animals use pigments and iridescence, which are often quite different from the ways that both are used on land. We then move on to the many functions and displays of self-generated bioluminescent light for offense, defense, camouflage, and communication. We end with fluorescence. Little is known about this, but the ocean provides a perfect optical habitat for it to serve important functions.

Although we feature certain animals in each of the five sections, most exhibit more than one optical phenomenon. In fact, many—like a transparent bioluminescent jellyfish with fluorescent tentacles and a pigmented gut—use four, even five, of these optical tricks at once! Conversely, just because we see an iridescent rainbow when we take a picture with a flash, or see fluorescence when we shine an ultraviolet light, doesn't mean that this has any biological relevance for the animal. Sometimes a flash in the pan turns out to be fool's gold.

Although the open ocean is distant from most of our daily lives, it is actually the most common habitat on the planet, filled with the most abundant organisms. We hope this tour of the visual diversity of ocean inhabitants—none of which is computer generated or enhanced—gives you a deeper appreciation of this vast underwater kingdom.

⇇

When you first jump into the open ocean, you may see nothing, or only the large animals like this coiled salp and the three ocean sunfish (*Mola mola*). Look closely, and above the rightmost *Mola* you can see the orange ink cloud left by a swimming worm. In fact, within this patch of water, thousands of tiny transparent organisms are drifting past unseen.

Depending on the amount of microscopic algae or sediment in the water, the color and clarity of transmitted light can vary greatly. Sometimes you can't see your dive buddy an arm's length away, and other times you can see the boat hovering 30 meters above you like a spaceship.

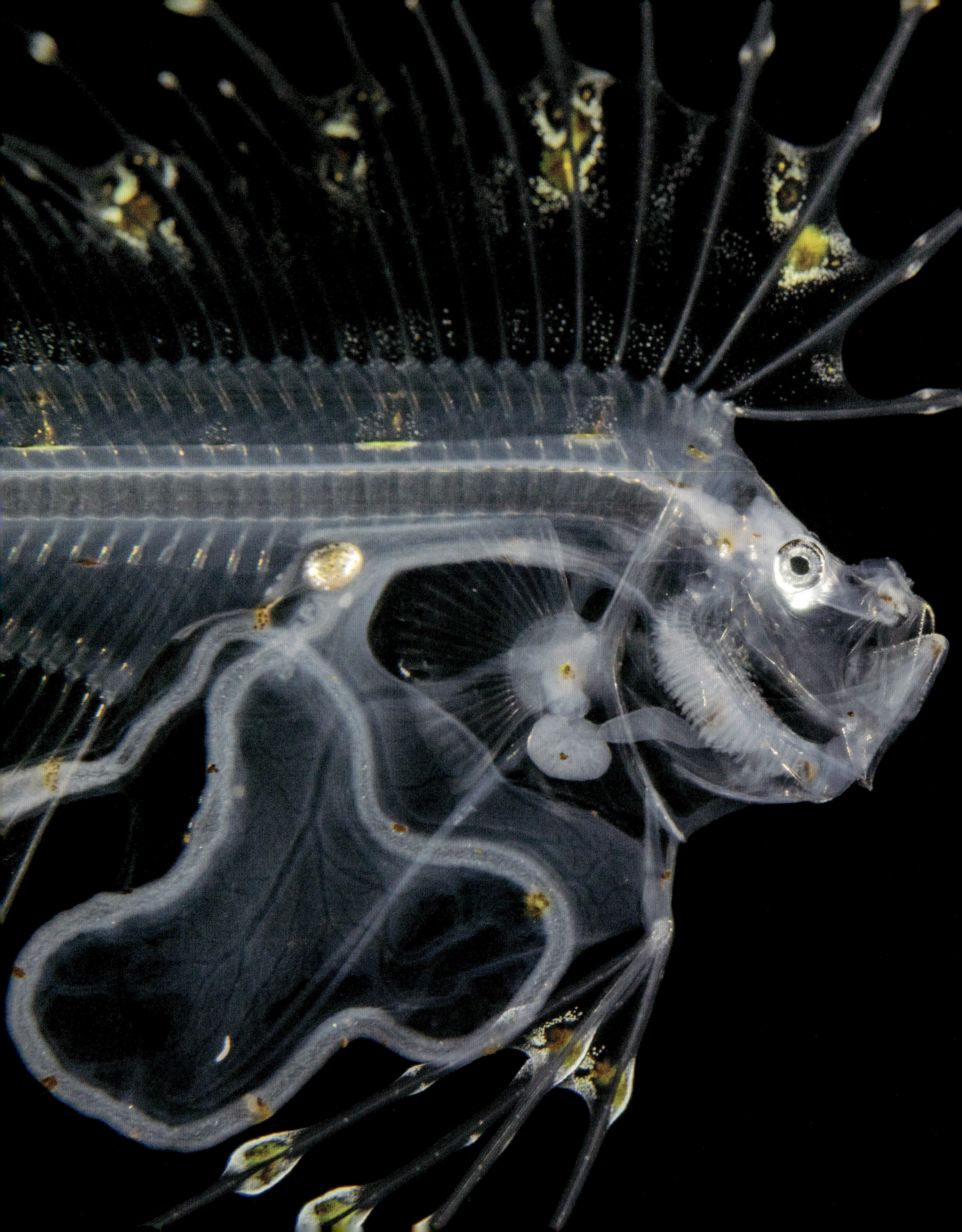

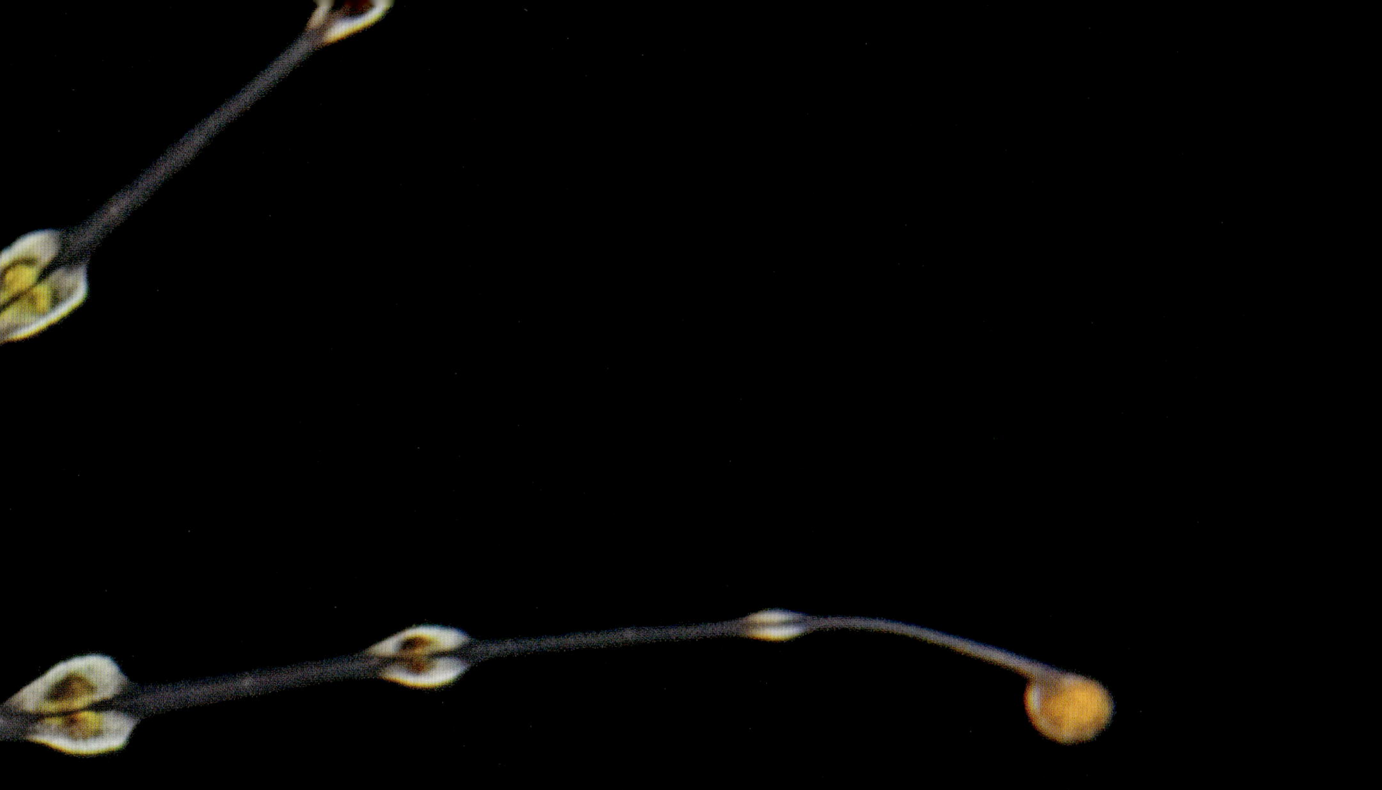

TRANSPARENCY

Life as kinetic glass

Both of us have been on many scuba dives in the open sea. Doing so requires training, followed by medical exams and piles of forms, and involves a lot of work, planning, and money. A big ship is needed to get out to sea, and then a small boat to travel a further couple of kilometers to dive safely. When we are finally beneath the surface, at first all we see is water. After a few minutes, if we're lucky, we start to notice faint objects slowly moving by. Some are as small as our fingernails, some are as large as our fists, and others are in chains that can be many meters long. A few are pinkish and more obvious, especially if they are on the larger side, but most are so clear that it's hard to see them at all. When we do manage to get a good look at them, the animals seldom look like anything we've seen on land or on the seafloor, even though they're related to these animals. Snails can look like little elephants, slugs can look like fish, and worms can look like swimming chandeliers. These animals, so rarely seen by humans, are actually some of the most abundant on the planet. But not only is it hard to get out to where they are, they can still be almost impossible to see, even when we're right in front of them.

While images of deep-sea fish and squid are fairly common, it's seldom appreciated that many of the animals in the open sea are essentially invisible. Whole-body transparency is perhaps the ultimate way to hide, because light simply passes through the animals as if they aren't there. A vast array of animals do this. The first set are often referred to as the gelatinous zooplankton (because much of their bodies appear to be made of a gelatinous material). The most familiar of these are the jellyfish, which we sometimes see (and dread to see) in waters near shore, or even washed up on the beach. Those we are most familiar with, known as the scyphozoans, are usually the least transparent. This group includes the moon jellies, sea nettles, cannonball jellies, and many others. However, there is a more diverse group of jellyfish, the hydromedusae, that tend to be smaller and far clearer. Closely related to this group are the siphonophores, which range from small animals being moved along by one or two pulsing chambers to large, complex animals, which can be seen as either a colony of different types of members or as one animal with organs that do different things.

In addition to the jellyfish, we have ctenophores (comb jellies) that superficially look like jellyfish but are less closely related to them than we are. Most of them are astonishingly clear, though a few are milkier or even red at depth (some of the red ones have clarity like a ruby). What unites them all are comb rows, which are made of stacks of little hairs (cilia) that row together to propel the animal through the water.

Then there are the wild swimming forms of animals we usually think of as earthbound: the snails and worms. As mentioned above, they look nothing like their terrestrial and seafloor relatives but are instead evolved to spend their lives invisibly wafting through the ocean. As with the jellyfish and ctenophores, many of these lose their transparency at depth, becoming red over at least part of their bodies. At the depths where scuba divers travel, however, they are gloriously clear. Some swimming snails have lost their protective shells (more like swimming slugs), while others retain them. As a result, they must continue to flap, with almost comical enthusiasm, to avoid sinking into the abyss.

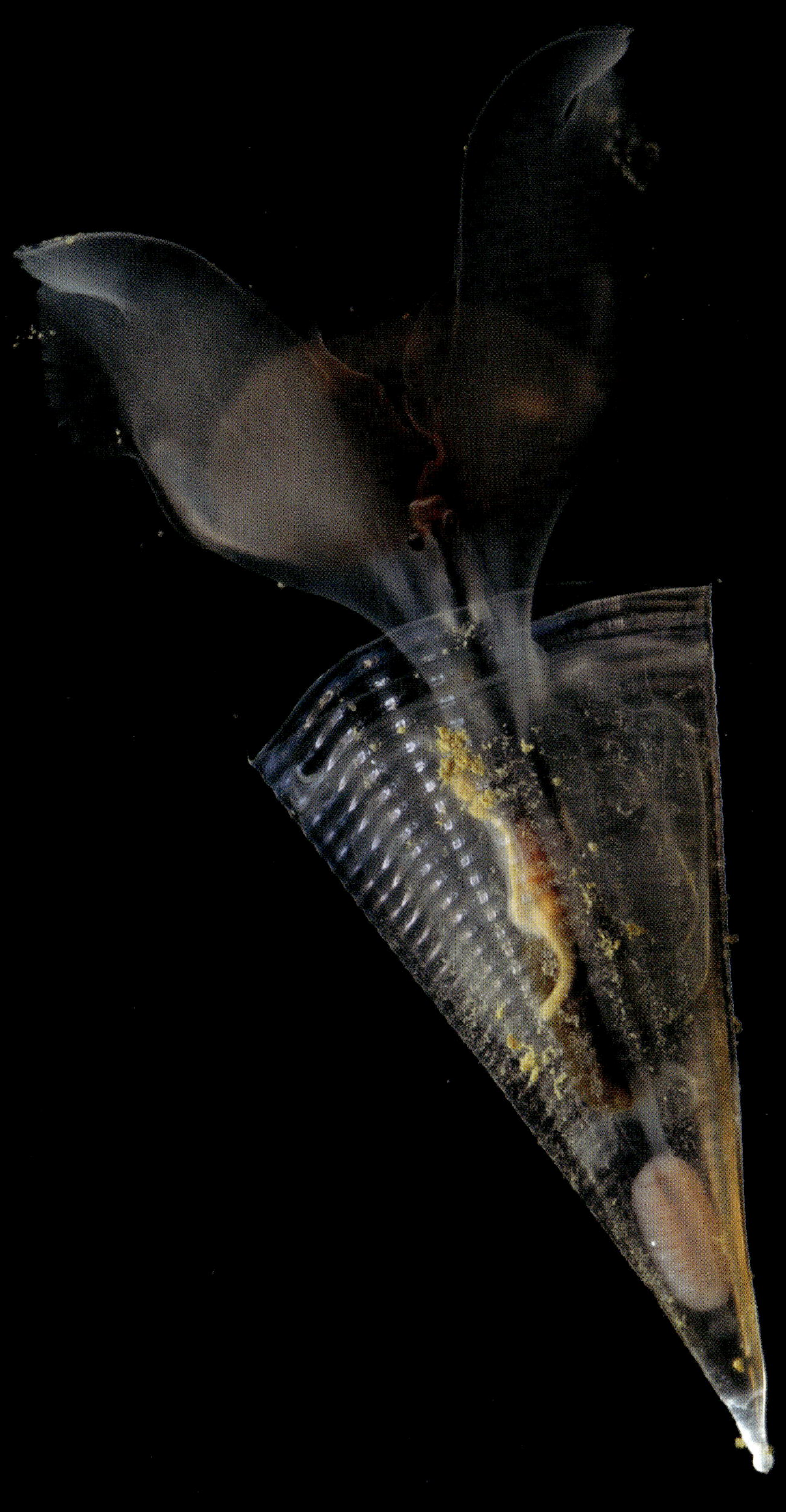

This pteropod is highly modified for permanently living up in the water column. Its foot has become flapping wings for swimming, and its shell has been reduced to be lightweight and transparent.

Not all transparent animals are gelatinous; the upper ocean is also full of transparent crustaceans of all kinds. These are related to crabs and lobsters, but spend their whole lives up in the water. In the shallower reaches of the ocean, they are often transparent. Some of the largest and most impressive are the hyperiid amphipods, which can be as big as the palm of your hand and so clear that you can barely tell they are there, even if they are in a bucket right in front of you. There are also many transparent fish, especially the young forms of opaque adults. Perhaps the most striking are the leptocephalus larvae of eels, which look like long, thin leaves and are so clear you can read a book through them.

Finally, we have transparent animals that are not closely related to anything most people have seen. The most numerous of these may be the chaetognaths (arrow worms), which are found in tremendous numbers in the open ocean, especially near the surface. As their common name implies, they look like short, thick arrows, complete with fins that resemble feathered wings. Again, while a couple of species have colored stomachs, in the sunlit parts of the sea they are usually extremely clear.

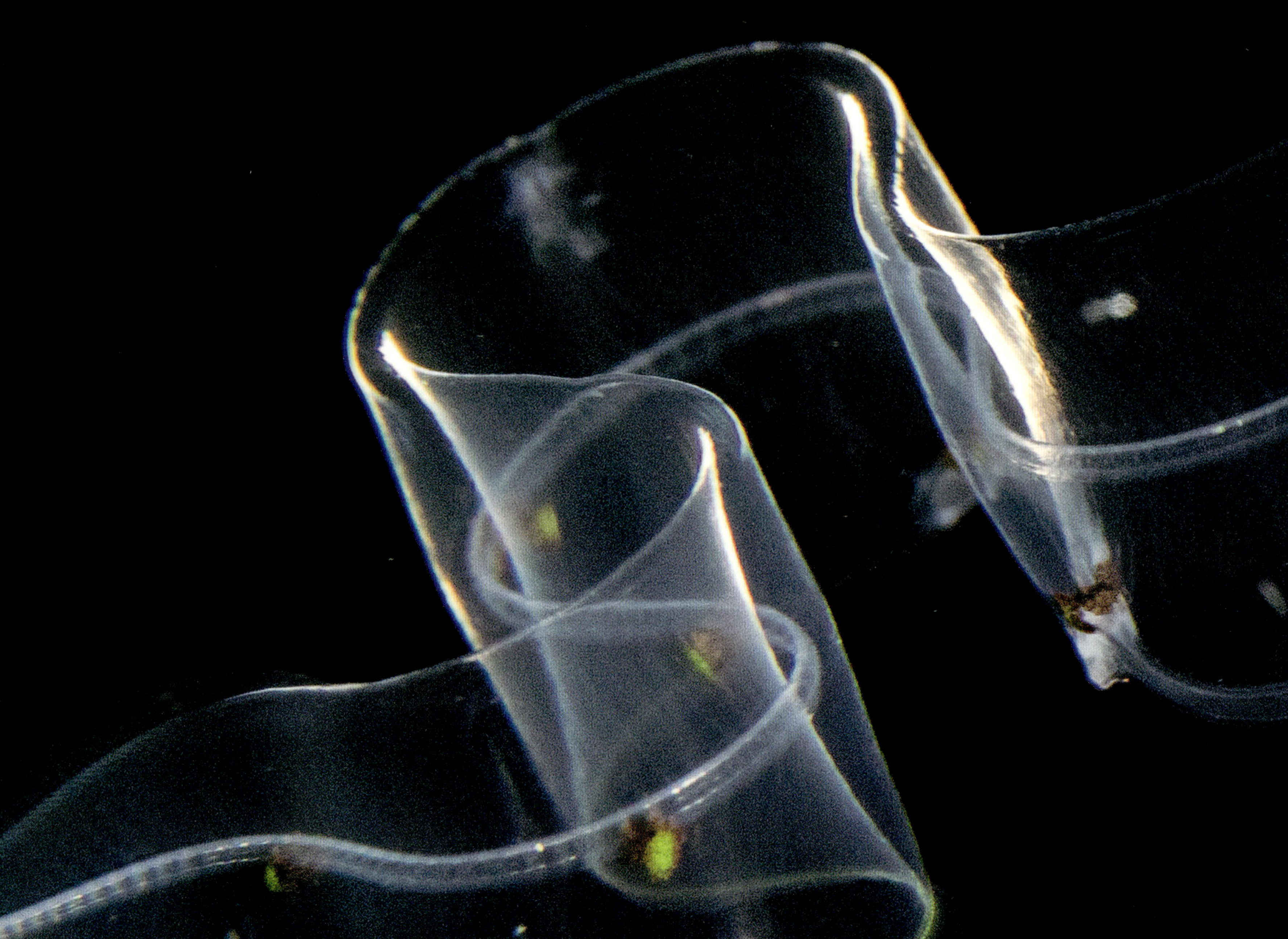

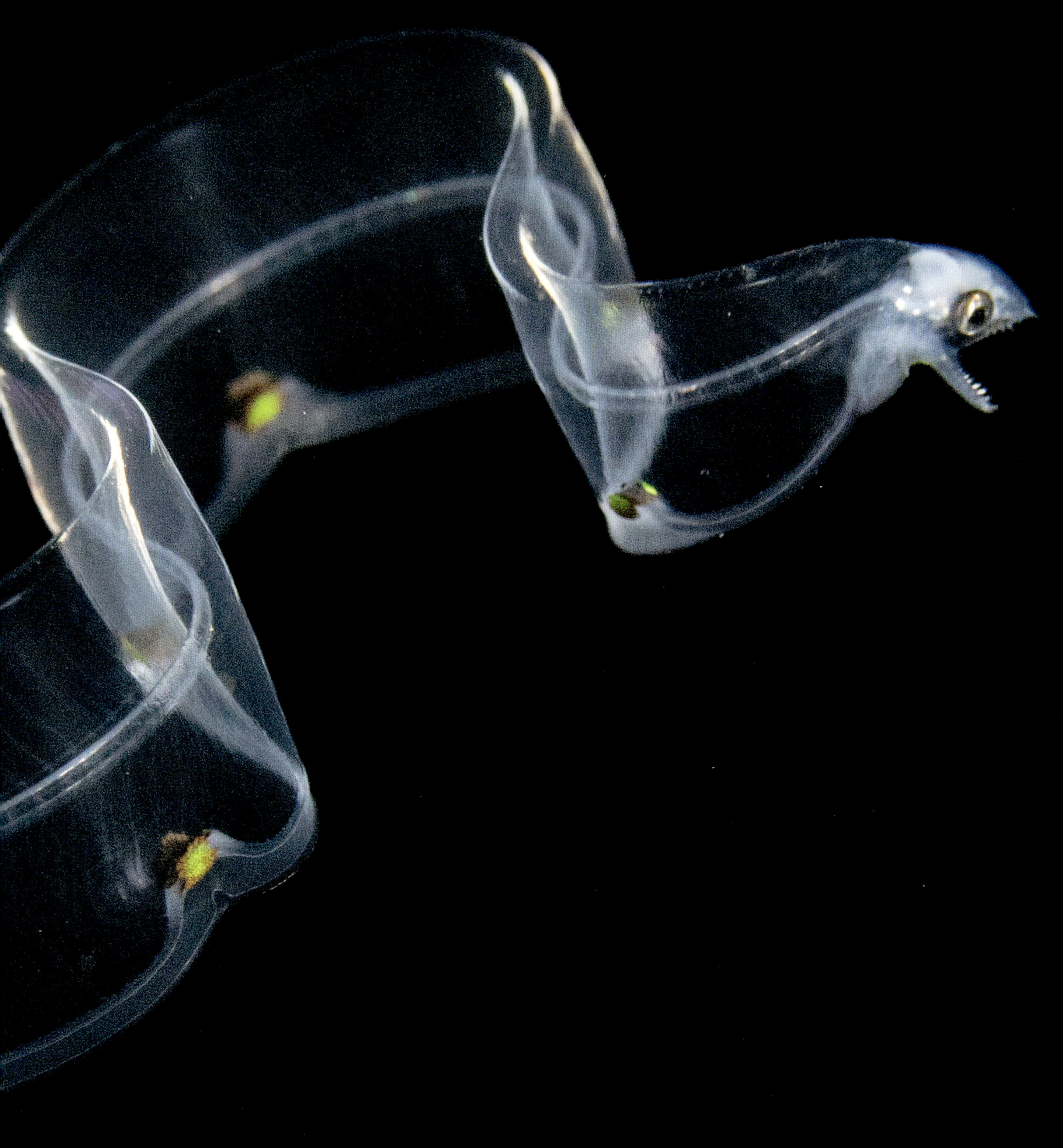

The upper reaches of the ocean house a diverse menagerie of swimming and floating animals that are remarkably clear, with many of these animals having red versions at depth. These animals were not well appreciated by earlier oceanographers that used nets, because the nets would shred them beyond recognition. But they are abundant in nearly all parts of the ocean. If you go into the water out there, they will be around you, even though you can't quite see them. When we scuba dive to collect them, we often have to get the sunlight just at the right angle, or use another diver's dark wetsuit as a backdrop, because against the diffuse blue they utterly vanish.

HOW TO BE INVISIBLE

In contrast to the abundance and diversity of transparency in the ocean, it's rare on land and on the seafloor. The most familiar transparent tissues to us are the corneas and lenses of our eyes. Both of these tissues are heavily modified to be clear. The cornea is a perfect alternating stack of tiny fibers that is so regular that light goes right through it, and the cells of the lens have lost nearly all their internal structures (organelles), leaving only a soup of almost identical proteins that don't interfere much with the passage of light. In both cases, there is no way these two tissues could survive without the rest of our body keeping them alive and in good condition. (The cornea isn't even really alive, but a fancy and very well cared-for sheet of tendon.) It is hard to imagine how an animal could modify the tissues of its whole body to be transparent when we can only manage to keep such a small part of our own bodies clear. When the clarity of our lenses starts to fail, we suffer from cataracts that cloud or even blind our vision. The lenses of deep-sea fish are also especially susceptible to clouding, in the same way that egg whites go from clear to opaque when hard-boiled. Nearly all the photos you see of deep-sea fish have milky white eyes, because their eyes don't stay happy for very long once brought out of their habitat.

The first trick to being a seagoing transparent organism is to make your insides clear. One might think that simply getting rid of colorful pigments should do the job, but getting the red pigment out of a steak would not make it clear, for example. There are many substances, such as snow, clouds, and milk, that don't have light-absorbing pigments but are still opaque. We can't see through them because light gets *scattered,* not absorbed, as it passes through. Think of a ball rolling across a pool table: it can bounce off the other balls and keep going (scattering) or at some point it may fall into a pocket (be absorbed). When light moves from one material to another, it usually scatters, which means it bends, bounces, or reflects. Any moderately sized animal has all sorts of internal complexity in the form of organs, nerves, and blood vessels, and so even without light-absorbing pigments, it will be milky or opaque. The light just bounces around inside and won't make it straight through.

In the same way that eye lenses and corneas are composed of simplified tissues, many transparent ocean animals also fill much of their insides with gelatinous material or fluid. Many also make their bodies very flat, so the path through the inside is short. Some, like many jellyfish and swimming snails, consist of a thin layer of cells wrapped around a large nonliving ball of gelatinous material. They likely employ other tricks as well—we just don't know yet.

Even the clearest object can be given away by reflections of light from the surface. An ice cube is a good example: the ice itself can be clear, but its shape is made obvious by all the light reflected from its surface. This is worse when a surface is complex—as you see in crustaceans—or in the air. This last issue is probably one of the main reasons why we don't see many transparent animals on land—it's easier for tissues to match the properties of water than it is to match air alone. Some transparent animals have solved this problem by covering their surfaces with anti-reflection coatings. These are much like the coatings on your sunglasses and camera lenses, except that they may be alive.

Eyes, stomachs, and reproductive cells present three other challenges to being completely transparent. Eyes need to absorb light to see, and stomachs contain food, which may either be opaque or glowing. A solution for both is to use mirrors to hide them. Despite how shiny a mirror may appear on land, in the open ocean it would mainly reflect more ocean, which looks pretty much the same in all directions. So, many transparent animals will put their digestive organs and eyes in iridescent mirrored bags that make them much harder to see, even though they're opaque. Reproductive cells, such as eggs, are already reduced to the essential components, so they may not be able to alter or simplify their composition. Some comb jellies are totally clear until they mature, and then the canals containing their sperm and eggs show up as a filigree of stark white lines across their bodies.

Together, all these adaptations allow animals to be transparent over much of their bodies, perhaps stuffing the few remaining opaque parts into mirrored bags. It's a remarkable feat of evolution.

NOW YOU DON'T SEE ME, NOW YOU DO

Finally, there are animals that are normally transparent, but become milky white in places only after you touch them. One of the best examples of this is the ctenophore *Cestum*, or the Venus girdle (p. 46). It is normally quite clear, but within a few seconds of it being touched, its surfaces become frosty white. Other animals, including certain siphonophores and swimming snails, do this as well, but only over parts of their bodies. For example, the swimming snail *Corolla* develops white spheres all over its normally transparent (and gelatinous) shell. Some boxy siphonophores only become opaque along their edges. Research has shown how some of these animals do this, but the question remains: why? The main idea is that the sudden appearance of the animal scares away any predator or unwanted interloper, much like the decloaking of a warship in a science fiction show. The truth—as is so often the case with the ocean—is that we don't know for sure.

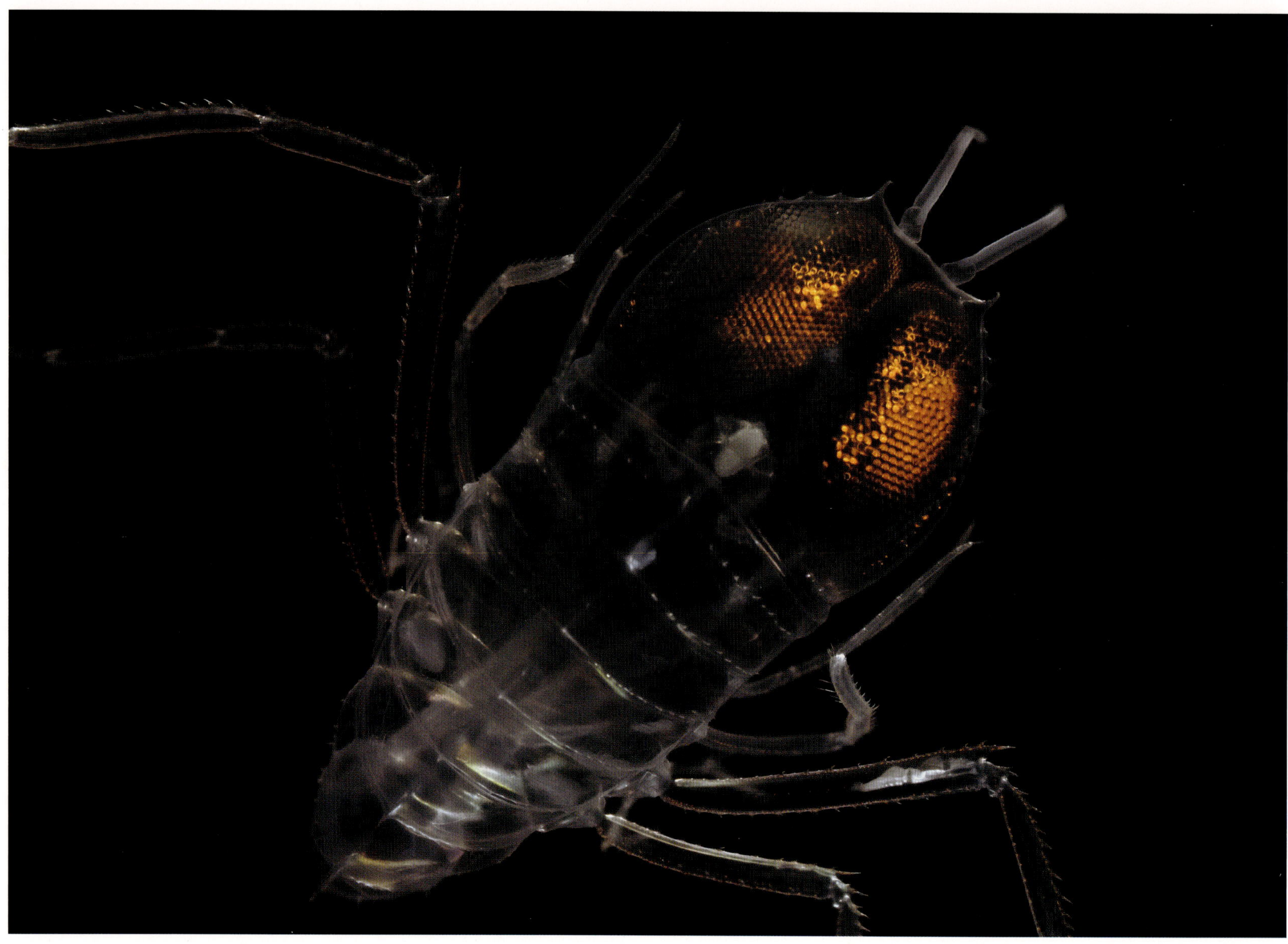

Cystisoma

This beautiful shrimp-like animal is a hyperiid amphipod. Not all members of this group are transparent, but those that are can take transparency to an expert level. The name *Cystisoma* means "body like a bag," which is accurate because almost the entire inside is empty of organs and instead is filled with a fluid that we know little about. They can be up to 10 cm long, and a big part of their length is made up of their enormous eyes, which, unlike many eyes, are nearly transparent and hard to see. If the light catches the eye at a certain angle, you can just make out a large, thin retina, tinted orange by the light-absorbing layer. We found that they use tiny bumps and a thin coating of bacteria on the surface of their bodies to reduce reflection. It's unknown whether this particular trick serves a purpose, but it does make these animals even harder to see.

In this head-on view, *Cystisoma* can be seen using a pair of its legs like "windshield wipers" to groom the outer surface of its massive eyes. Just like with an automobile, this process probably keeps its view unimpeded by splatters of marine snow falling from above. The clear dome above the orange retina is actually an array of fiber-optic guides, which direct the light onto the retina from across the visual field.

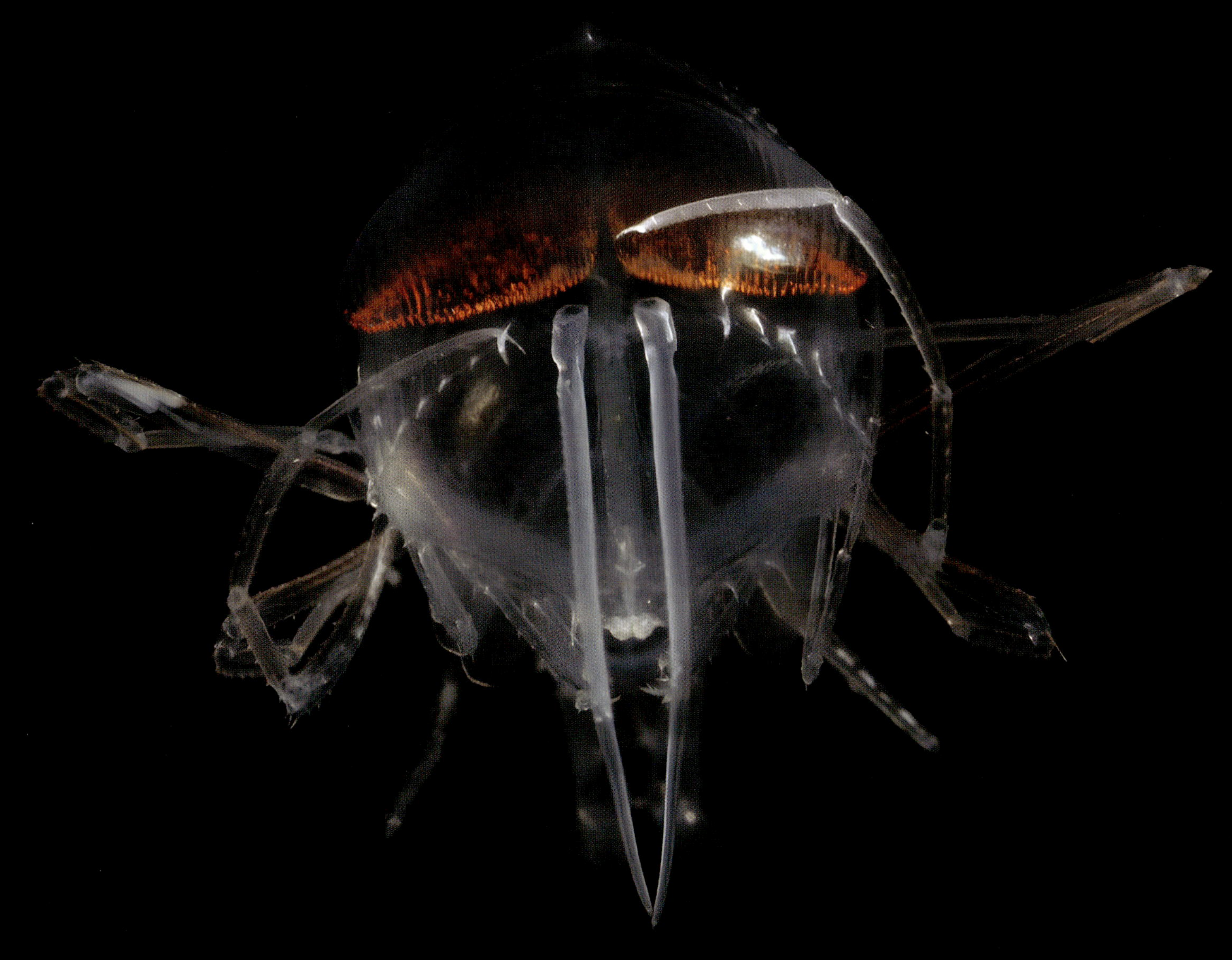

Leptocephalus larva

The so-called leptocephalus larva of an eel mystified marine biologists until recently because their stages of life are found in very different places than the adults (rivers versus open ocean) and they change appearance. Eel larvae can be at least 30 cm long, but are nearly always only about 1–2 mm thick. This—combined with their lack of red blood vessels and apparent lack of organs—makes them completely clear. The stripes down the body reveal the segmented nature of fish muscle. Although many transparent animals can be given away by the food they eat, leptocephalus larvae are not, perhaps because they eat transparent food such as marine snow.

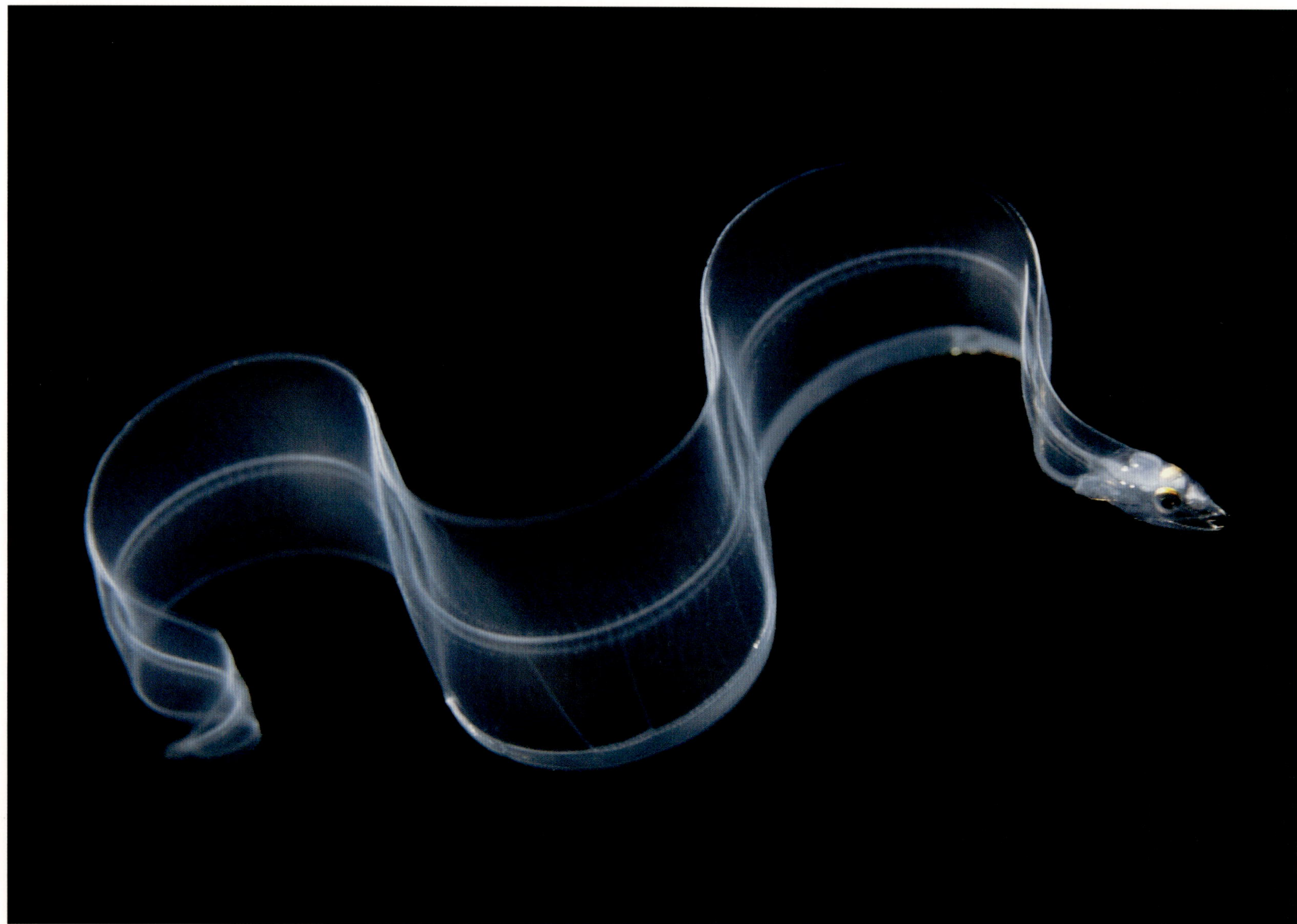

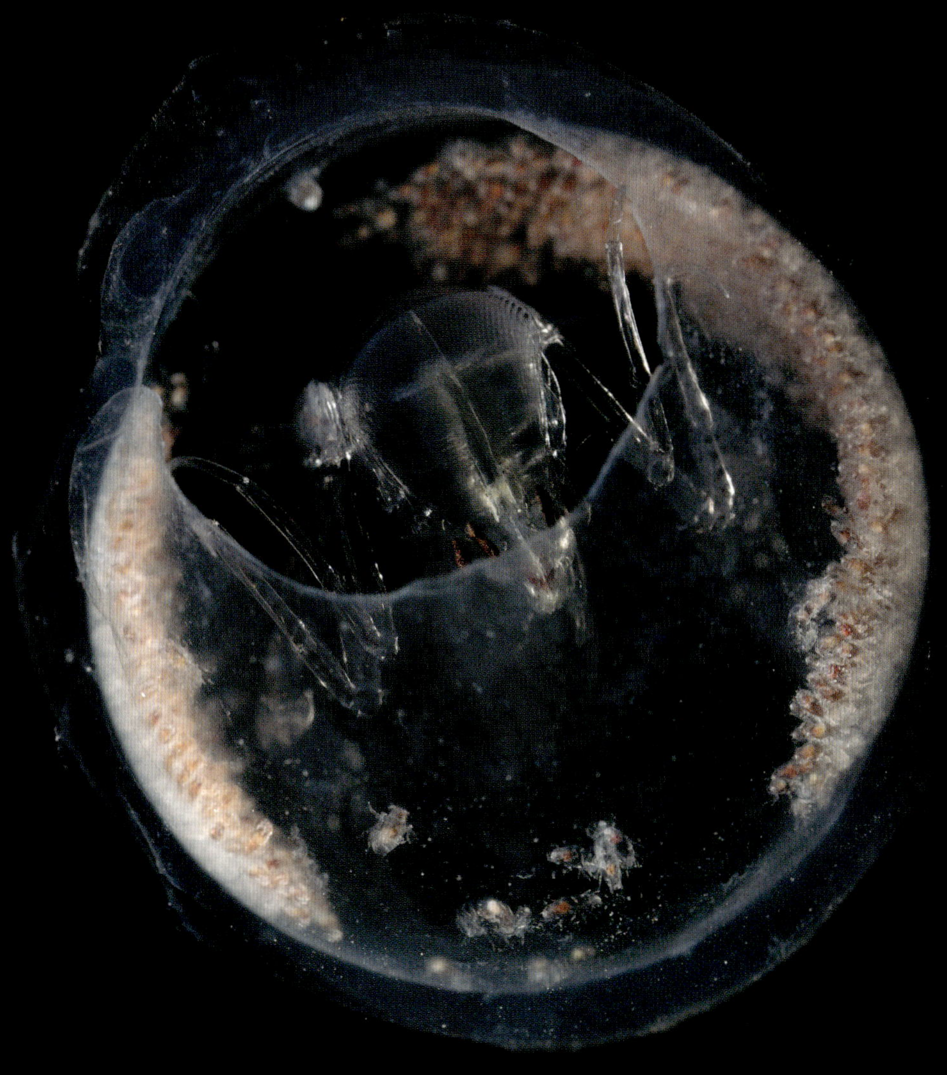

Phronima in a salp

Phronima, another hyperiid amphipod, is much smaller than *Cystisoma* (pp. 24–25) but still highly transparent. It has four large, transparent eyes that almost entirely fill its head. You can see the four sets of photoreceptors as brown dots at the bottom of the fiber-optic dome. Its transparency is striking, but it is better known for the way it parasitizes another animal to reproduce. It finds a salp—a short, transparent tubular animal—eats all the nutritious parts out of it, and then changes the animal from a gelatinous tube to a thicker plastic-like cavity that it uses to lay its eggs. It swims around with these eggs until they hatch, then continues to swim the young around as if the former salp were now a stroller. Here, you can see the hundreds of babies lining the walls of the barrel. Some in the lower middle are already crawling around, looking like miniatures of their mother.

Cranchiid

Cranchiids, known as glass squid, are a beautiful group that is highly transparent because—like *Cystisoma* (pp. 24–25)—their body wall is thin and surrounds a cavity that is mostly filled with fluid. Unlike *Cystisoma*, however, the cavity can also be used for jet propulsion. The fluid is also used to help with flotation because it has been altered by the animal to be slightly lighter than the seawater around it. However, some parts of the animal are harder to make clear, especially the pigmented arms, the eyes on transparent stalks, the brain between them, the stomach, and the pair of white gills.

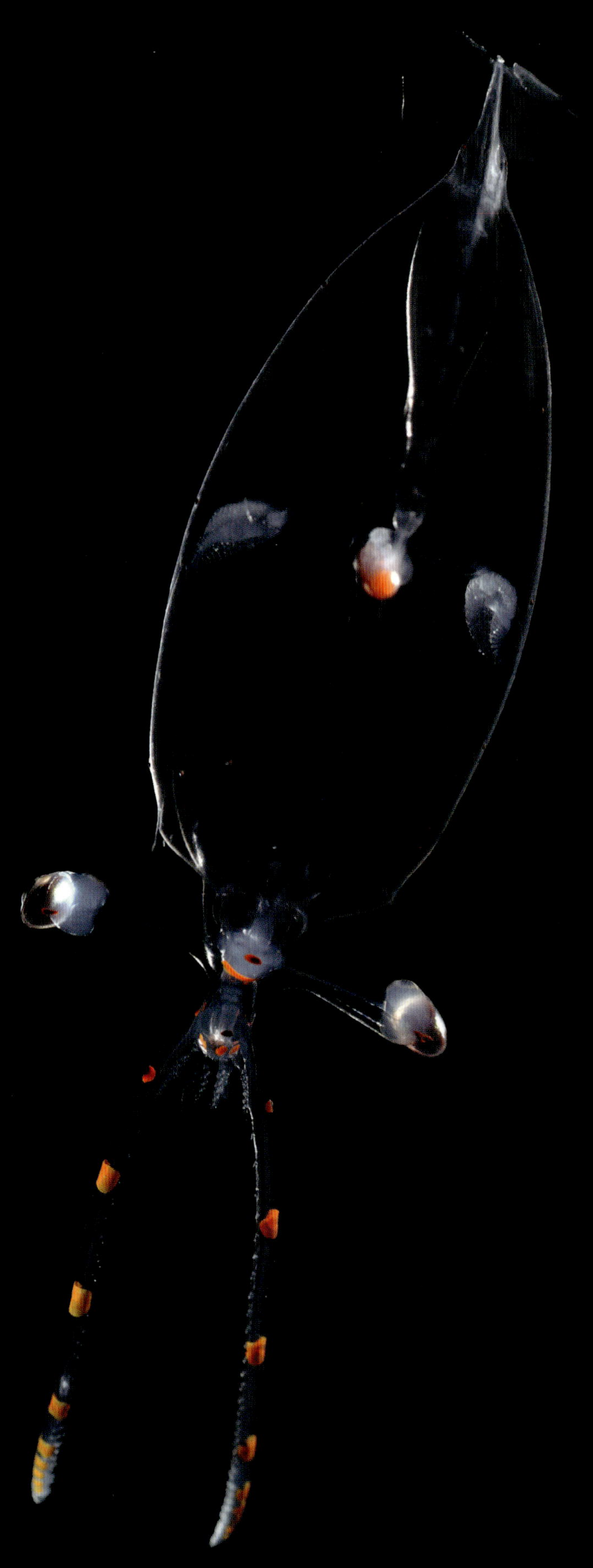

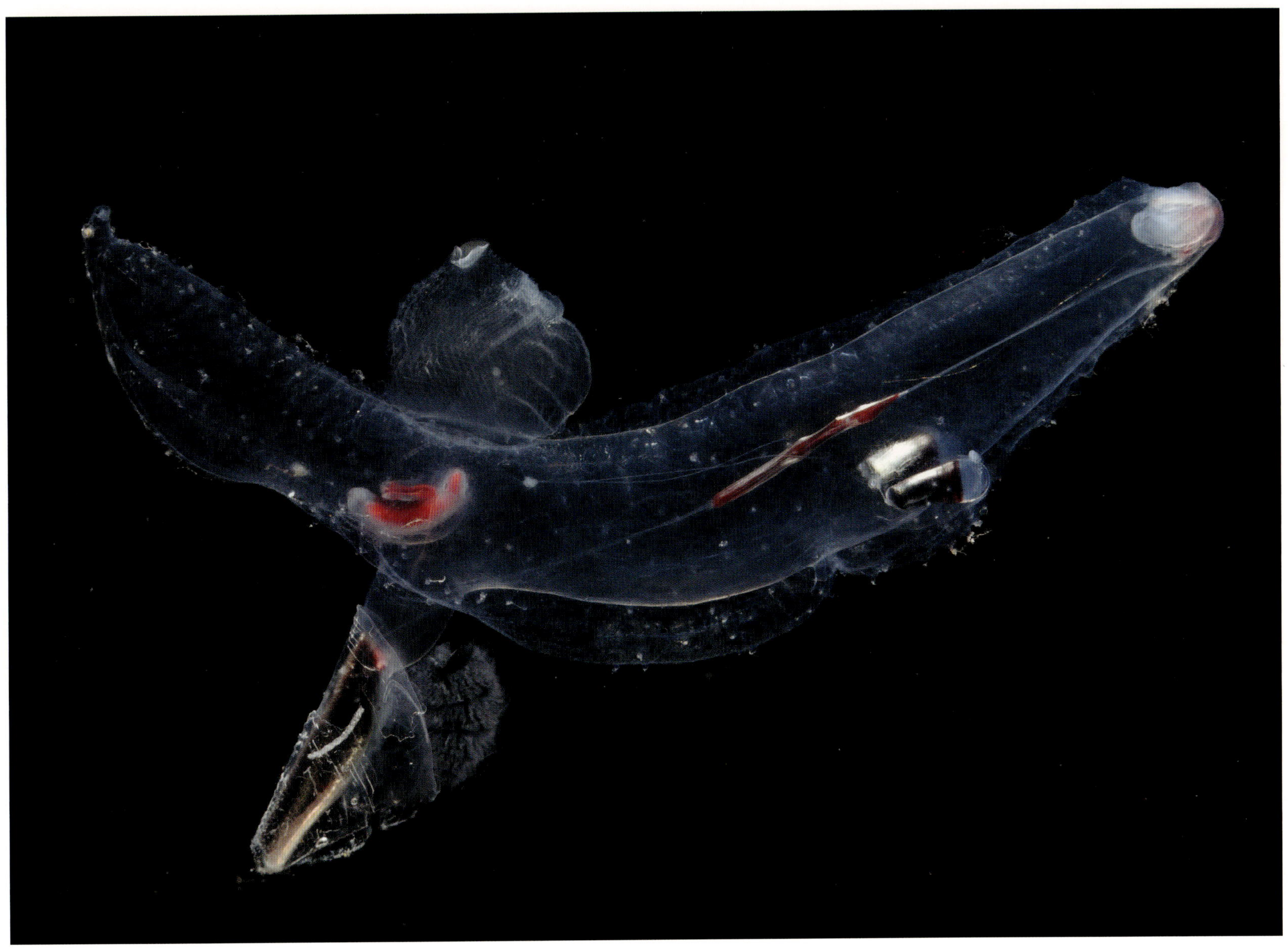

Carinaria

Heteropods, like *Carinaria*, are actually gastropod snails that have adapted to live in the water column. Because a normal snail shell would be hard to keep afloat, *Carinaria*—along with most other swimming snails—have a much smaller shell with thinner, lighter walls. The smaller shell not only makes the animal harder to see but also provides a tiny refuge for the tuft of gills. The typical foot of the snail is still used for movement, but now as a winglike fin that undulates to move the animal through the water. All the related heteropods have well-developed eyes, but this species has extremely large lenses, balanced like a scoop of transparent ice cream on their silvery eyes.

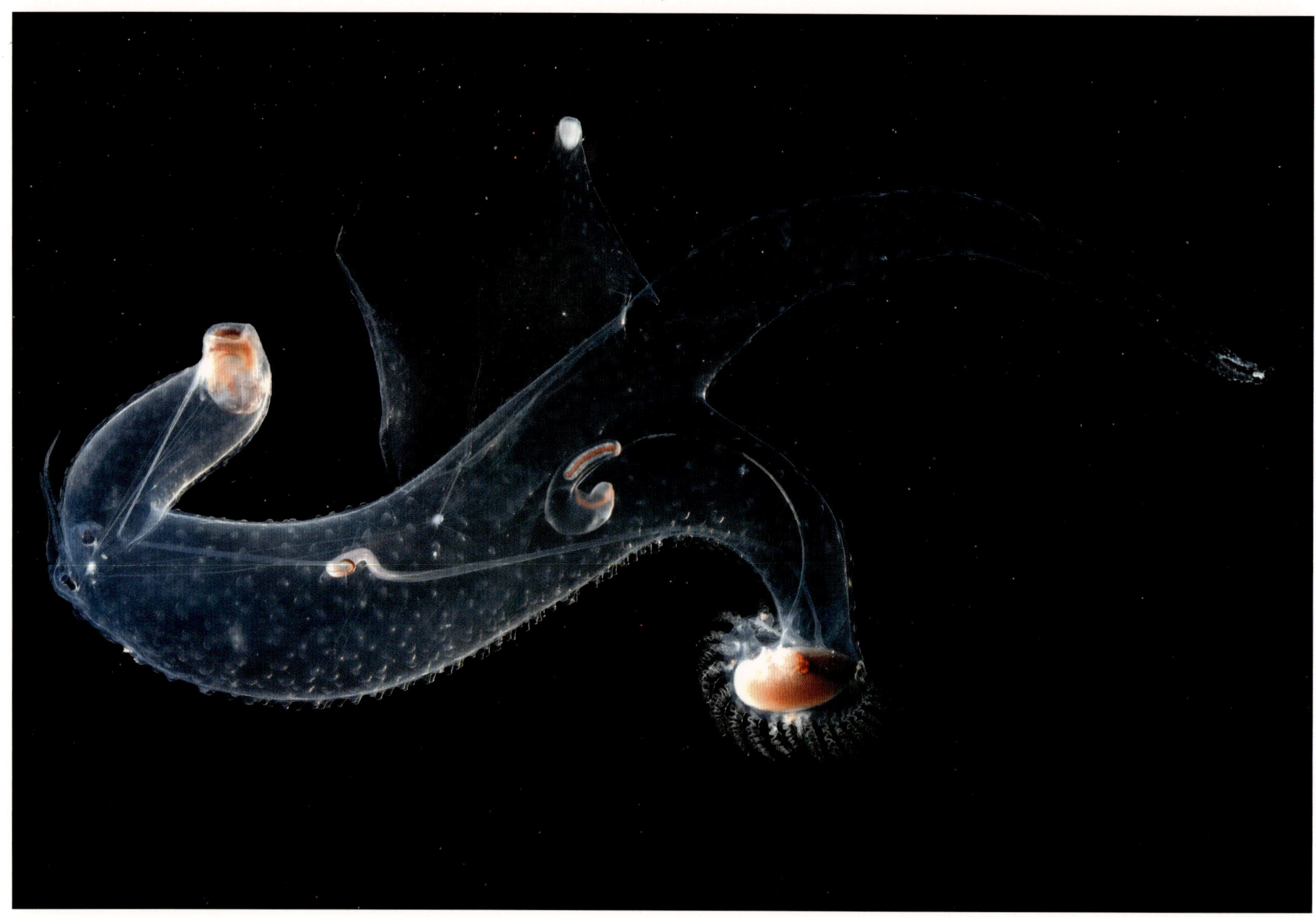

Cardiopoda

These heteropod snails look much like their close relative *Carinaria* (p. 29), except that they have gone one step further and completely lost their shells. Like other heteropods, they feed on various gelatinous animals, such as jellyfish and other swimming snails. They perform an unusual trick, which is especially uncommon for a transparent animal that seems strongly adapted for camouflage: when disturbed, these animals spread a multilobed red "flag" near the end of their bodies. This red object appears to have evolved to be quite visible in the open ocean, although at most depths it will appear black, not red. We have seen it unfurl this flag both when it strikes prey and when we prod it with a finger. This is probably acting as a "flash of darkness," in the same way that sudden bioluminescent flashes can be used to startle.

Vitreledonella richardi

The glass octopus goes to extraordinary lengths to remain clear. In addition to its transparent body, it has also modified its eyes and gut for further camouflage. Instead of the eyes being approximately spherical as usual, they are tubular. This allows them to have large lenses, which makes them more sensitive to light, while at the same time minimizing their visibility. On the downside, the tubular design limits their field of view, so their vision is a bit like looking through a telescope. They also minimize the appearance of their gut using the same mirrored-needle trick that the heteropods and other squid employ.

Together, these traits make the animal truly hard to find, which leads us to question: how do rare deep-sea animals find each other? We know from deep-sea video that these animals don't just spawn into the water but that they find each other to mate. The open ocean is vast, and these animals are uncommon, so—as with so many deep-sea animals—how they find each other remains a mystery.

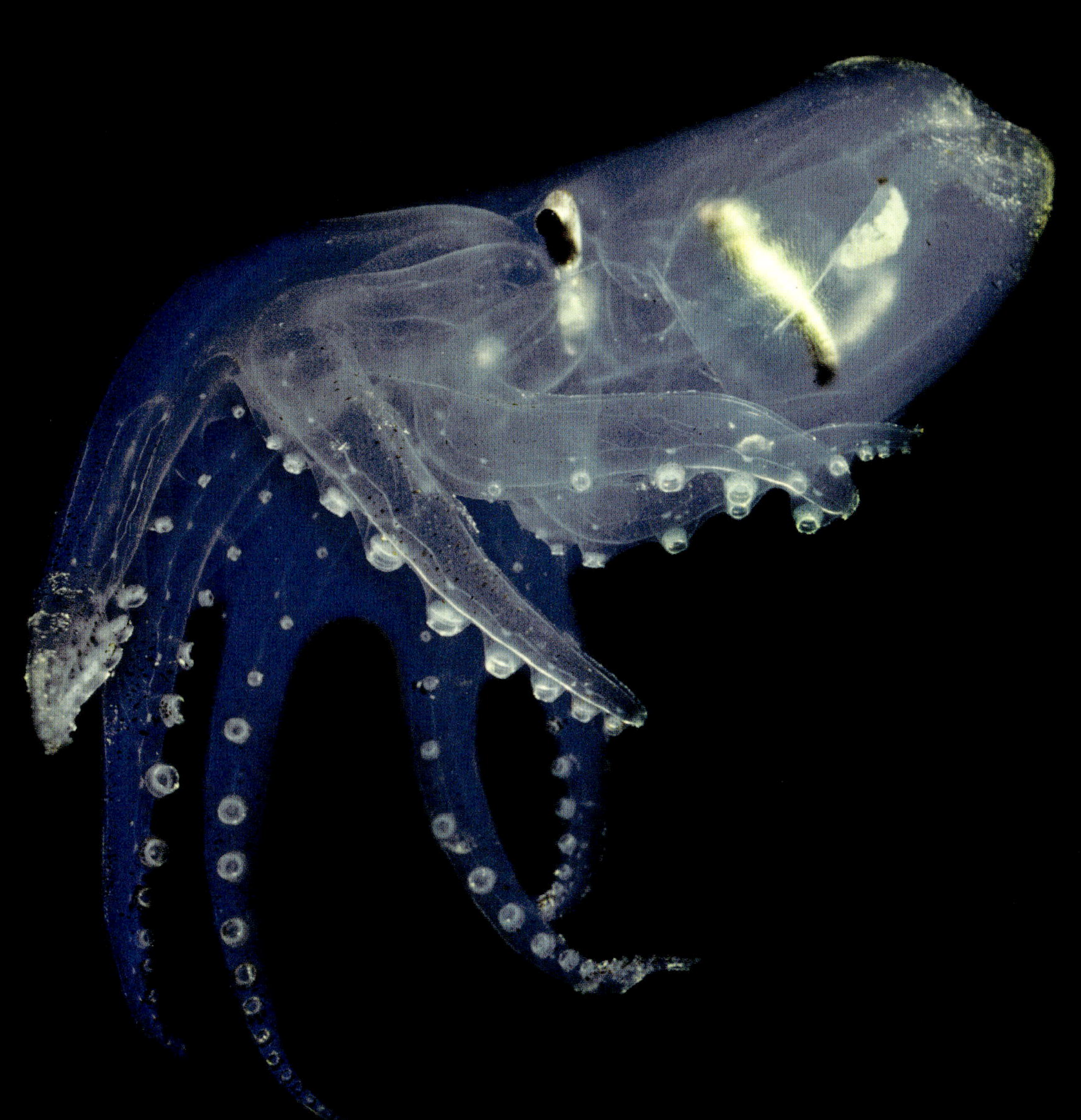

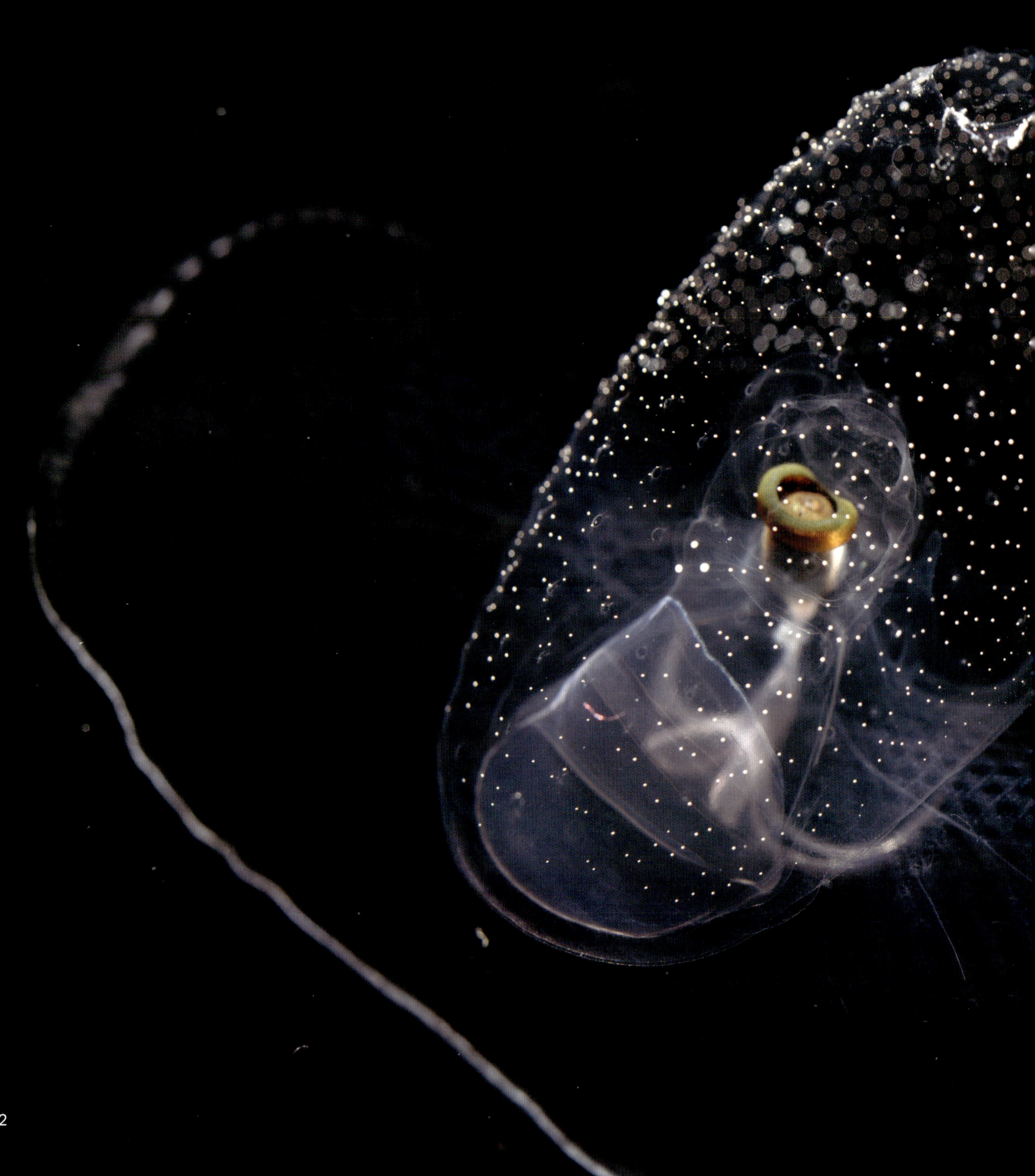

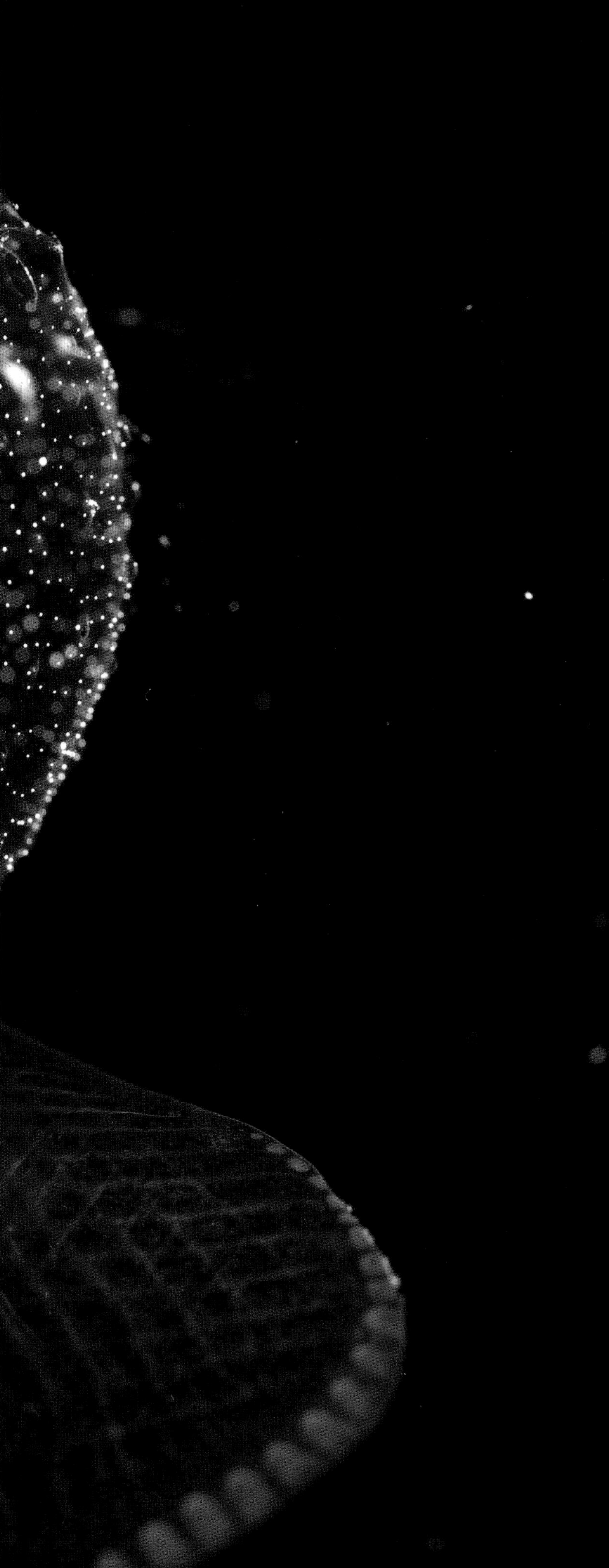

Corolla

This pteropod is another type of pelagic snail that evolved a swimming lifestyle independently of its heteropod cousins. In these pseudothecosomes ("false shell bearers"), the shell has become gelatinous and transparent. The crosshatched muscle fibers can be seen in the modified foot, which flaps like a pair of wings, giving these animals the nickname "sea butterflies." The white spots dotting the outside of the body are usually transparent too, but they can suddenly turn opaque when the animal is disturbed. This "blanching" is thought to be a daytime equivalent of a bioluminescent flash to make the animal suddenly appear larger.

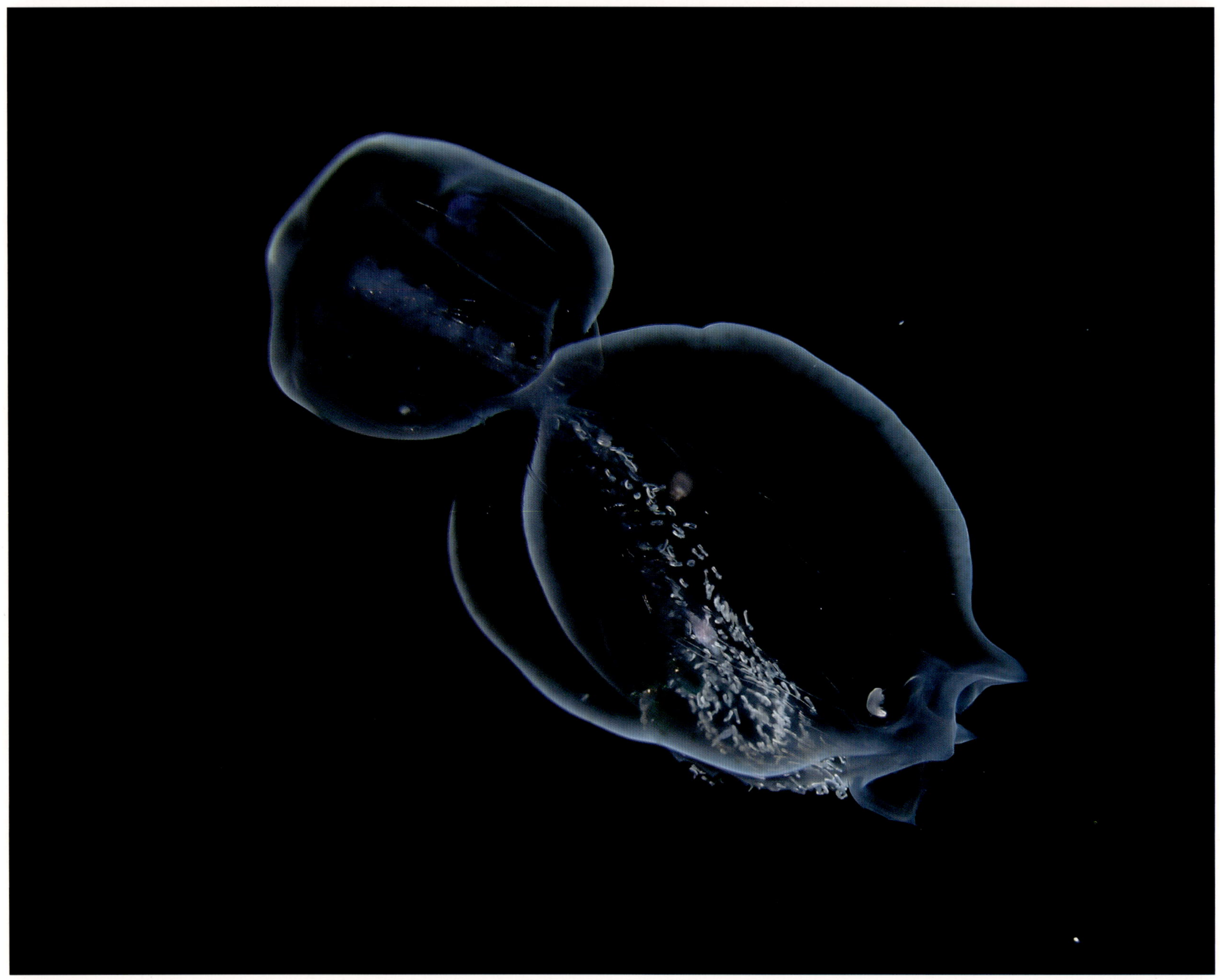

Abyla

Abyla is a type of siphonophore—a relative of hydromedusae, which are a type of jellyfish. They are usually completely transparent, but when touched they and other related siphonophores may take on a frosted appearance. It was shown that another siphonophore, *Hippopodius*, becomes opaque by "un-dissolving" a protein in their tissues. This process (known as precipitation) makes their bodies milky in the same way that precipitating water makes fog. Like most related siphonophores, *Abyla* can do both blanching and bioluminescence, so it has optical distractions ready for both day and night.

Pseudosagitta

Unlike many other animals in this book that have relatives on land or in freshwater, chaetognaths are only found in the ocean. For this reason, they are far less familiar to most, despite being abundant in their own habitat. They are commonly called "arrow worms," but are a distinct evolutionary group, not closely related to earthworms or other worm-shaped animals. Turns out worminess is one of evolution's most popular body shapes. Although they're only a few centimeters long, you wouldn't want to bump into one if you're a tiny drifter. They pounce and grab their prey with an imposing array of hooks around their head, and pass them into their mouth where they are paralyzed with tetrodotoxin—the same venom found in pufferfish. Their bodies are essentially invisible, but like most animals, they are betrayed by the opacity of their reproductive cells. On this *Pseudosagitta*, you can clearly see the clusters of eggs and the white strands are packets of sperm along the sides of its body.

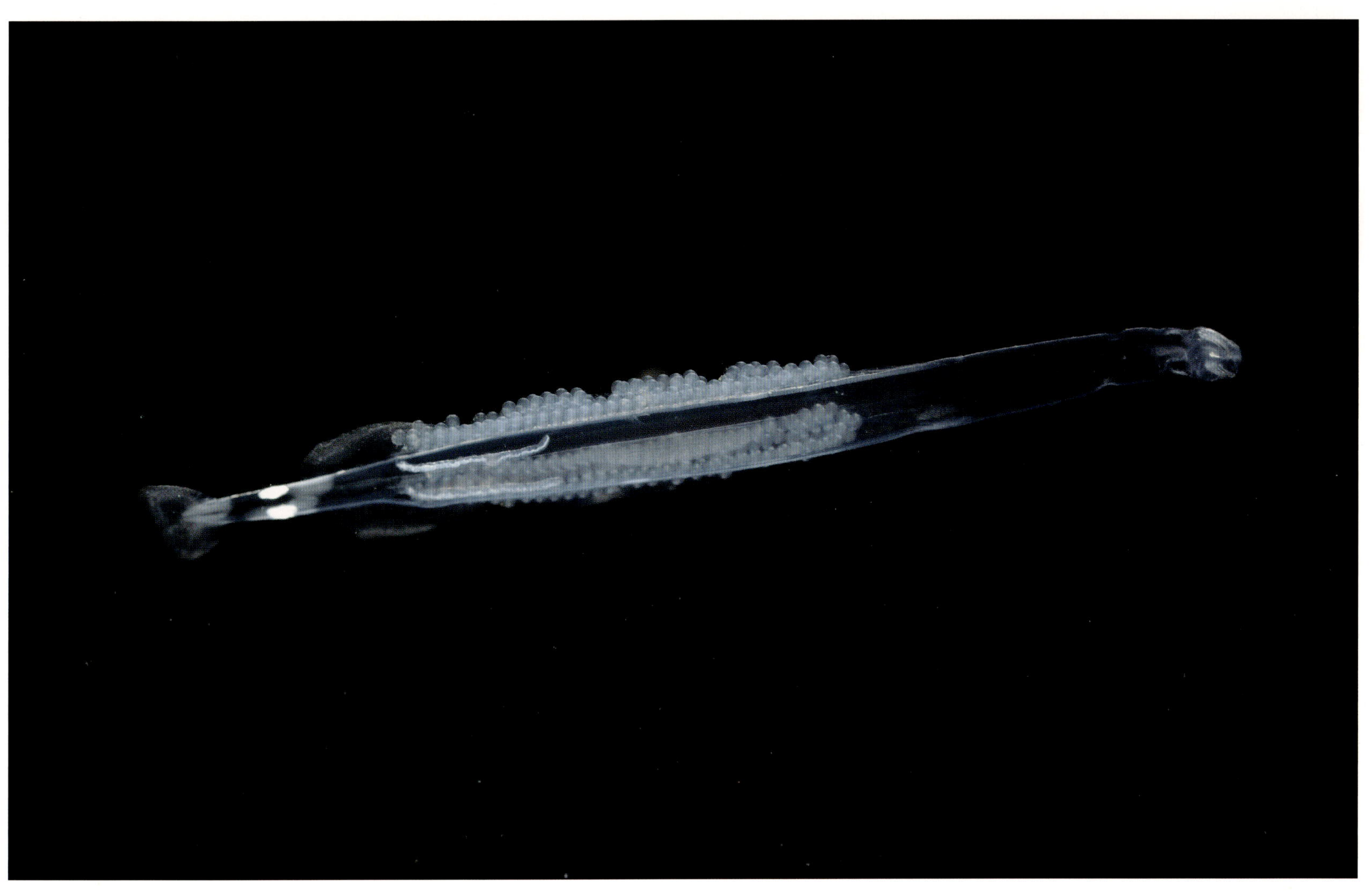

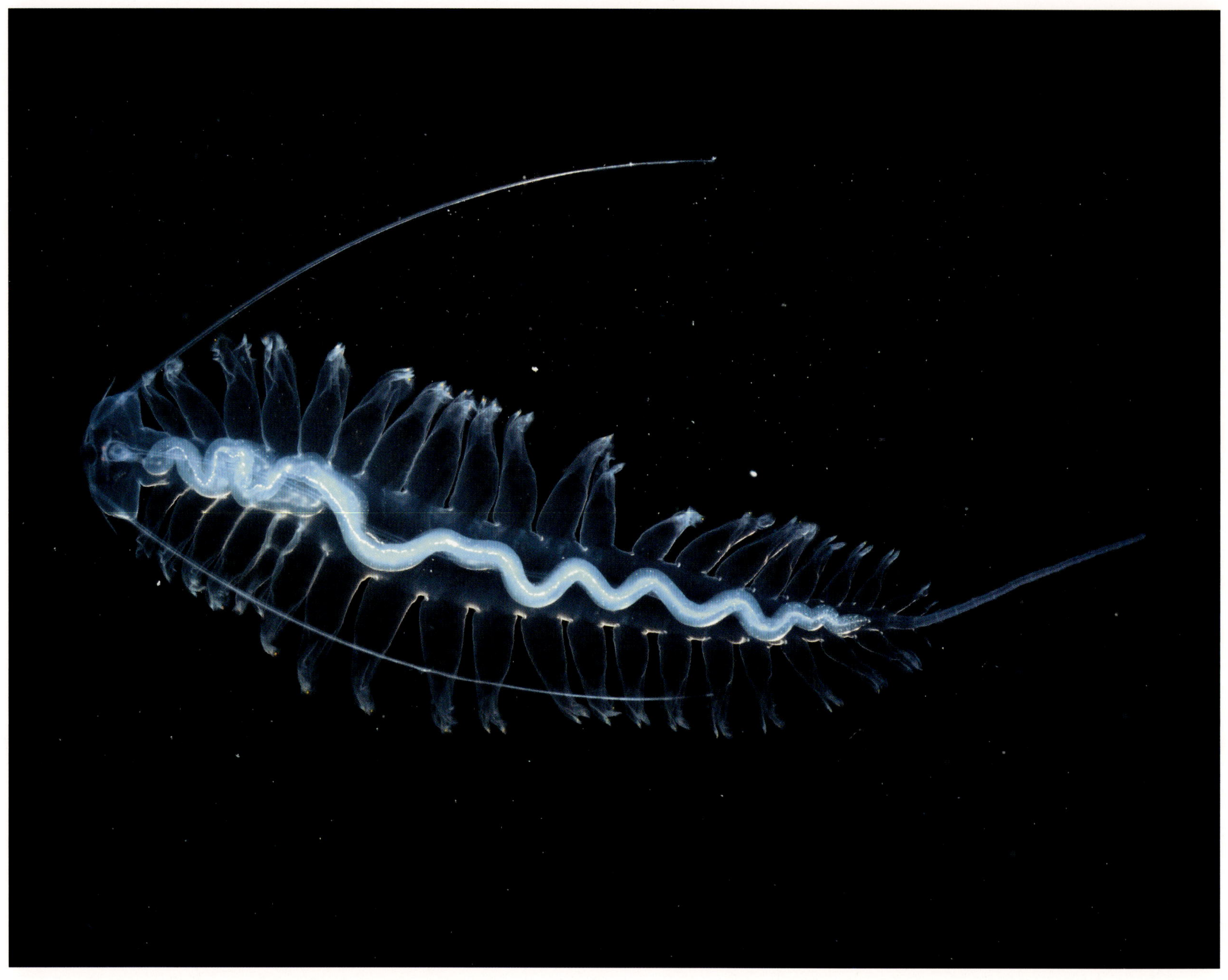

Tomopteris

The evocatively named gossamer worms are some of the most beautiful of the swimming oceanic worms. Some are small and clear; others can be over 30 cm long or have red guts. We have seen them consuming gelatinous prey; in this photo, the animal's transparency is compromised by the opacity of the meal that is crammed within the length of its gut. Although they are shown here for their striking transparency, they're also unusual in being one of the few animals that can emit yellow light (more about this in the Bioluminescence section).

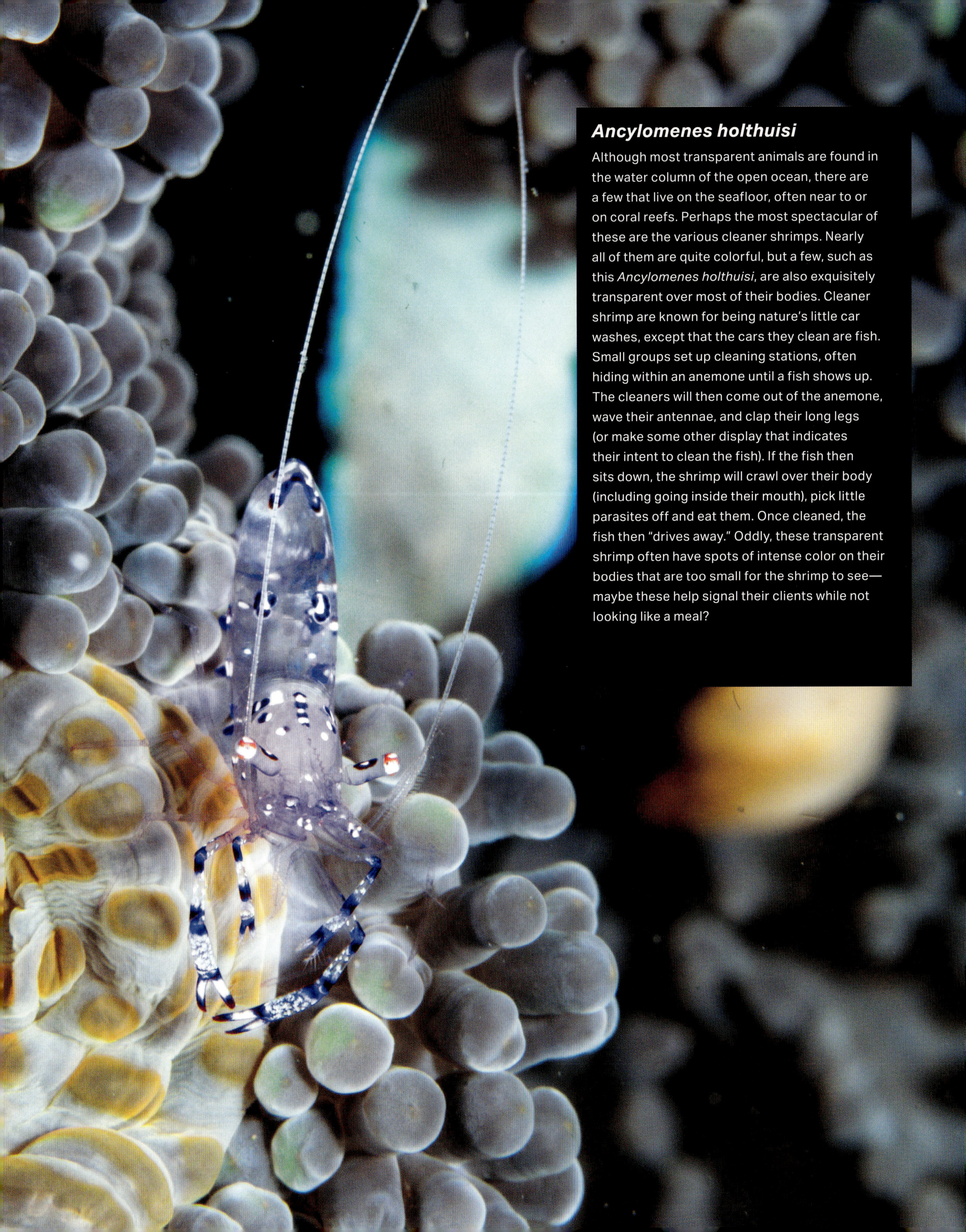

Ancylomenes holthuisi

Although most transparent animals are found in the water column of the open ocean, there are a few that live on the seafloor, often near to or on coral reefs. Perhaps the most spectacular of these are the various cleaner shrimps. Nearly all of them are quite colorful, but a few, such as this *Ancylomenes holthuisi*, are also exquisitely transparent over most of their bodies. Cleaner shrimp are known for being nature's little car washes, except that the cars they clean are fish. Small groups set up cleaning stations, often hiding within an anemone until a fish shows up. The cleaners will then come out of the anemone, wave their antennae, and clap their long legs (or make some other display that indicates their intent to clean the fish). If the fish then sits down, the shrimp will crawl over their body (including going inside their mouth), pick little parasites off and eat them. Once cleaned, the fish then "drives away." Oddly, these transparent shrimp often have spots of intense color on their bodies that are too small for the shrimp to see—maybe these help signal their clients while not looking like a meal?

Acorn worm larva

One of the most remarkable facts about animals that live in the ocean—even in tide pools—is that most of them have young (larvae) that don't look anything like the adults. In fact, the larvae of starfish, barnacles, snails, sea urchins, lobsters, and countless worms typically don't look like any animals in our normal experience, often being a tiny conglomeration of hairs, spikes, and gelatinous forms, sometimes with giant eyes. These larvae may float around in the sea for months until some environmental signal tells them to metamorphose, and then—like caterpillars becoming butterflies—they settle down and turn into their adult form. The species shown here is the drifting larva of an acorn worm, a little over 1 cm long. Early in life, they drift in the water but then turn into opaque yellow worms that burrow into the sediment.

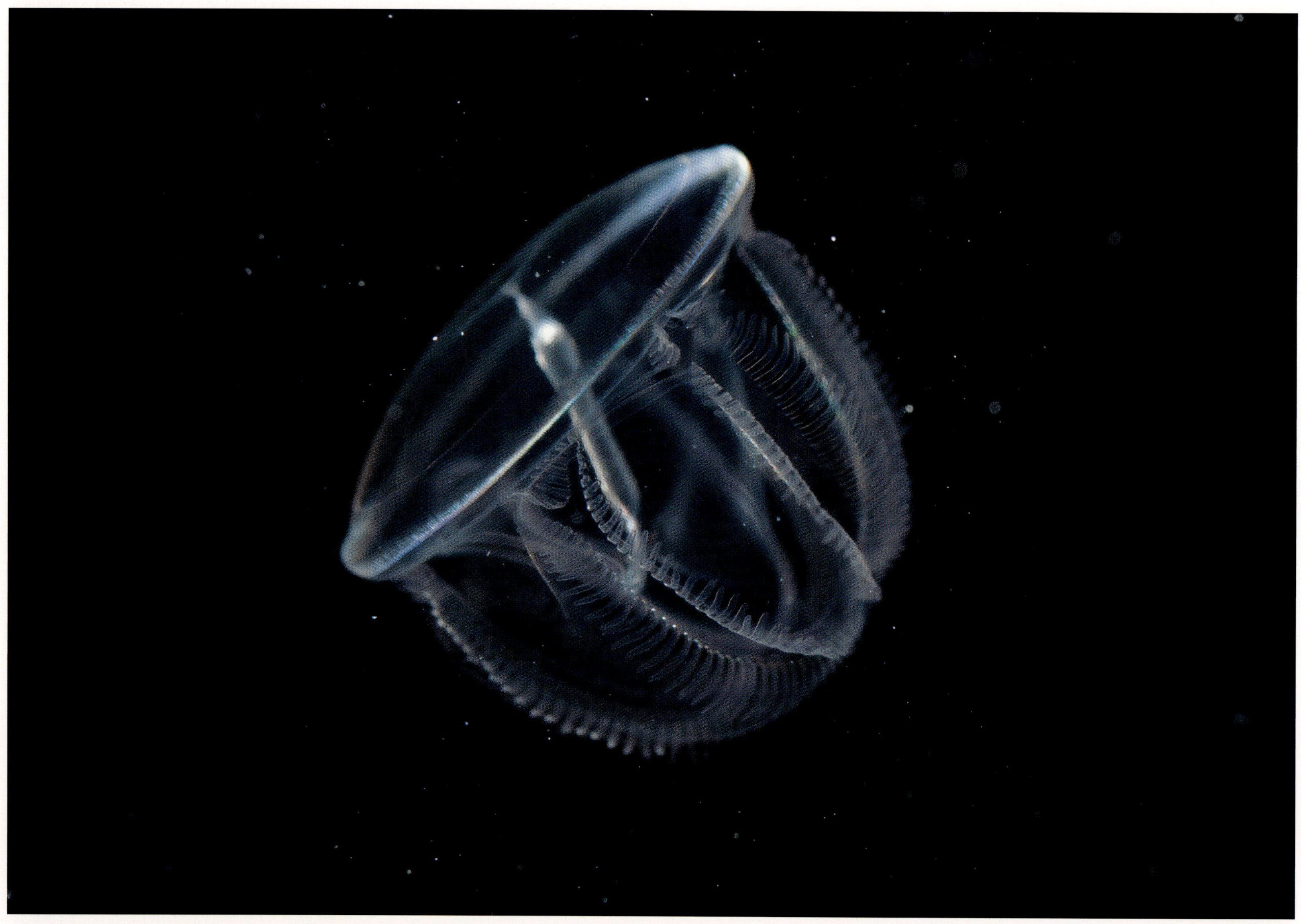

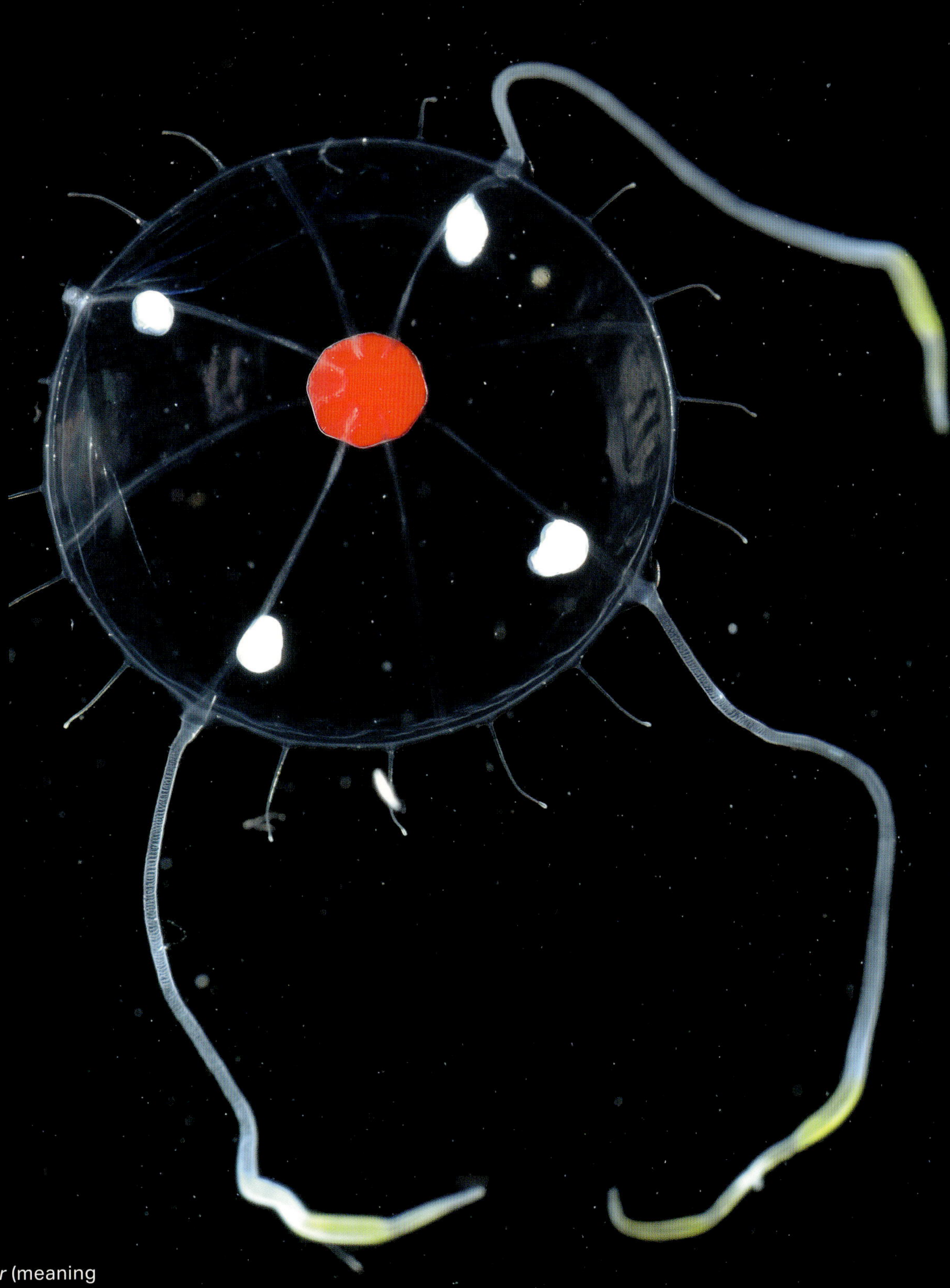

Tetrorchis erythrogaster

This hydromedusa *Tetrorchis erythrogaster* (meaning four gonads and red stomach) illustrates many of the challenges to achieving full transparency. The stomach is pigmented red to mask glowing prey, the gonads are opaque because they are packed with irreducible cells, and the tentacles are frosted with the complex stinging capsules that cnidarians use. This species vibrates those yellow tentacle tips, so there is likely some prey attraction going on as well. At least the bell is clear!

Fish larvae

We often see animals that are transparent except for a few strategically placed pigmented spots. These can sometimes be used to mask bioluminescent prey in the gut or to function as photosensitive pigments, but at times it appears that the pigmentation is used to mimic the appearance of another—perhaps less palatable—organism. For example, the extended chin barbel of this cusk-eel larva *Lamprogrammus* looks exactly like a strongly stinging siphonophore.

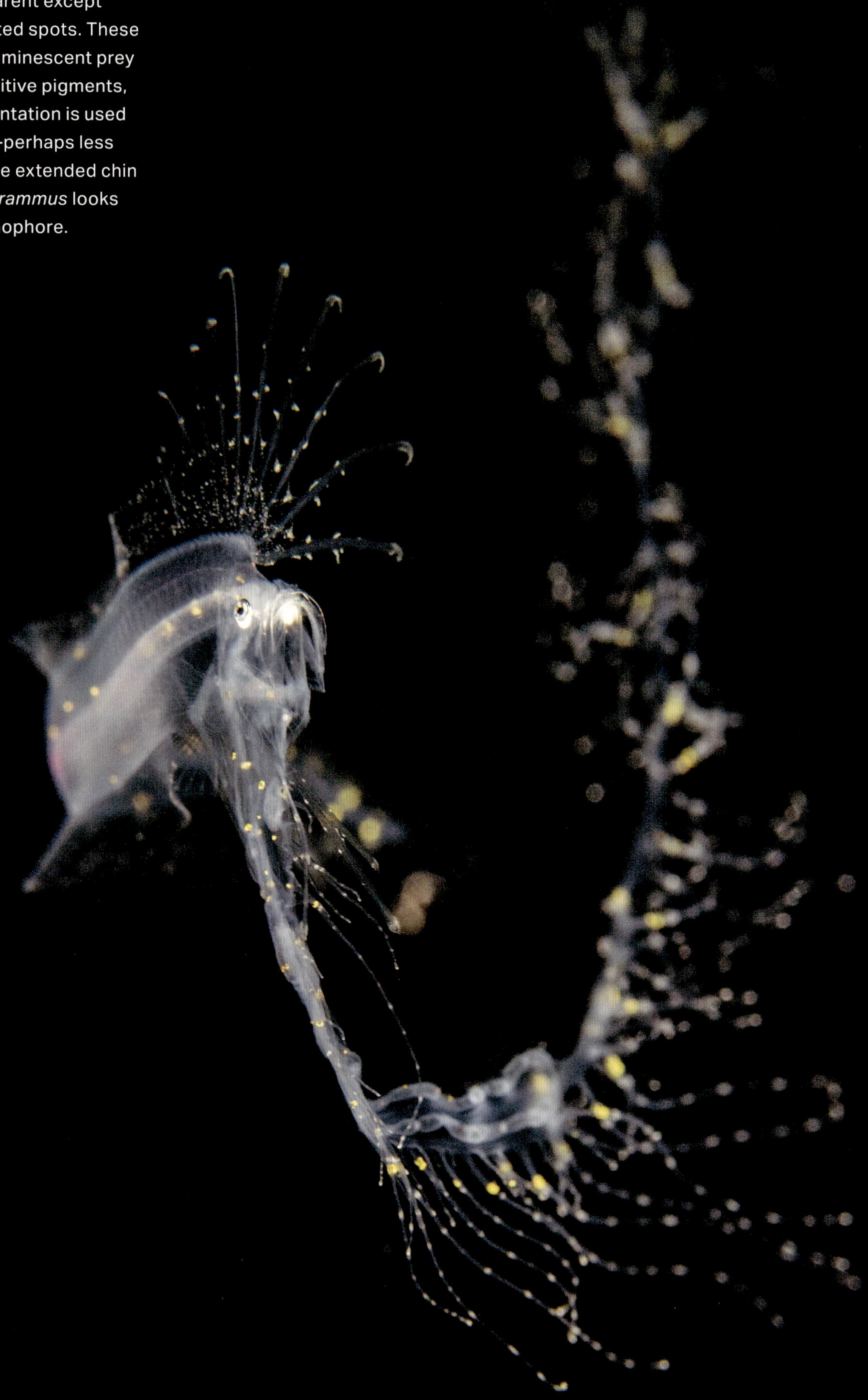

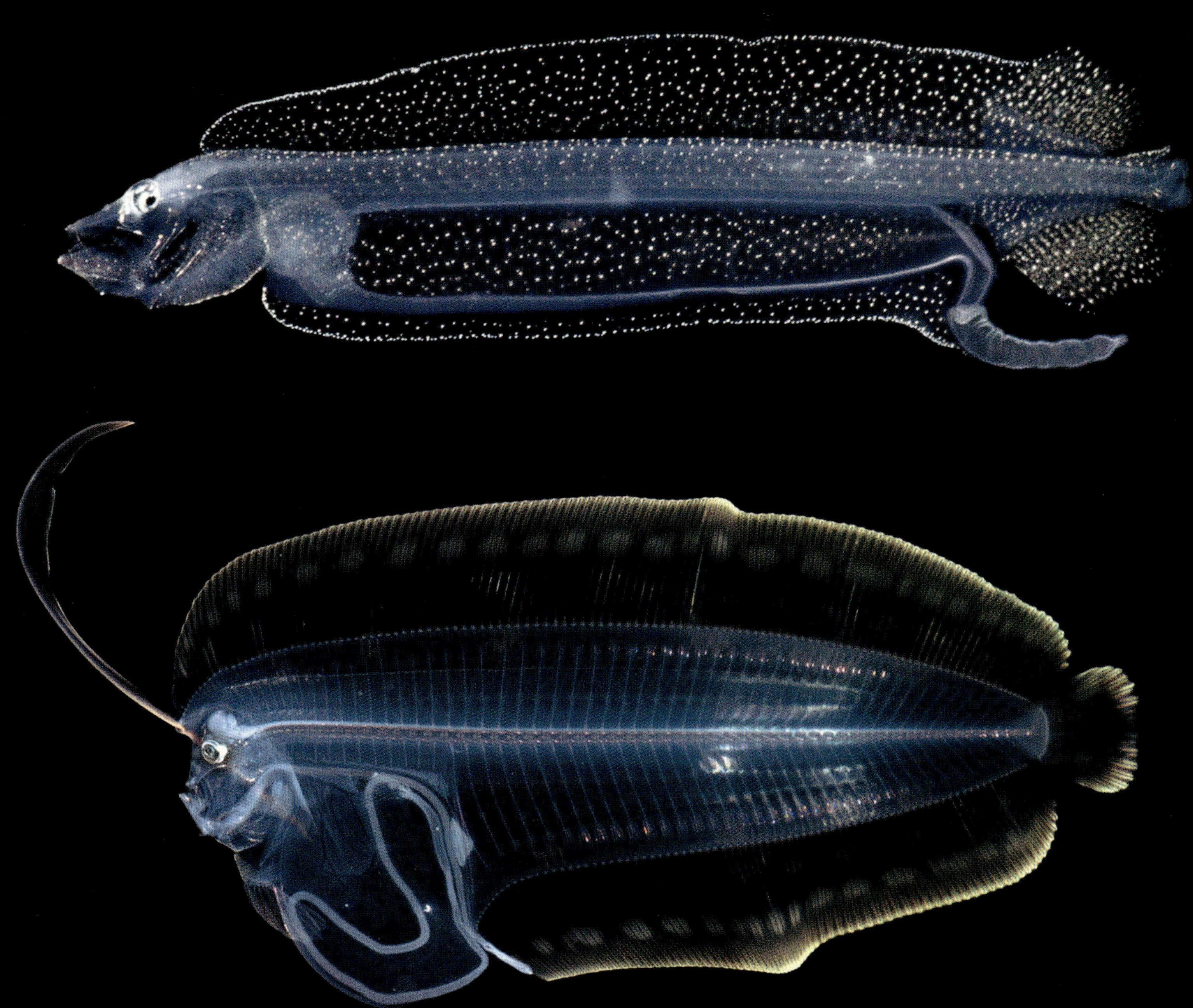

Larvae of many types of fish, all the way up to the giant oarfishes, look like an anatomical sketch of basic fish parts. You can see the layers of muscles, the meanderings of the digestive tract, the eyes, the gills, and some of the bones. During this vulnerable period of their lives, everyone is out for a meaty snack, so hiding in plain sight becomes a popular strategy. Later in the book, you will see what this juvenile *Bathophilus* (above) grows to be, and this young flounder *Chascanopsetta* (right) will transform both its body shape and pigmentation as it develops.

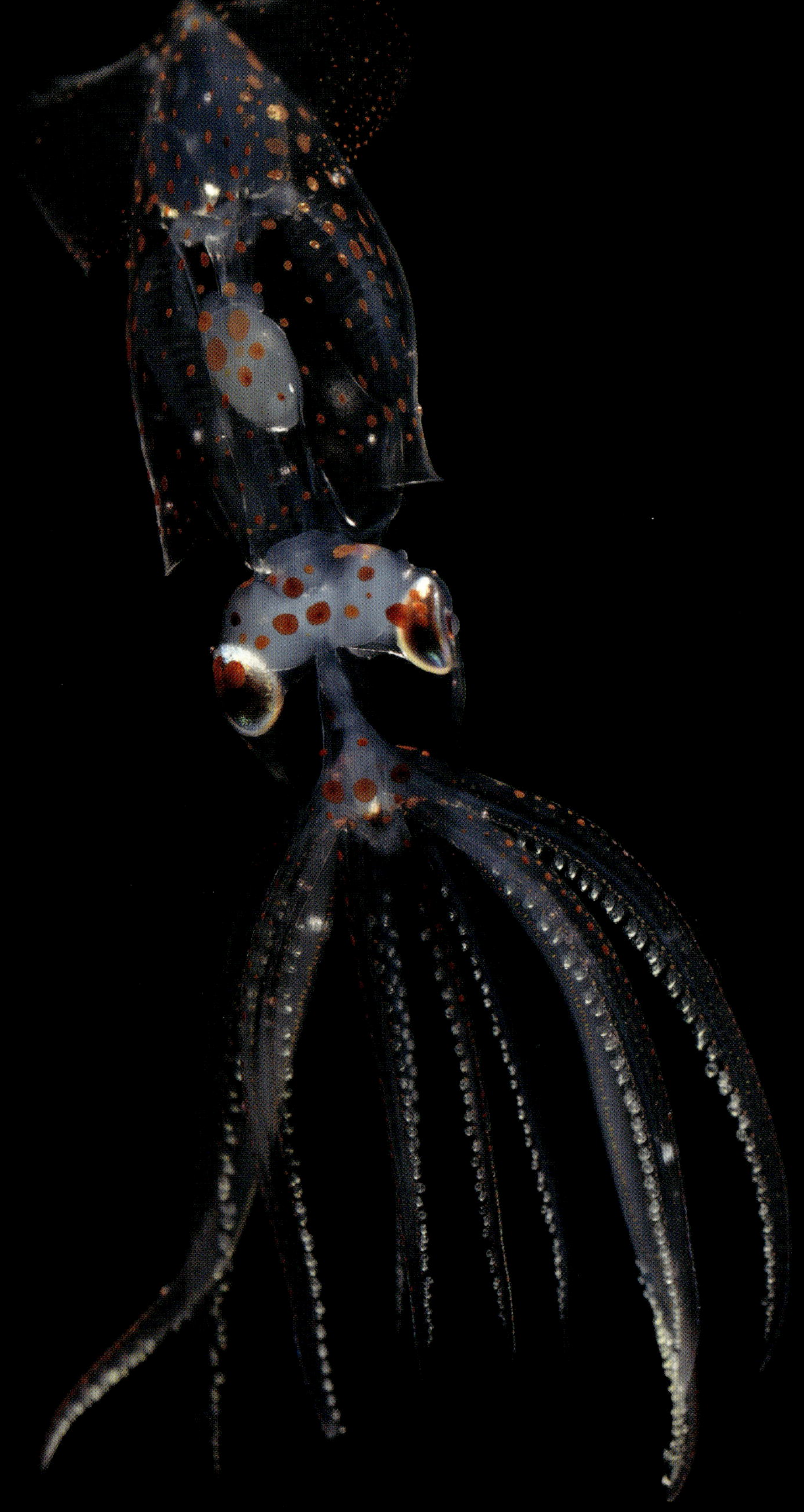

Ancistrocheirus lesueurii

Although the adult of this gorgeous sharp-eared enope squid is opaque, younger animals are crystal clear. They do still have the colored circles (chromatophores) that allow squid and octopuses to rapidly change color, but in the young these are few and far between. These can rapidly alter their transparency—we observed the chromatophores expanding (as in the photo) and contracting in sync with the pulses of their conical mantle. Like many oceanic squid, these animals have hooks instead of suckers on their arms and are covered with light organs. Lots of animals in the open ocean would love to eat these muscular squid, so being inconspicuous is one of their best defenses. At least they have a couple of years before they have to take on their adult responsibilities, like avoiding being eaten by sperm whales.

Helicocranchia

It's hard not to project cuteness onto this squid and interpret that the opening of its siphon is a little smiling mouth. In fact, its mouth is hidden at the base of its short arms, and the two "horns" are its long tentacles. This squid illustrates two of the strategies that animals use to hide their non-transparent parts as well as they can. Because eyes need to absorb light to function, squid will either wrap them in a mirrored bag or replace the absorbed light with a bioluminescent light of their own. It's also safer to have an opaque stomach than to swim around with a glowing "Eat here" sign, so to minimize the silhouette they keep the stomach (here orangey brown) oriented vertically. It may also be swathed with a reflective coating.

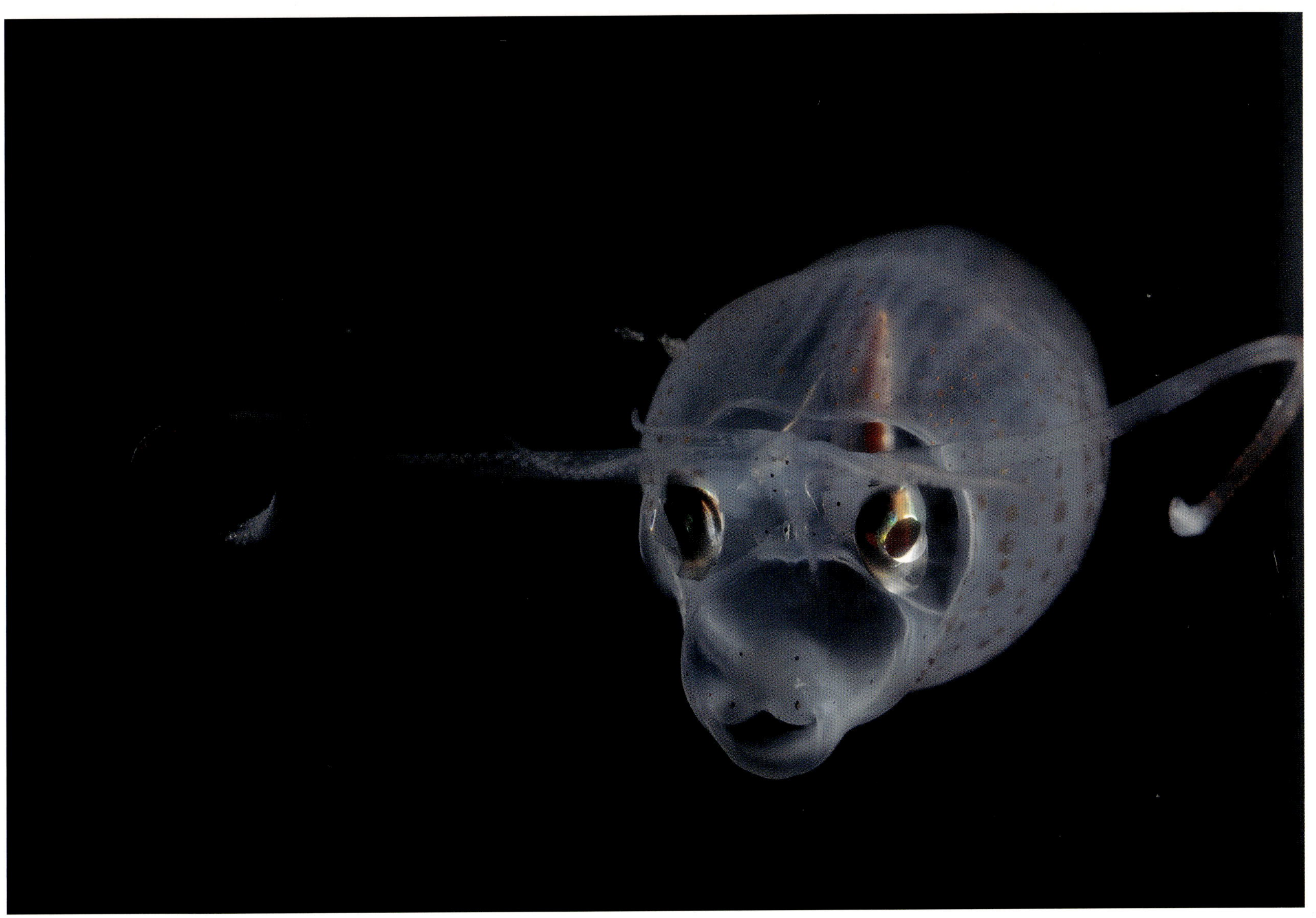

Bargmannia

Siphonophores divide the functions of life—feeding, swimming, reproduction, and defense—between different "zooids" along the length of a central tube. This deep-sea species is turbocharged for propulsion, with an array of seventeen swimming bells that contract in coordinated pulses to spread out the feeding tentacles and get the animal moving through the water. In some siphonophores, the trailing portion can be at least 20 meters long, but those species usually have *fewer* swimming bells. It's safer to gradually get up to speed; a sudden yank could tear the long stem in half, leaving the swimming section without a way to feed. In the deep sea, the pace of life is generally slow—unless you're escaping a predator who has seen through your camouflage.

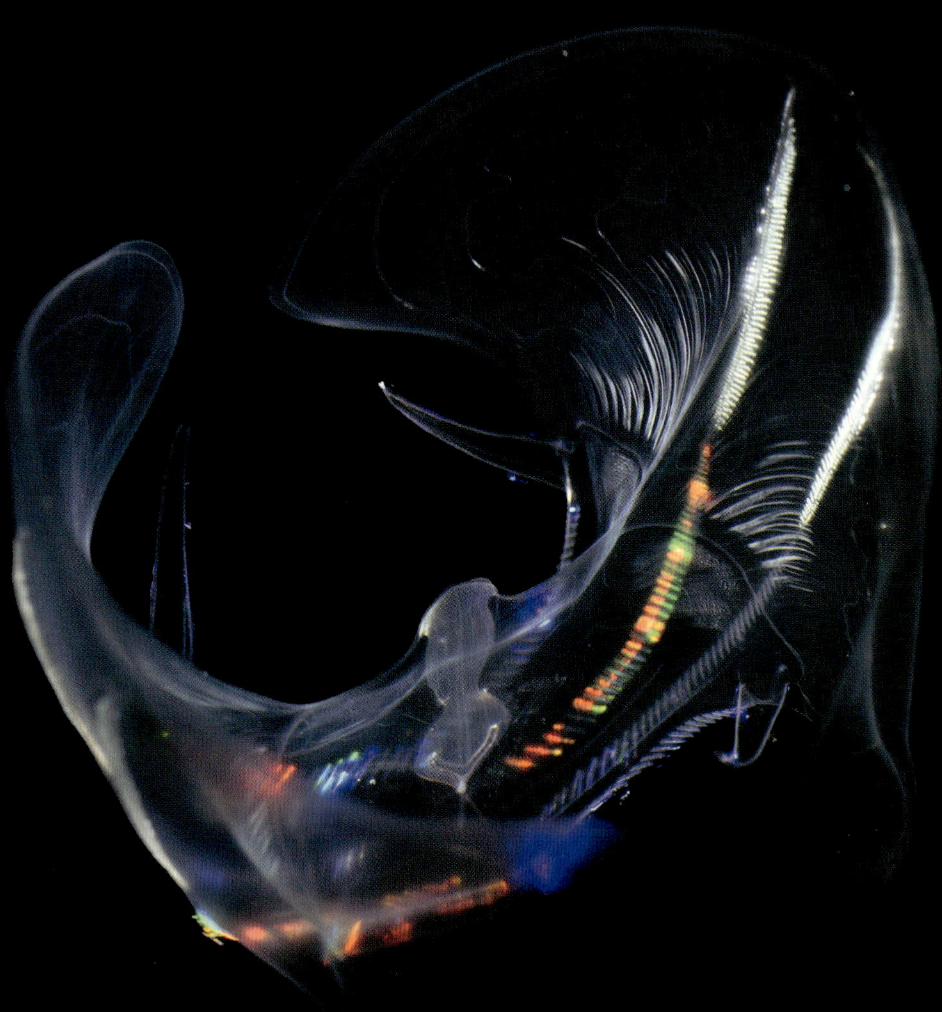

Ocyropsis maculata

Ctenophores are so named for their eight rows of combs, which are in turn made of rows of tiny hairs called cilia. By beating in synchrony, this whole system can move the animal through the water. In fact, these are the largest animals to use ciliary propulsion.
The ctenophore *Ocyropsis* is a good illustration of the challenges to being transparent. It tries extremely hard to vanish, but can't mask the frosty tubes containing its reproductive cells. When disturbed, it also moves by flapping its muscular lobes, and the sturdy mesh of muscle fibers give it a frosty appearance. One species is named *O. crystallina* for its crystal clear appearance, while another species is called *O. maculata* for the dark spots on the tips of its lobes. The reason for these dark patches, also found in the ctenophore *Cestum*, are still unknown.

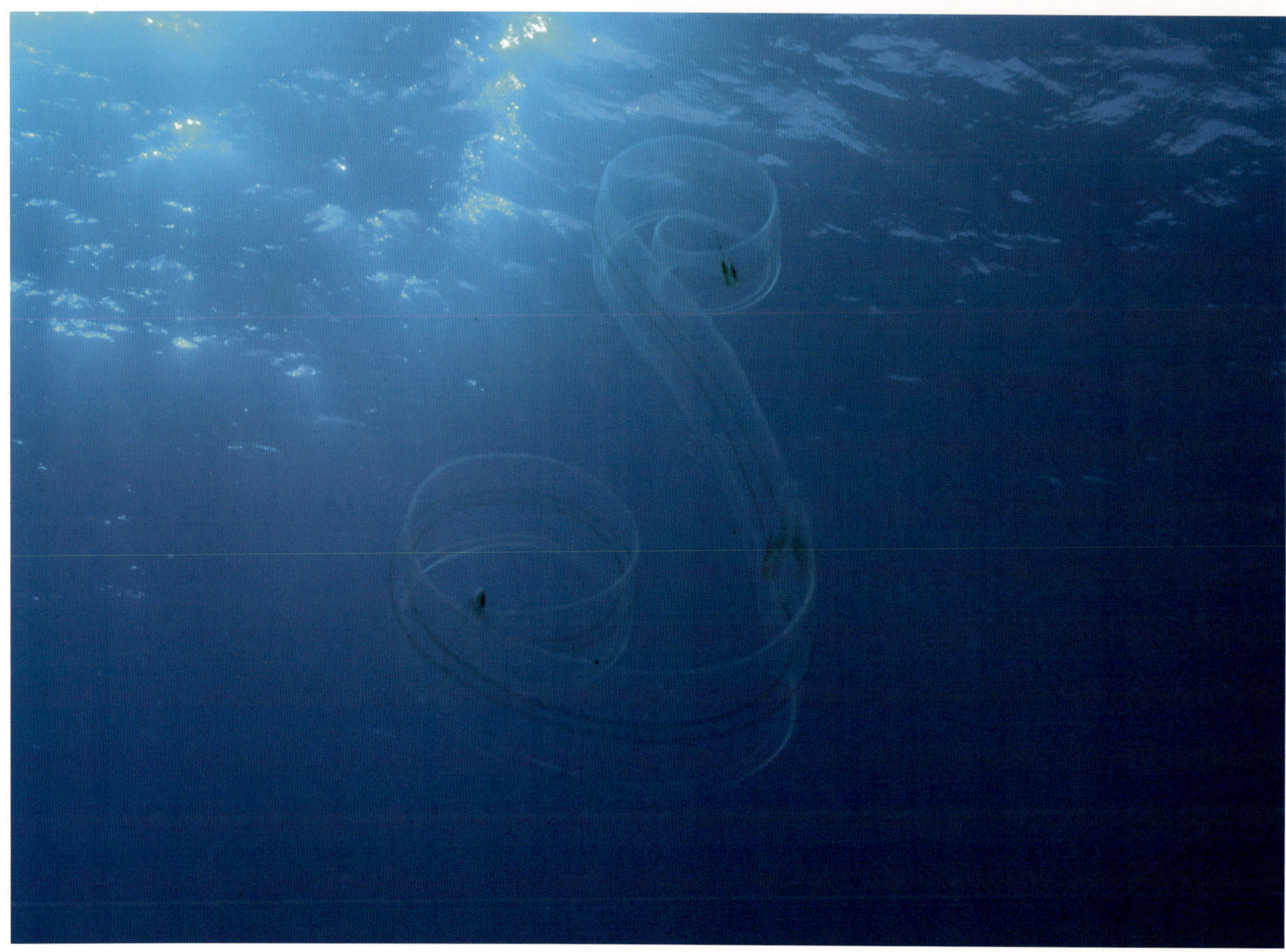

Cestum

To get most of the images in this book, we carefully captured specimens and then photographed them in the lab under controlled conditions. This photo shows the ribbon-like ctenophore *Cestum* in its natural habitat, bathed in sunbeams and surrounded by infinite blue. This upward-looking angle is often how we find transparent animals, by getting the lighting at just the right angle to highlight the edges.

Thalassocalyce

Steve: On first seeing a *Thalassocalyce*, collected from a submarine, I wrote in my notebook: "Frilly, brainy looking." That description has stood the test of time. If not for the sinuous canals and eggs, there would be almost nothing visible with this comb jelly. Its name has a nice double meaning of "sea cup." *Thalassa* is another Greek name for Thetys, goddess of the sea, and *Calyce* is the name of the water nymph from mythology. Unlike most ctenophores, which feed with tentacles, lobes, or by chomping, *Thalassocalyce* is almost like a Venus flytrap. It creates an open cup shape, and if a shrimp ventures inside, the ctenophore snaps the opening closed. Rather than tangling or overpowering its prey, *Thalassocalyce* smothers it by remaining pliable and damping out the impact of the escape responses. We've found this adaptation in many unrelated midwater animals, including a doliolid, a jellyfish, and even a swimming nudibranch (a type of sea slug).

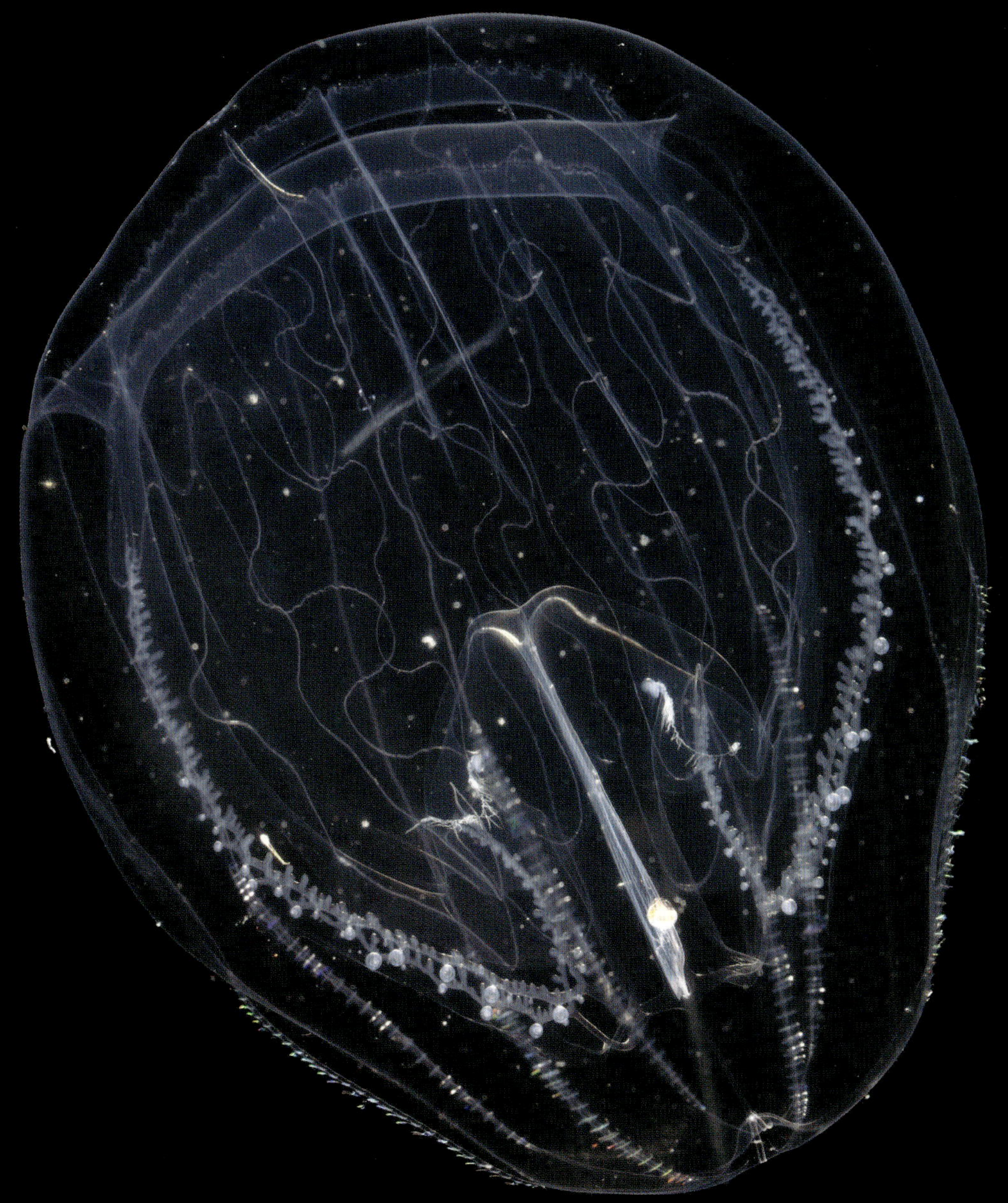

Ocyropsis canals

One of the most enjoyable things about working on transparent gelatinous animals is that we can see their inner workings—the twistings and turnings of their canals—despite the fact that their fragility can make them challenging to observe intact. We've even done a study based on the prey items that we could see inside transparent stomachs. The base of this ctenophore *Ocyropsis* is a hub of digestive tubes and ciliary conduits, all surrounding the central white dot of a gravity-sensing organ that tells the animal up from down in its fully 3D environment.

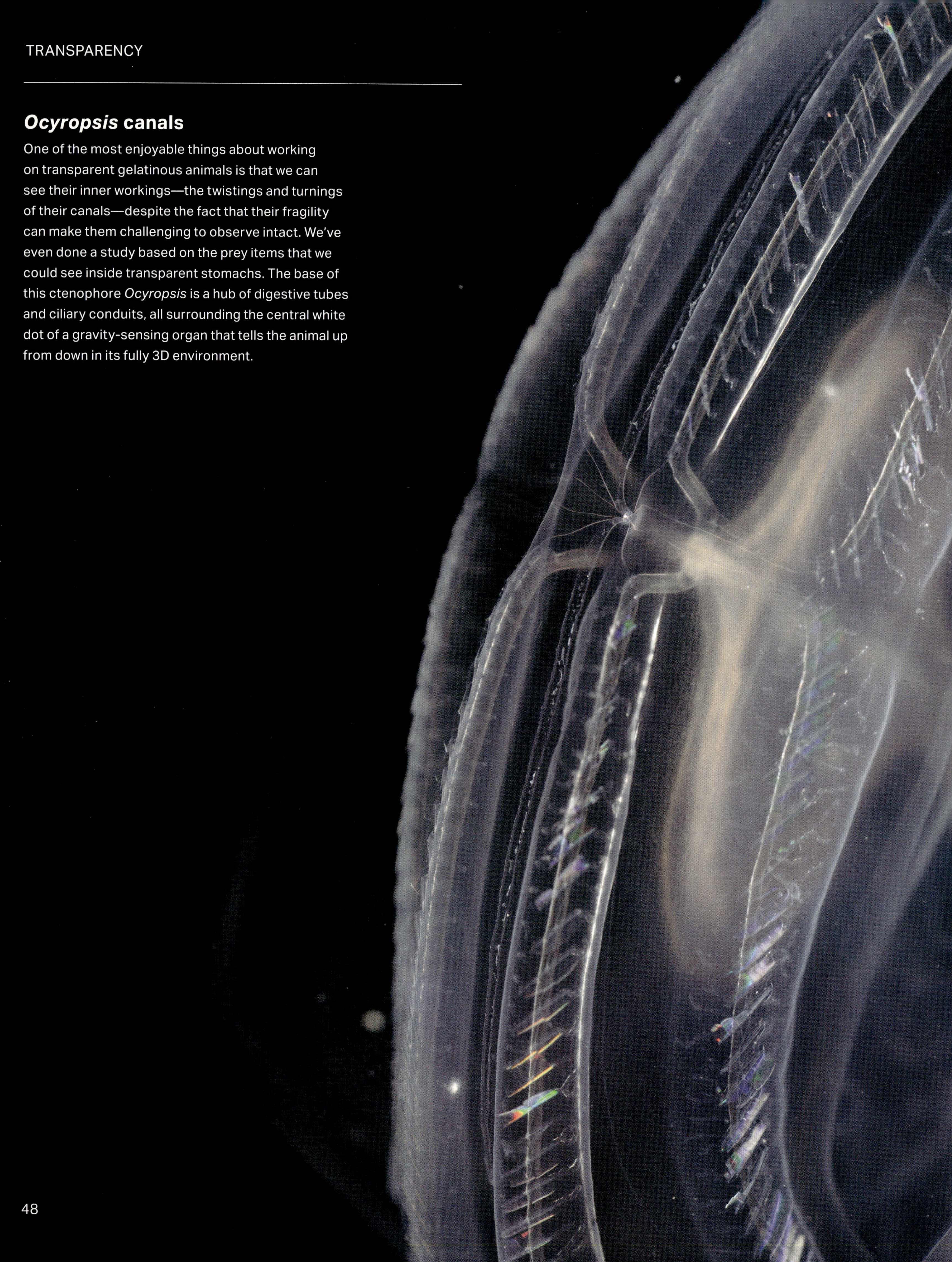

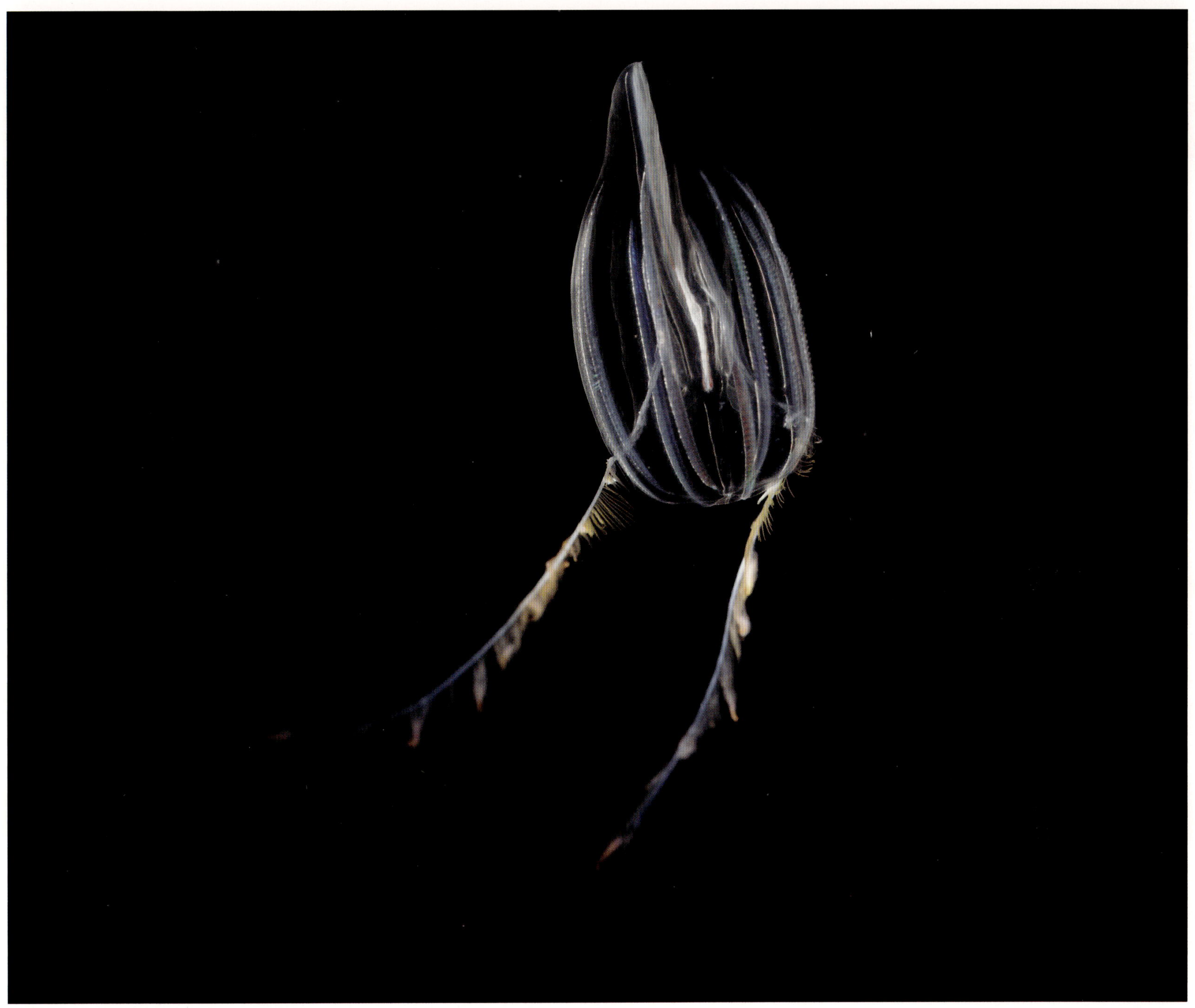

Hormiphora palmata

At times organisms might break their transparency "on purpose." This ctenophore extends a net of sticky tentacles to trap prey, much like a spiderweb. But some of the side branches of the tentacles have an unusual shape. In other animals, conspicuous appendages are used to attract prey, and maybe the distinctive tentacles of this ctenophore also enhance its visual appeal to potential prey.

which we named for molecular biologist Lynne Christianson, has a transparent swimming bell with a few fine digestive canals running across it. These canals provide nutrients to the thin sheet of muscle that pulses to propel the animal through water. Nutrients are provided by prey captured by the orange stomachs strung along behind.

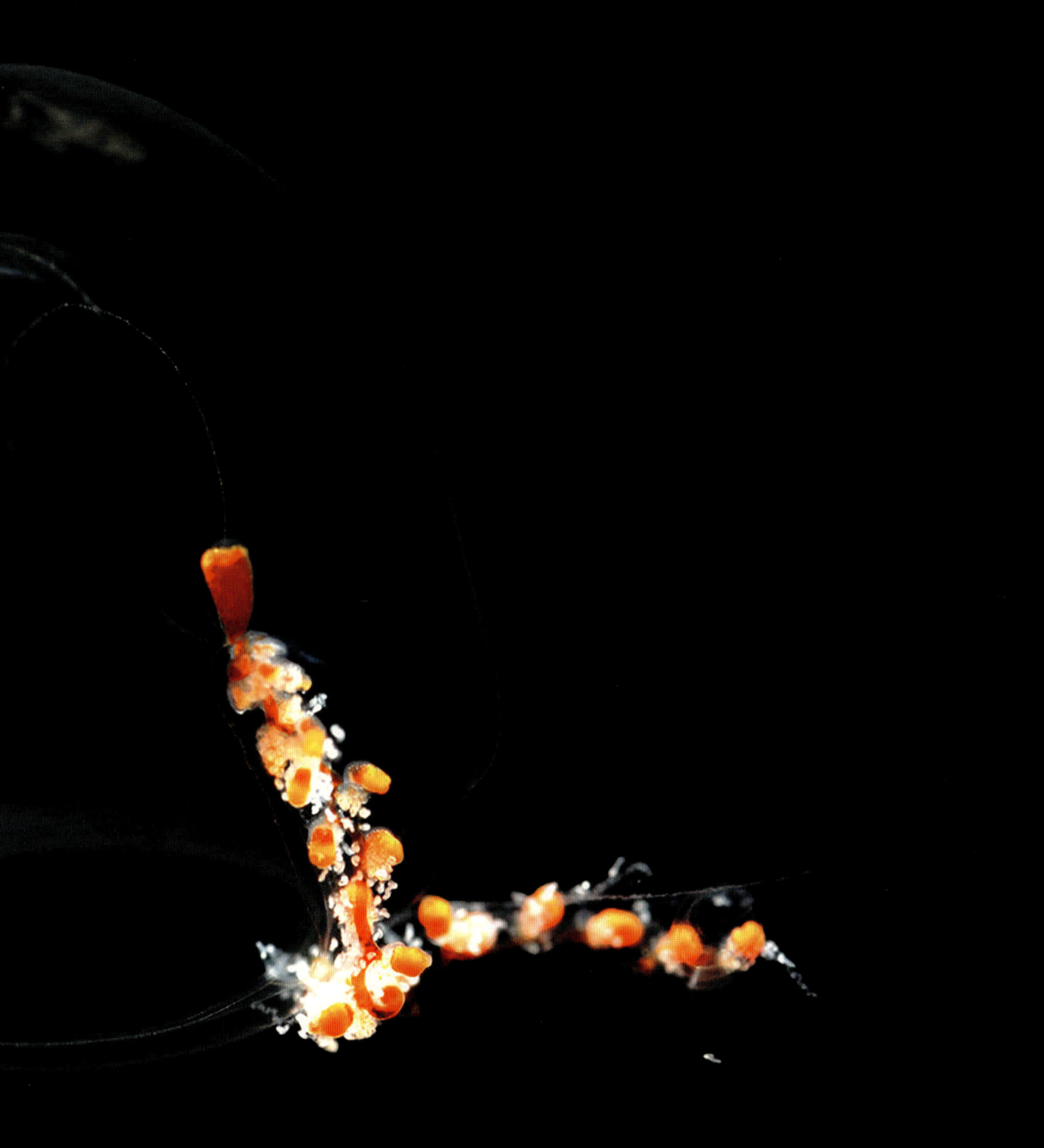

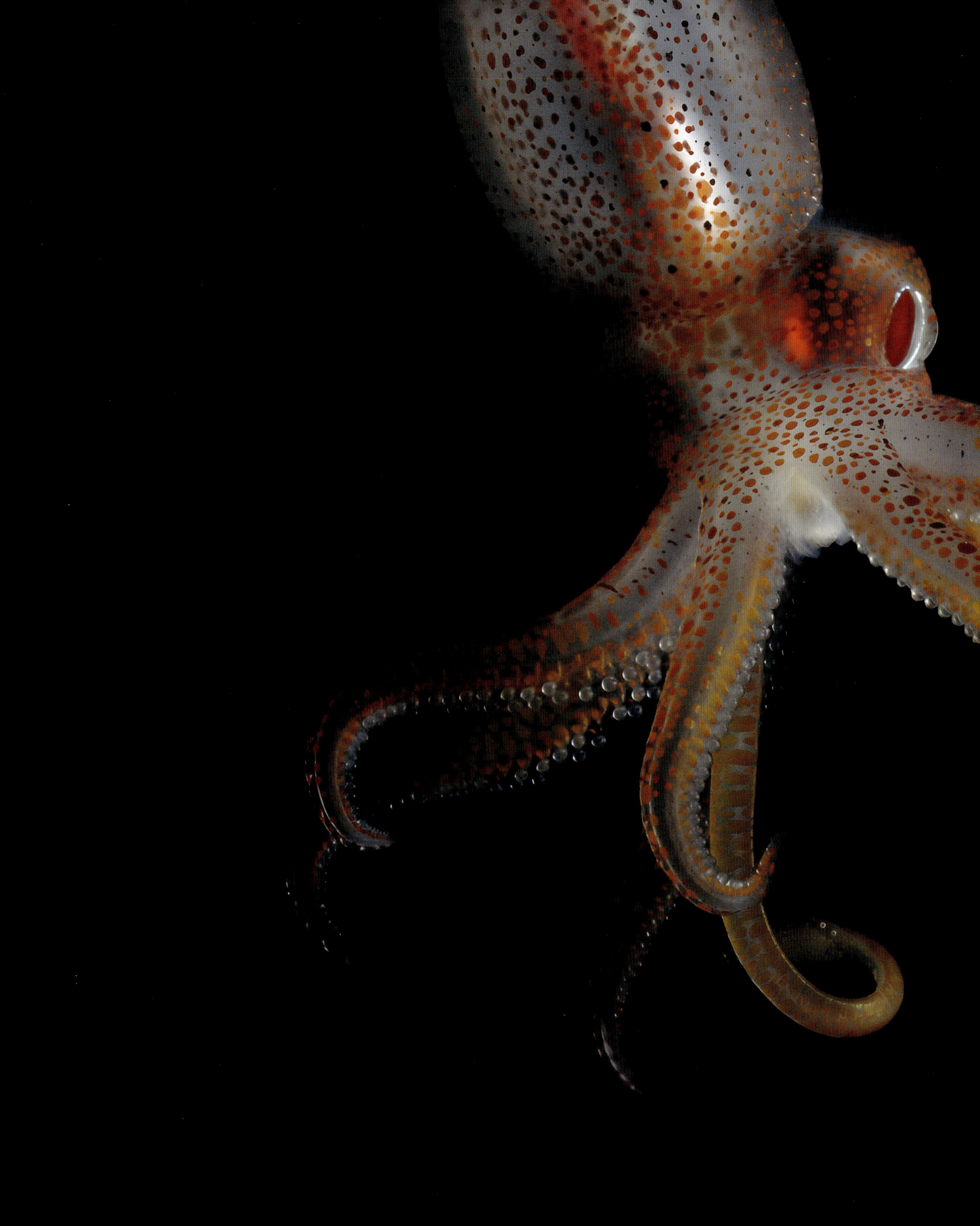

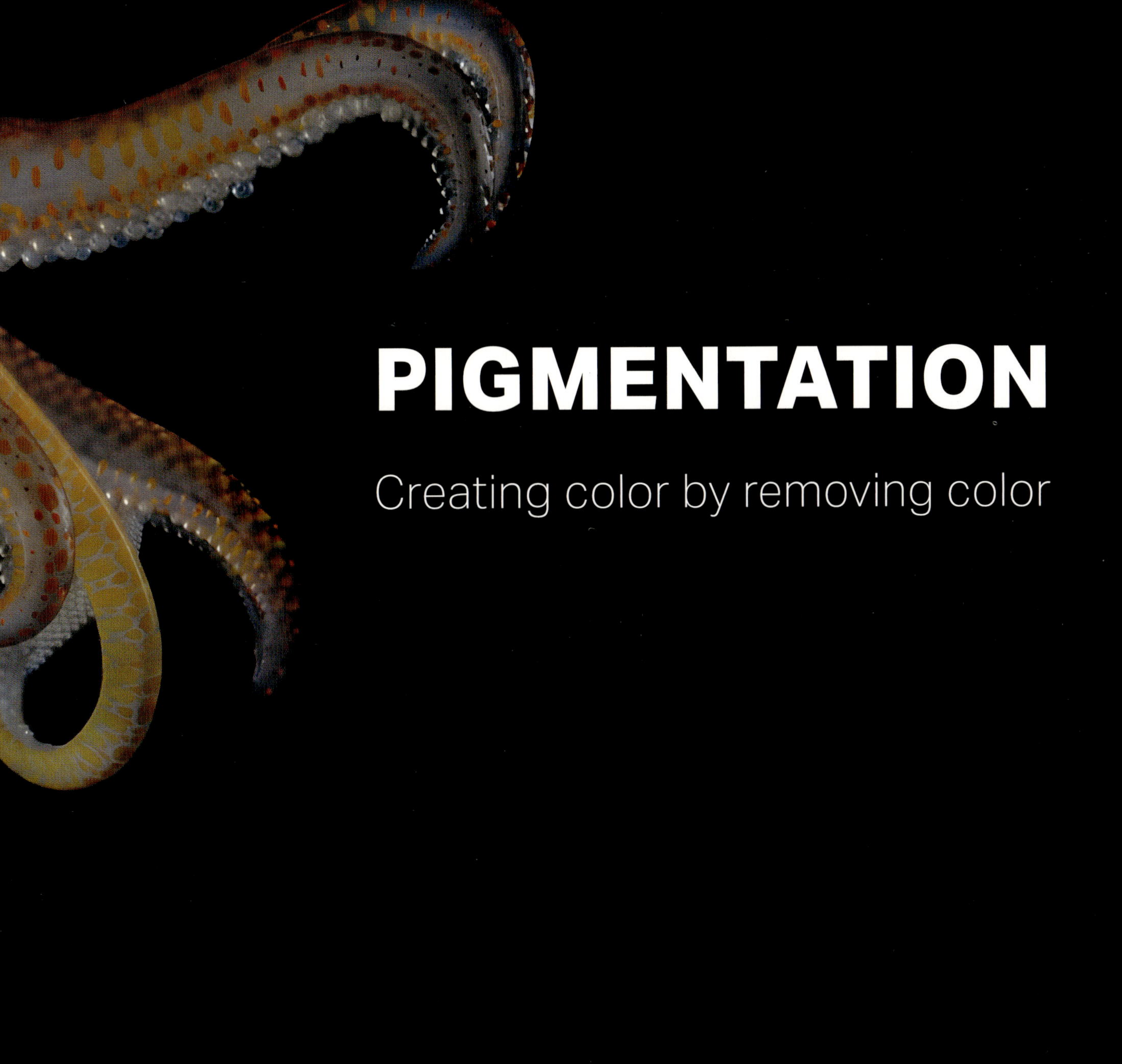

PIGMENTATION

Creating color by removing color

There is no such thing as a white photon. What we perceive as white light is actually a blend of brilliant single-color photons, like a bowl of multicolored candy. If you could zoom way, way in (and still perceive color) and slow time way, way down, when you looked at a white surface you would see it being pelted by a technicolor hailstorm, with much of the hail bouncing back up into the air. If you turned your attention to a blue object, you would see all the colors striking it, but most of those that bounced back to your eyes would look blue to you. Many of the colors we see in our daily lives—from our clothes to magazines, the trees, and the bees (though not our smartphones)—are the result of pigments: chemical compounds that absorb certain colors of light more than others. Photons interact with materials in different ways depending on their wavelength, which we perceive as color.

For pigments to function best, the starting point needs to appear to be white light, but there is no absolute white. The white from a clear sky, a foggy day, or the smiling face of a full moon all have different relative amounts of colors across the spectrum. Your computer screen tricks you into perceiving white by serving up carefully calculated portions of vivid green, red, and blue photons. Different tints of white light—like the offerings you get when buying a light bulb—are described as "warm" (if they appear more orange) or "cool" (if they appear more blue). Paradoxically, these terms are backward from the actual temperature of a warm or cool light source: the blue stars you see in the night sky are far hotter than the red stars. The reason that the properties of white light are important is that pigmentation is a *subtractive* process. It can only reveal a subset of the colors that are illuminating a scene.

WITH PIGMENTS, LESS IS MORE

To understand pigmentation you must know that the colors you see are those that are *not* absorbed by the pigments. Green is the very symbol of growth and vitality. Fresh green leaves indicate the process of photosynthesis, where the power of light turns water and carbon dioxide into sugar and oxygen. The poetic twist is that chlorophyll—the pigment that captures light to drive the reaction—doesn't care a whit about green light. Most plant chlorophylls absorb blue and red light from the ends of the spectrum, and the green light we see is what is leftover in between. (Other photosynthetic organisms in the sea have pigments that do absorb green light, but in land plants, not so much.)

Similarly, in the ocean, red and orange pigments are indicative of the colors that are *not* absorbed. A pigment that absorbs blue light and removes it from the scene reflects back any longer and shorter wavelengths of light. Marine animals that appear bright orange or red when we photograph them with artificial white lights are actually brown or black under most natural conditions. To hide, they don't have to develop a black pigment that absorbs

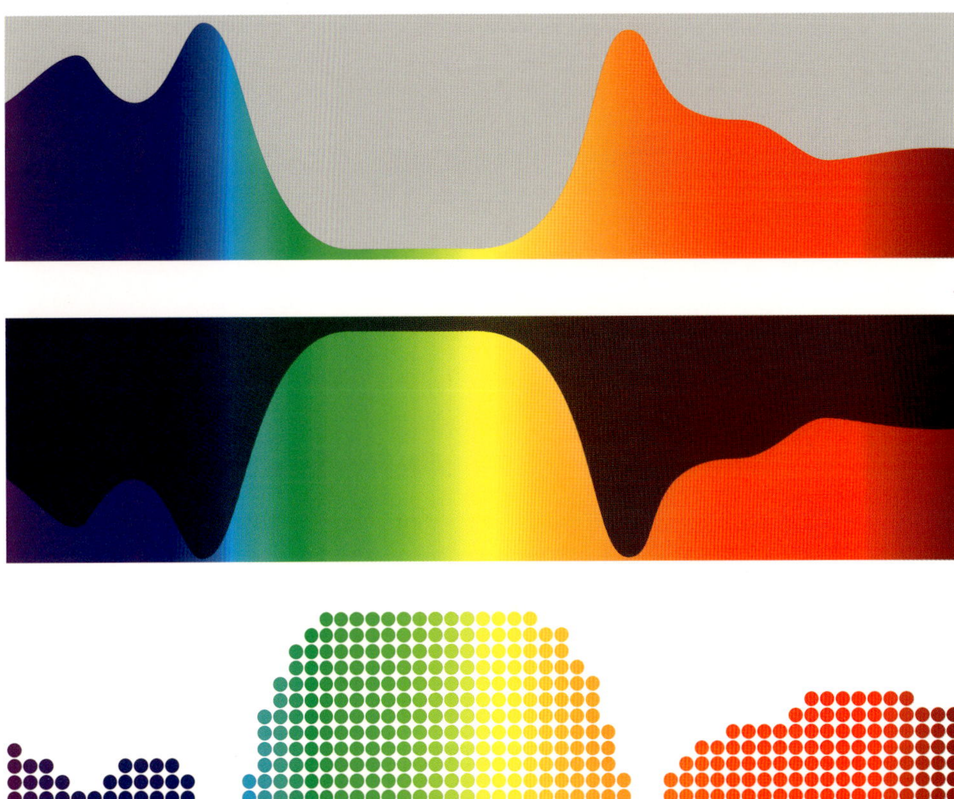

Typically, when scientists discuss the properties of a pigment, they show a graph of its absorbency, where higher peaks correspond to more loss of that wavelength.

To understand the visual appearance of the pigment, we have to flip it around and visualize it being subtracted from white light. Finally, we can visualize the reflected colors; here, mainly bright green.

To underscore that light is a distribution of discrete, colored photons, we've converted the continuous spectrum into stacks of same-colored dots whose height indicates their relative abundance.

every color of light, just one pigment that absorbs blue light effectively. As we have expressed elsewhere, one of the most important differences between the visual environment underwater and on land is that undersea light has already had the non-blue wavelengths filtered out. Unless they can perform some optical alchemy, which you will encounter in later sections, animals in the sea mainly use pigments to create brighter or dimmer shades of blue, and can only slightly shift the color between varieties of blue and green.

THE BASIS OF PIGMENTATION

Walk into any hardware store and you will encounter displays featuring thousands of paint swatches with evocative names like "Bahama Blue," "Essence of Fall," "Treasure Map," or "Romantic Moment" (a shade of purple). This variety is achieved using blends of only thirteen concentrated pigments, counting black and white. Instruments in the store can scan any colored object, then figure out how much of each stock pigment is needed to add up (subtract out) to the same reflectance. Just like in primary school, pigments can be combined together, with blue plus yellow equaling green. But, also just like in primary school, once you add a pigment you can't get that light back, so your paint-mixing experiments begin an inexorable slide toward brown.

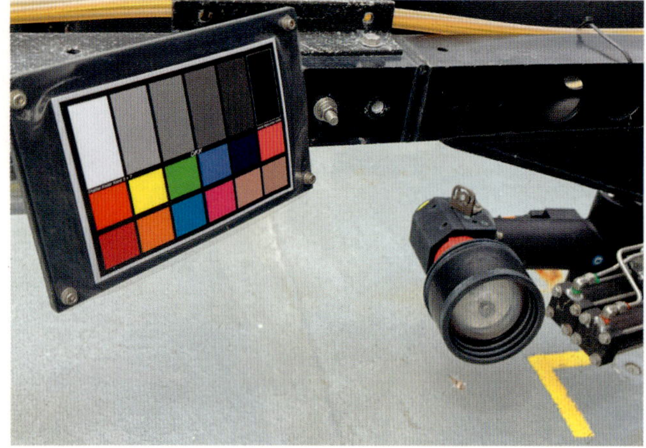

Surface

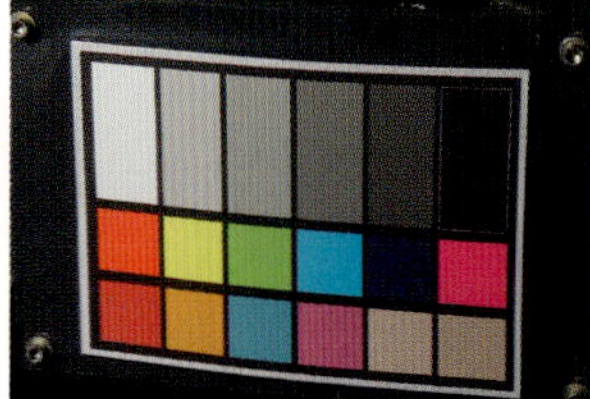

5 meters depth

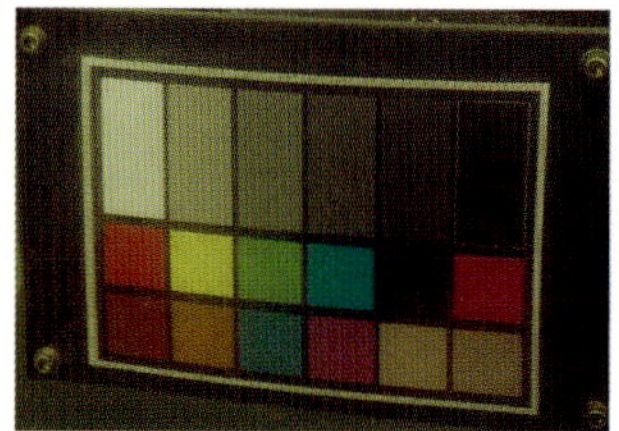

25 meters depth

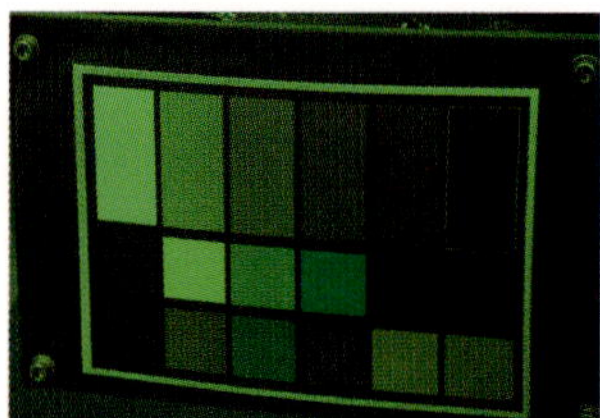

The appearance of pigment-based colors changes as you descend into the ocean because the available light changes rapidly with depth. To illustrate this effect, we mounted a color chart on our submersible and turned off the lights to film it during the descent. On this particular day, the ambient light was quite green due to an algal bloom. When illuminated with the white lights, all the colors were easily visible. As we descended, by 5 meters depth you start to see the decrease in the red and magenta squares. At 25 meters, the red and magenta squares have faded to brown, and the yellow, green, and blue squares have converged to dim shades of green.

In nature, animals act like the paint salesperson: they mix and match a relatively limited number of chemical types to achieve a diversity of colors. Operating via a subtractive process, pigments can't make colors brighter than the original available light. The range of colors you can create with pigments, even under bright white light, is a limited subset of a full range of visible wavelengths—violet to deep red—that are present in a beam of sunshine.

The same classes of natural pigments are often employed by very different, unrelated organisms, in another example of convergence (and in some cases, indicative of dietary links). Some of the common pigment types are porphyrins, which give our blood hemoglobin and jellyfish their red color; carotenoids, which give both carrots and deep-shrimp their orange color; ommochromes, found in squid chromatophores; and melanins, found in fish and other black-colored animals.

PATTERNS THROUGH SPACE

The prevalence of transparency and different colors of pigmentation change with depth and habitat type. We have classified these optical regimes based on predominant coloration used. At the surface of the water, many animals have blue or lavender pigmentation. Transparency is most common in shallow waters below the surface, but it continues to be important down into the upper parts of the deep sea, where vertically migrating organisms dwell. This zone is also home to orange or partially red animals, which try to be as see-through as possible but strategically employ pigment to help hide bioluminescent prey. Deeper still, you find the fully black and red animals, which are beyond the twilight regions so do not have to worry about creating a silhouette when viewed from below. As you approach the bottom, animals are more likely to be yellow, ochre, or pale salmon in color. These seem odd

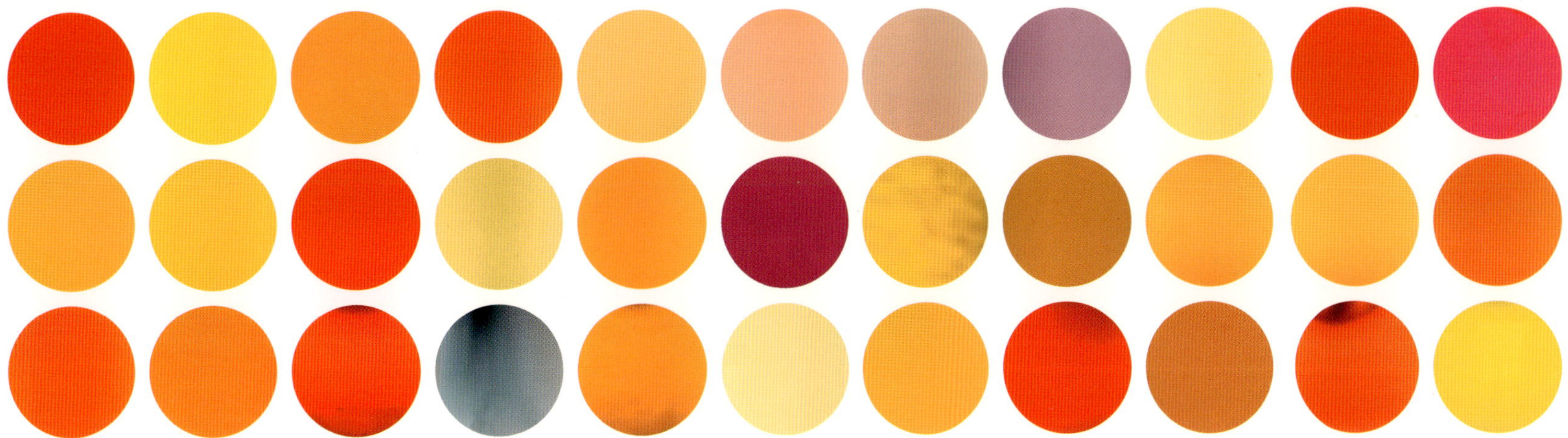

Isolated pigments of a variety of oceanic animals, photographed in test tubes against a white background.

at first, but their reflectance in the blue helps the animals blend in with the seafloor. Throughout these habitat ranges you can find numerous examples of gut pigmentation.

Some animals can adjust their pigmentation, either behaviorally or in response to changes in the illumination, like sunglasses that darken when you go outdoors. Many animals—even large ones such as sharks, dolphins, and whales—are light-colored below and dark-colored above to better match the environmental light from different perspectives. Cephalopods are incredibly adept at rapidly changing their coloration and patterns (stripes and spots) with precise control. Octopuses even roughen their surface texture to match their surroundings. Cuttlefish produce radiating stripes for mating displays, and squid have an entire visual vocabulary of colorful displays indicating whether they are aggressive, afraid, or curious. Squid also squirt pigmented ink clouds into the water. These can either be a billowing cloud to hide behind or they can take the form of "pseudomorphs"—decoy stunt doubles—with skinny strands for skinny squid and a series of round puffs for globular squid. Some squid will rapidly change from transparent to pigmented, or even inject ink within their clear mantle cavity, to make their bodies suddenly appear larger.

In the following pages, we will survey some of the diverse forms and functions of pigments found in marine animals.

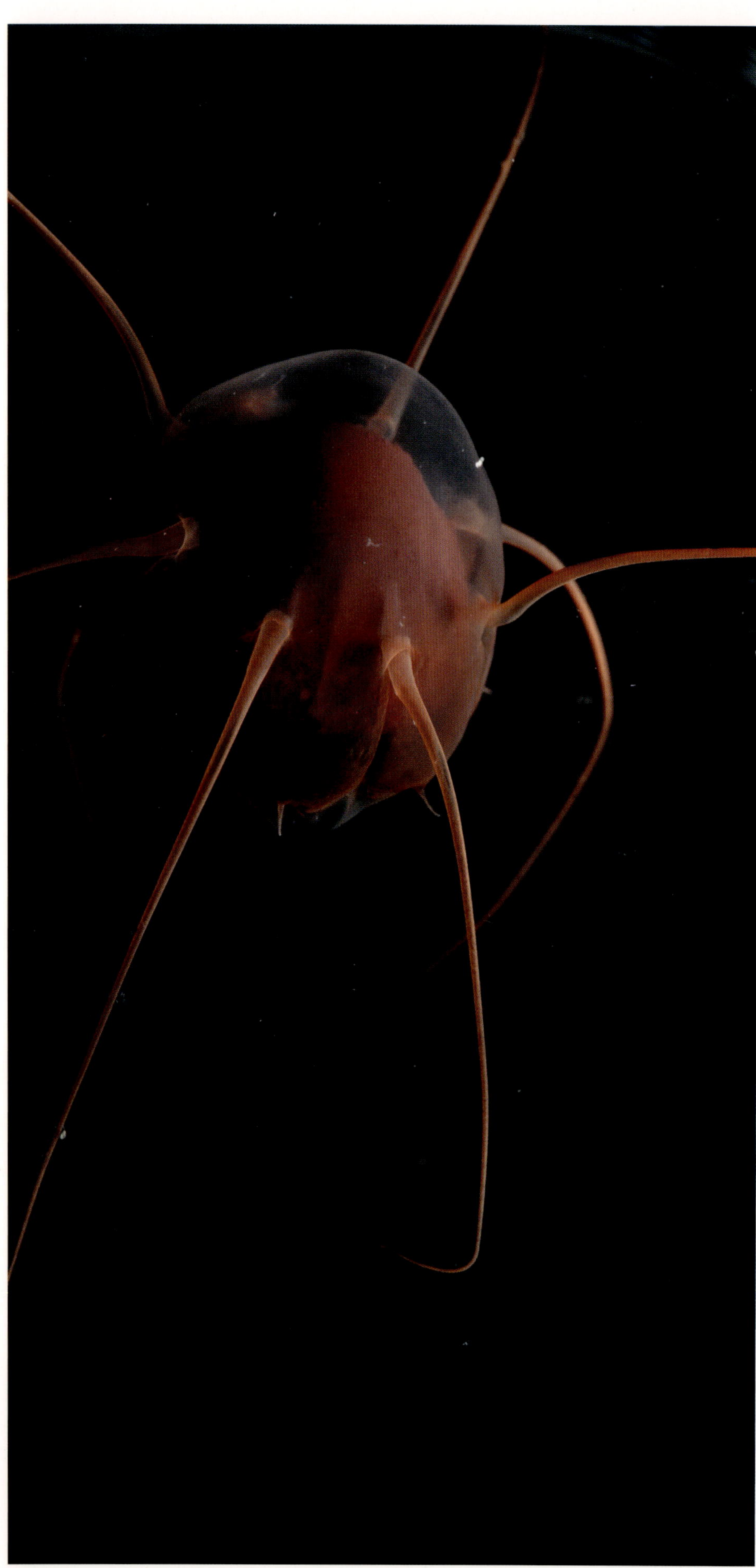

Aeginura

Aeginura is in a group of jellyfish called Narcomedusae, which are known to eat other jellies rather than more typical fare, such as crustaceans. While some species only bother cloaking their stomach with pigmentation, *Aeginura* extends the pigmentation along the length of its tentacles. Most of these red-colored jellies are also bioluminescent; how their pigmentation might affect their bioluminescent displays requires further study.

Atolla

One of the most famous deep-sea jellyfish (if such a thing exists) is the coronate medusa, *Atolla*. Nearly all members of this group gain their reddish-brown coloration from pigments called porphyrins. These types of chemicals have an elegant ring-like molecular structure and are also found in our blood (hemoglobin) and in plants (chlorophyll). We always say that red pigments absorb blue light, making them black in the depths. Two meddlesome facts make this simplified story a little more complicated. First, if you look closely at what colors porphyrins absorb, blue wavelengths are not absorbed strongly. Second, when exposed to bright light, porphyrins can become harmful oxidants. So, the range of conditions where they act as good light-screening pigments is limited.

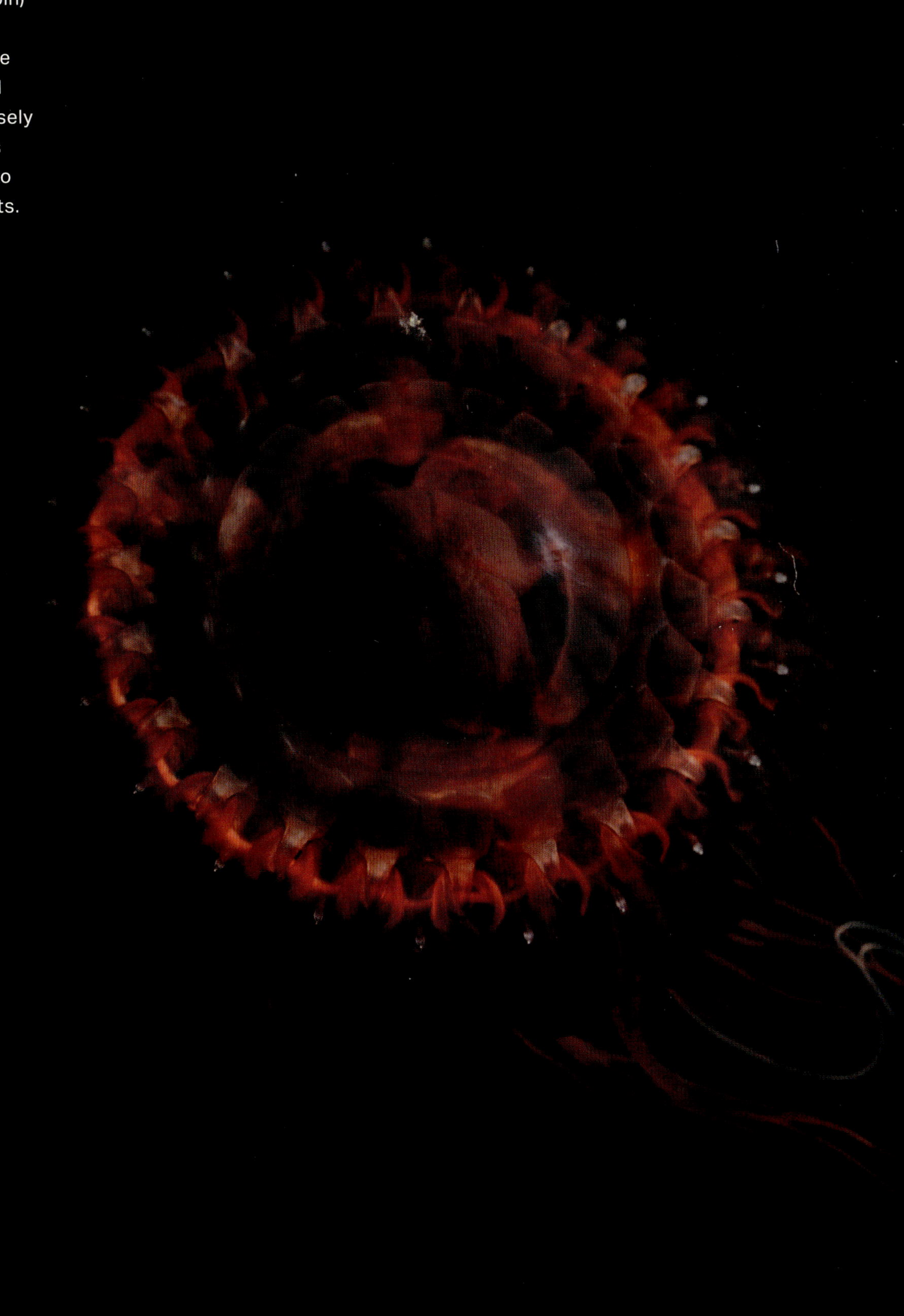

Mimonectes on a calycopsid medusa

Although this jellyfish has cloaked the inside of its bell with pigmentation, its hitchhiking parasitic amphipod (another type of crustacean) does not seem to mind being out in the open. Amphipods will often stroll around in and on jellies, allowing their juveniles to develop while attached to the flesh of their host.

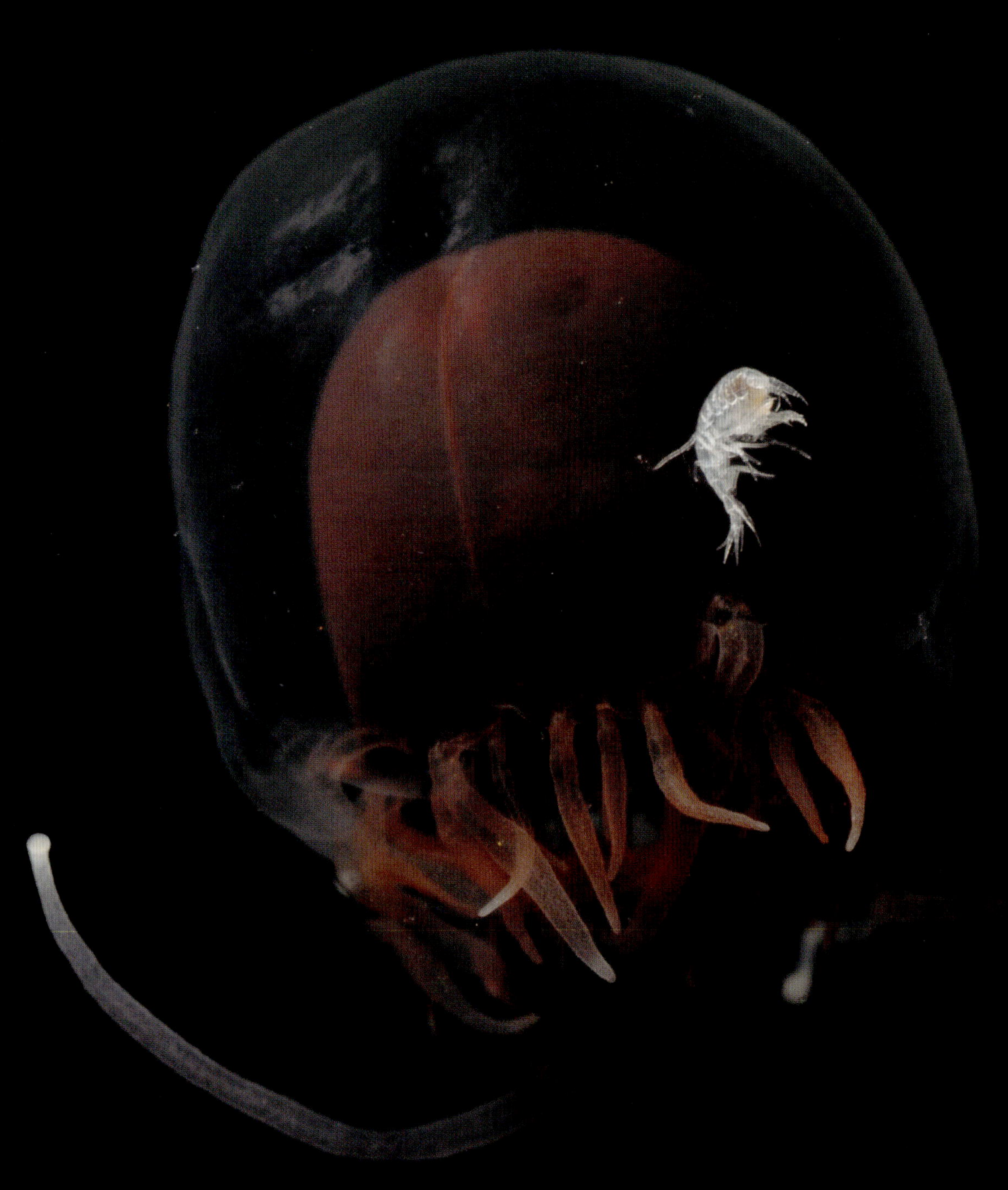

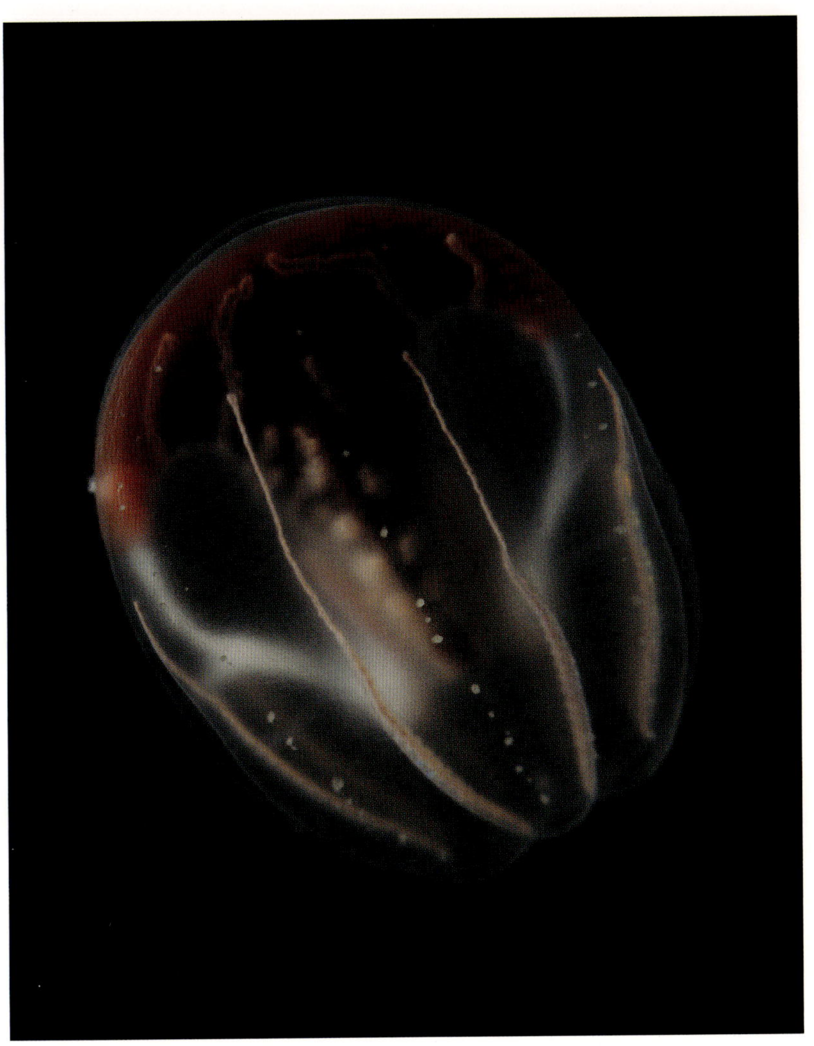

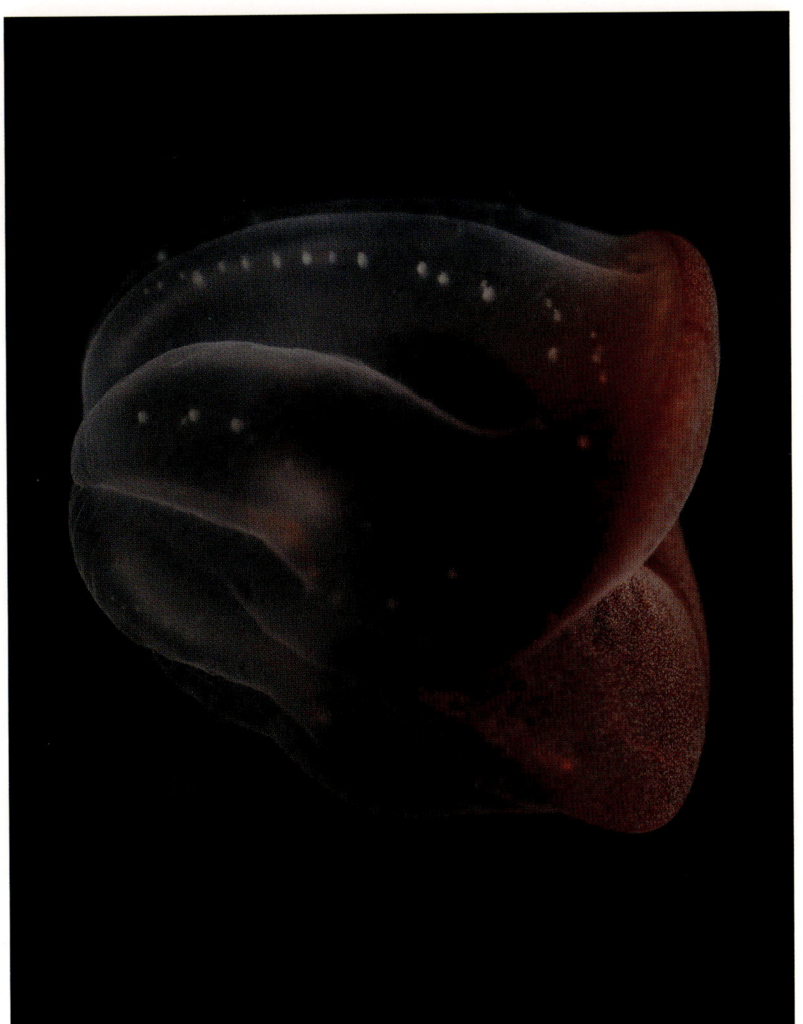

Bathyctena chuni

The comb jelly, *Bathyctena chuni,* was one of the first gelatinous deep-sea species discovered and named. Most ctenophores are too fragile to be collected by nets, but the comb jelly's muscle-infused body holds up well, even when collected from a depth of over 2,000 meters. The biologist who first described it in 1909 wrote that you could tell it was adapted to the deep sea because its lips were pulled tightly together, in order to resist the extreme pressure (spoiler, not true)! This is another member of the red-bellied club, which makes sense considering it lives in permanent darkness and eats prey that have a high probability of glowing, even after their consumption. The columns of yellow spots in between the comb rows are actually concentrations of bioluminescent chemicals, which are emitted to make a cloud of sparkling light.

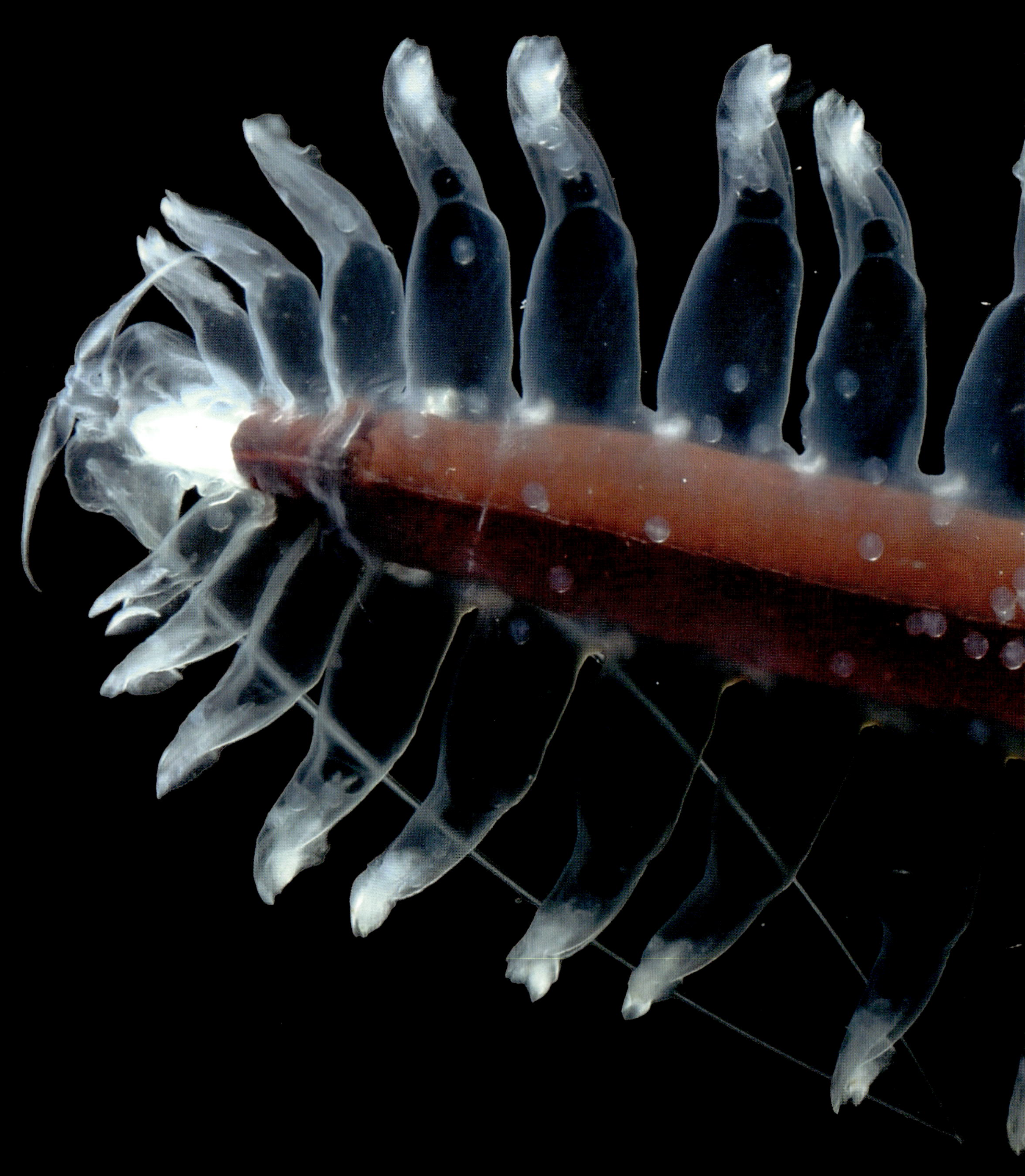

Tomopteris polychaete

Tomopteris has been featured in the Transparency section (p. 36) for its clarity, but some species lose this clarity by cloaking their gut in deep-red pigment. This pigmentation seems more prevalent in larger, deeper species. In the deep, the risk of being exposed by glowing prey must outweigh the risk of having their silhouette seen against the increasingly dim background. The white spheres are its eggs, which can circulate freely within the body cavity.

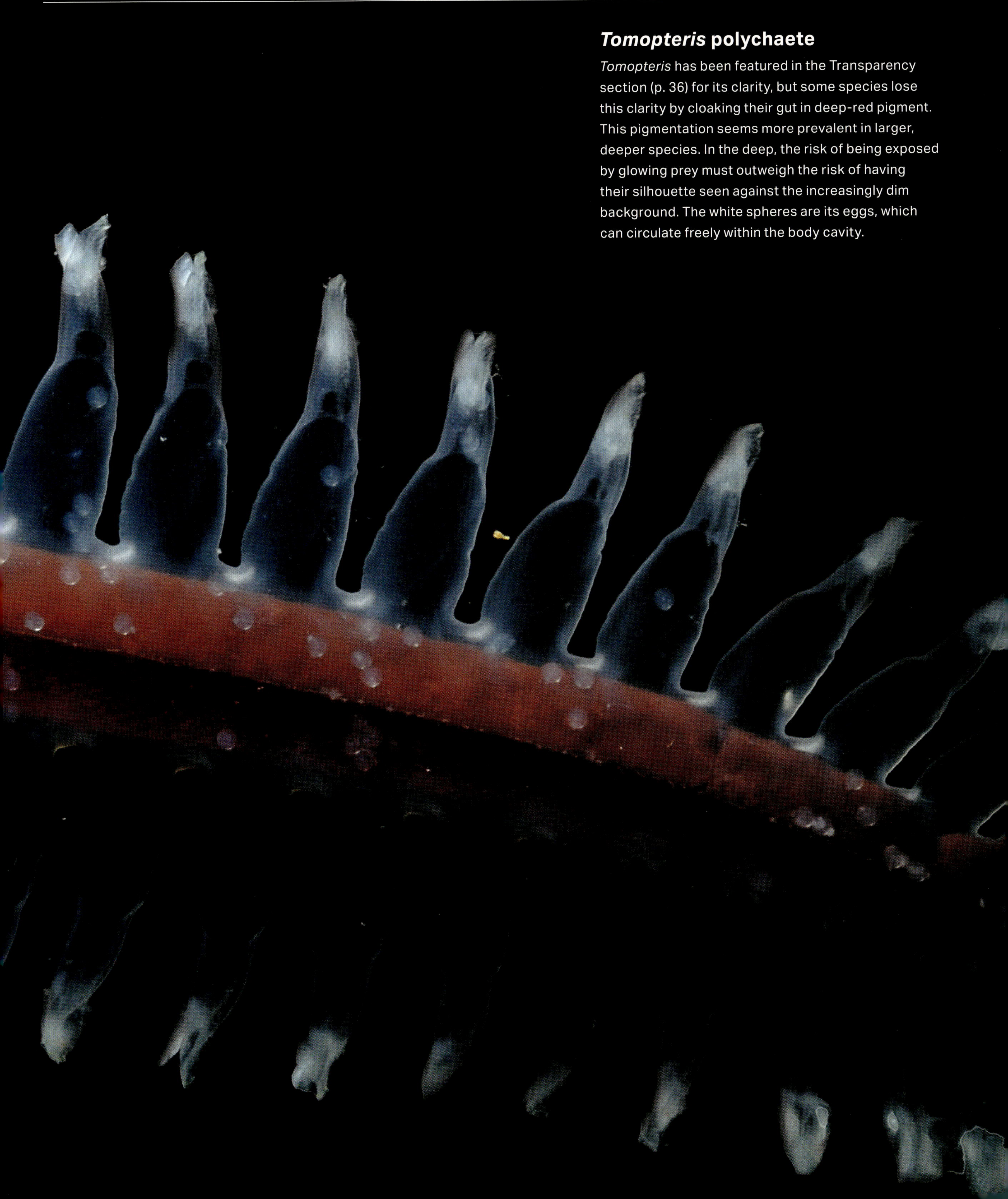

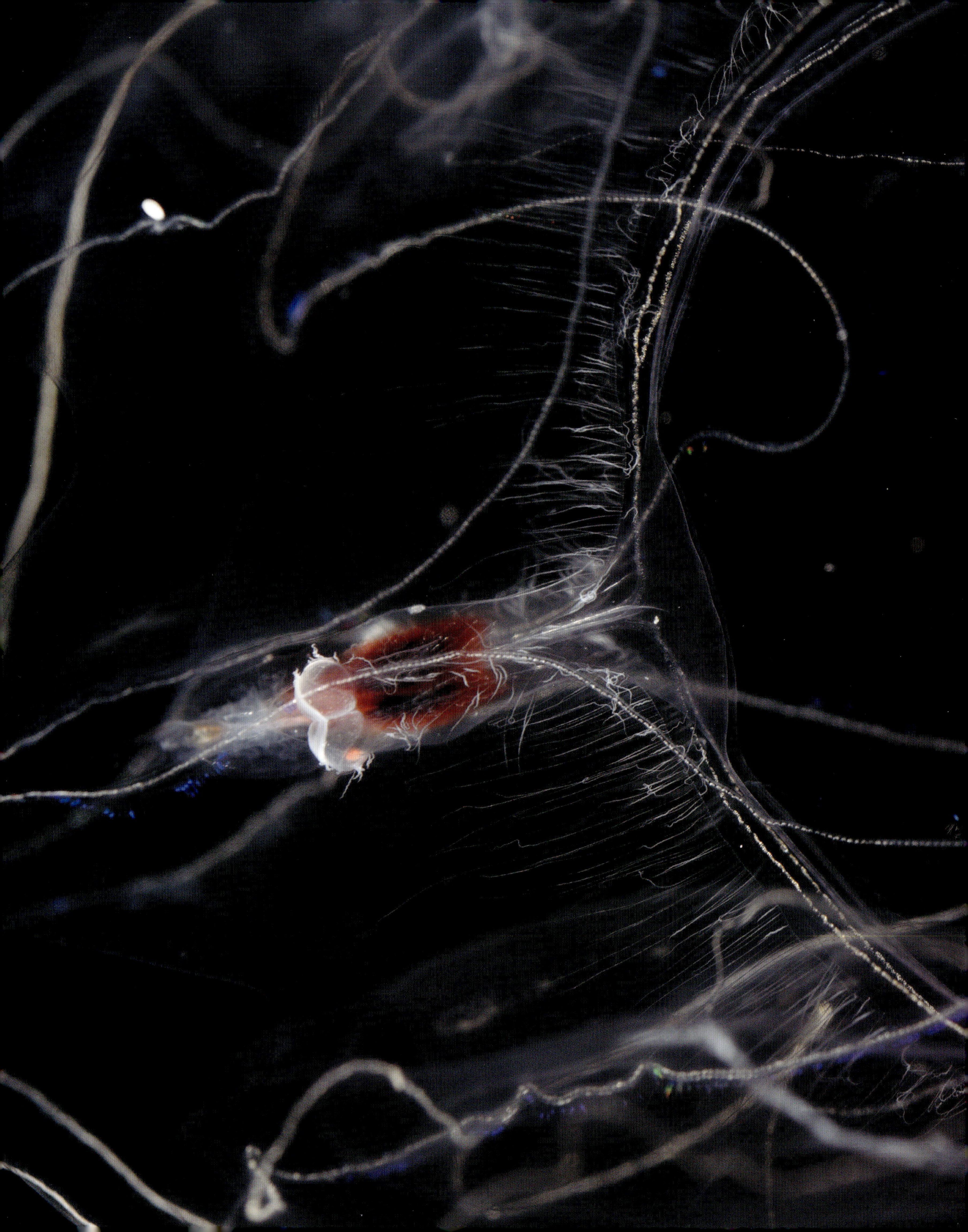

←

Bathocyroe

If any comb jelly has a flaw in its transparency, it will almost certainly be the pigment used to mask the gut. In this fragile deep-sea ctenophore, the stomach is cloaked in dark-red pigment, because even tiny copepods often emit bioluminescent light.

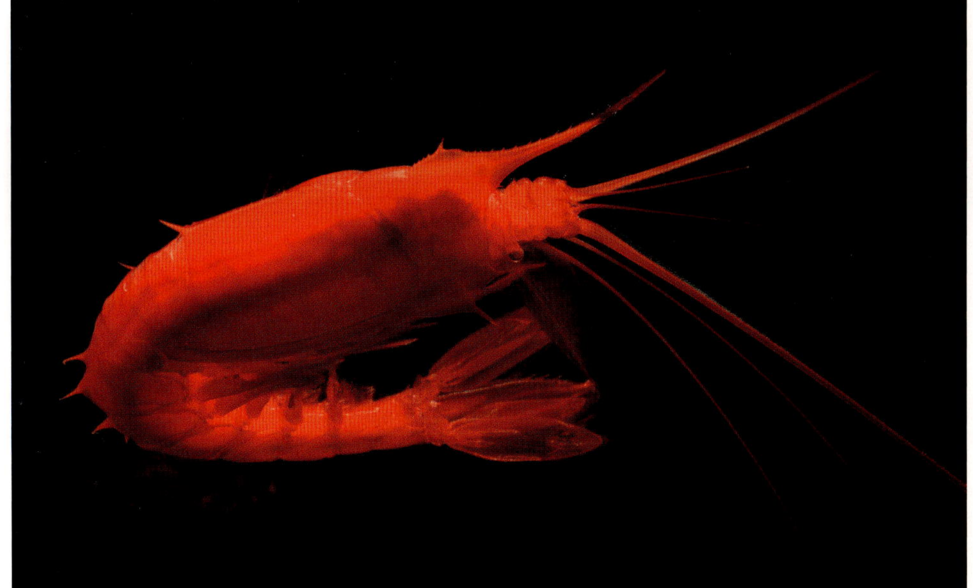

→

Gnathophausia

This large (12 cm) shrimp was one of the first deep-sea animals whose physiology was studied using modern techniques, revealing its impressive ability to survive in waters with very low oxygen levels. In addition to having an explosive bioluminescent display, its deep-red coloration makes it essentially black in the deep.

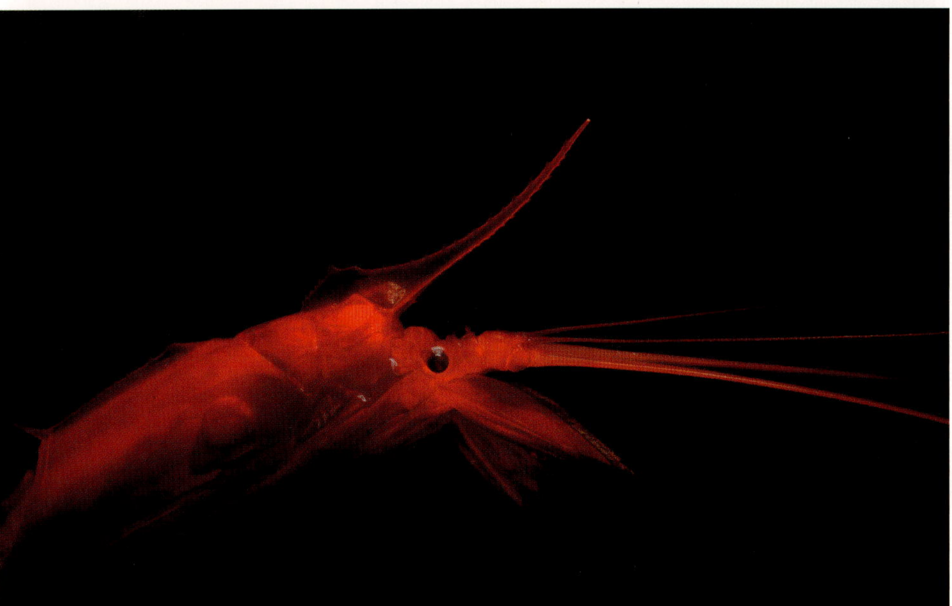

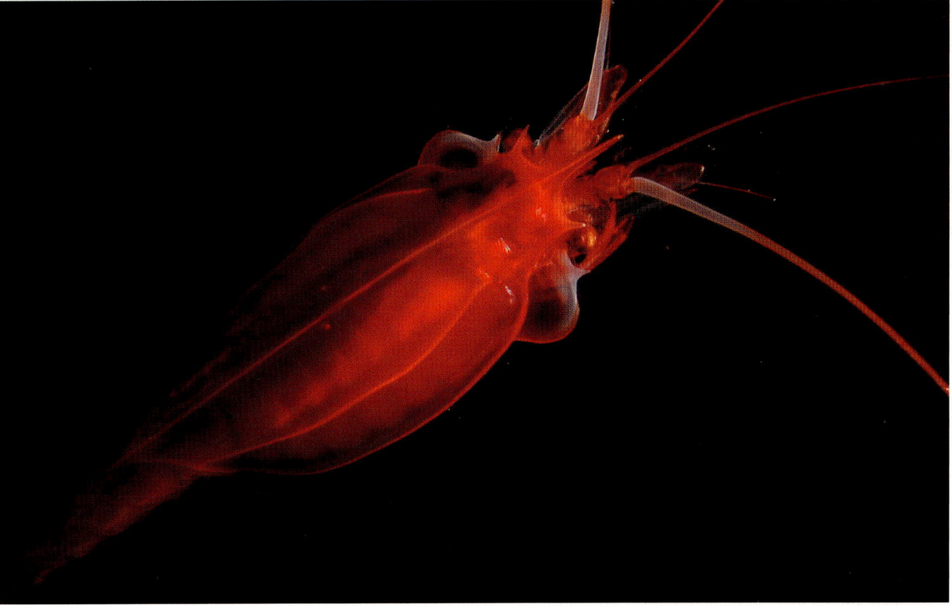

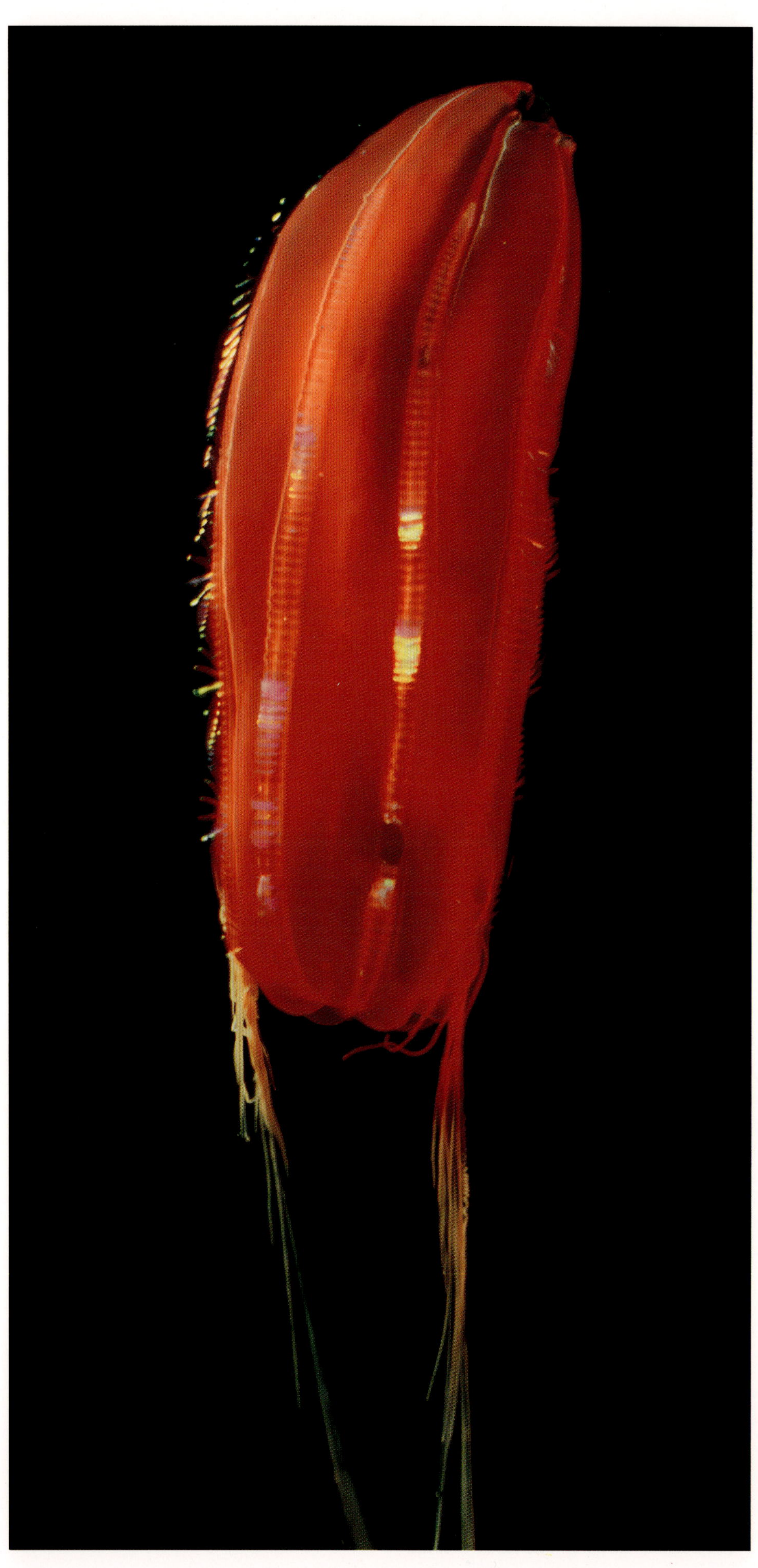

Large red ctenophore

This comb jelly can reach the size of a baked potato, but despite being large and conspicuous, it still hasn't been given a formal name. Its body is fully red, and just to be extra safe its stomach is lined with black pigment to mask its consumed prey.

Mastigotragus

Many squid have pigmented chromatophores, which they can dilate and contract to change their overall coloration. The whiplash squid *Mastigotragus* (formerly *Mastigoteuthis*) is fully invested in its red coloration. Despite the common use of the term "tentacles" to describe all cephalopod appendages, squid typically have eight arms and two extensible tentacles. In the whiplash squid, the tentacles, which look like white ropes in this photo, can be elongated to great lengths, inspiring both its scientific name (*mastigo* = whip) and its common name.

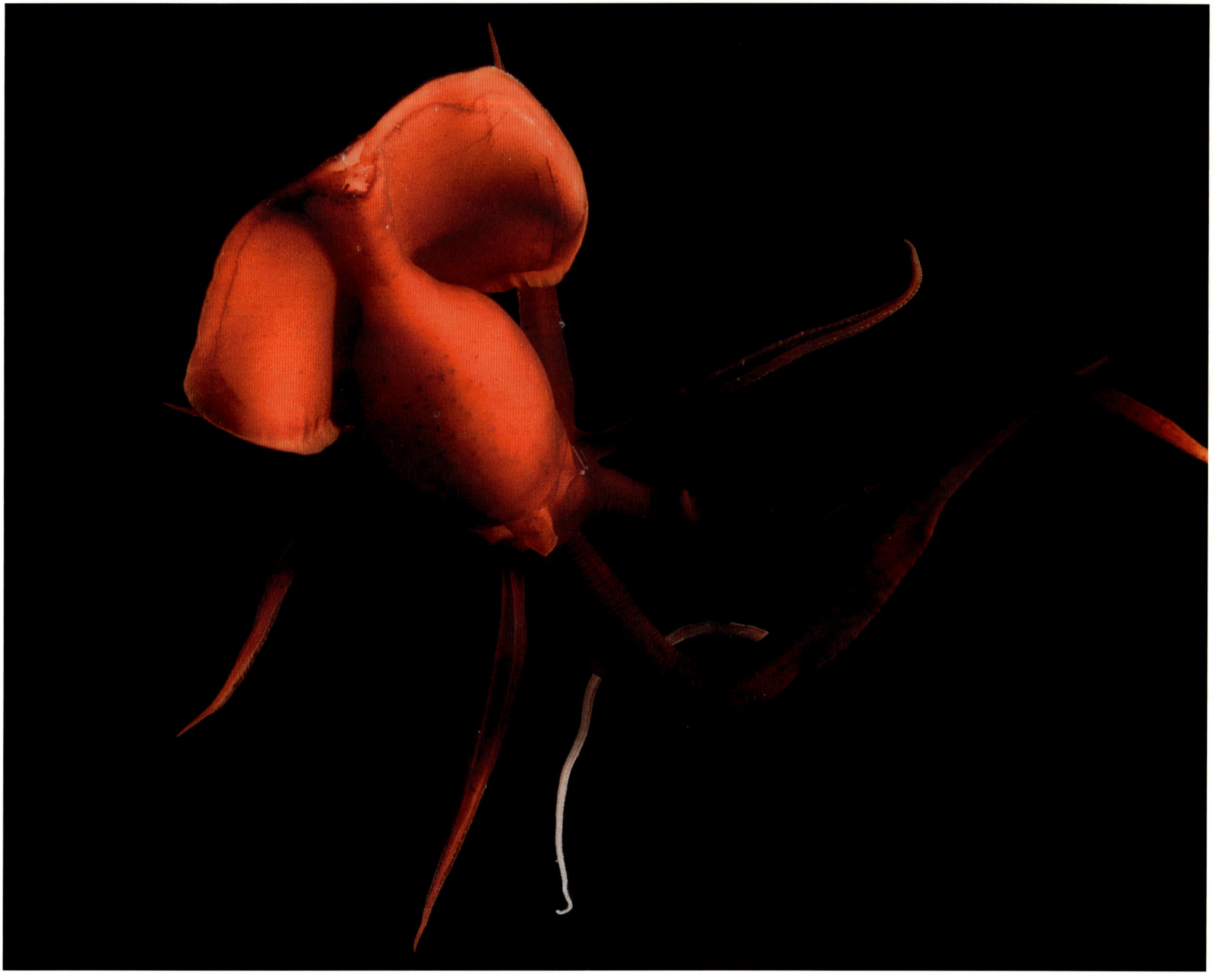

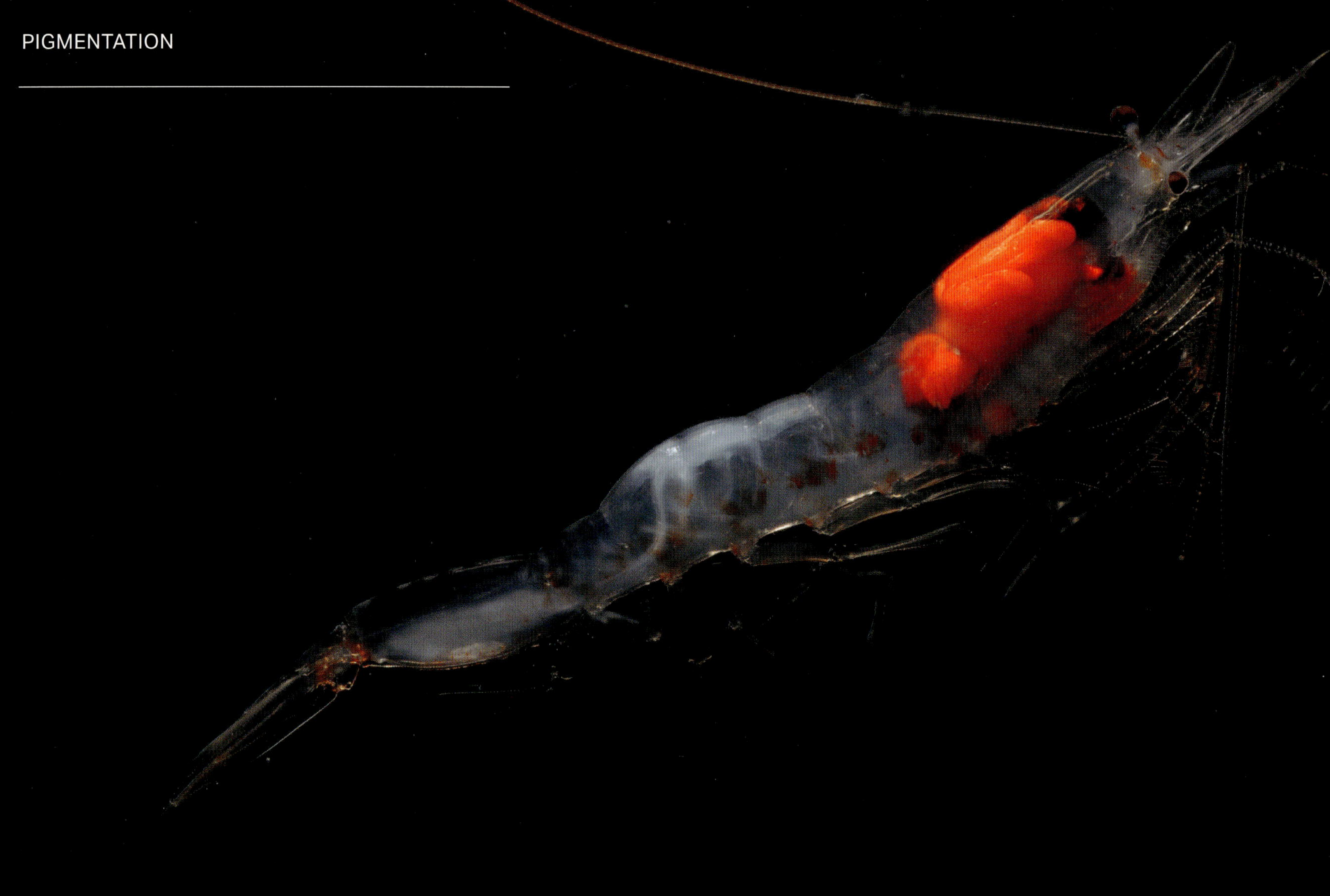

Eusergestes similis

Shrimp make some of the best snacks for ocean inhabitants. This species must be extremely concerned about being captured, because it uses several adaptations to hide from hunters. Its internal organs are masked in red pigment, but because this could create a shadow when viewed from below, it replaces the absorbed light with dim bioluminescent light of its own that shines just below its belly. As it swims up and down in the water column, it carefully tilts this light organ so it is always shining straight down where the shadow would be cast. If light were to shine out to the side, this would serve as a beacon to predators at the same depth.

Phylliroe

This unique nudibranch sea slug swims like a fish, with side-to-side undulations of its body, and consumes jellyfish. The function of its orange chromatophores is unknown, but they could serve a photoprotective function; they live in shallow open-ocean waters with plenty of sunlight. After a time in the dark, the spots contract to make the animal mostly transparent.

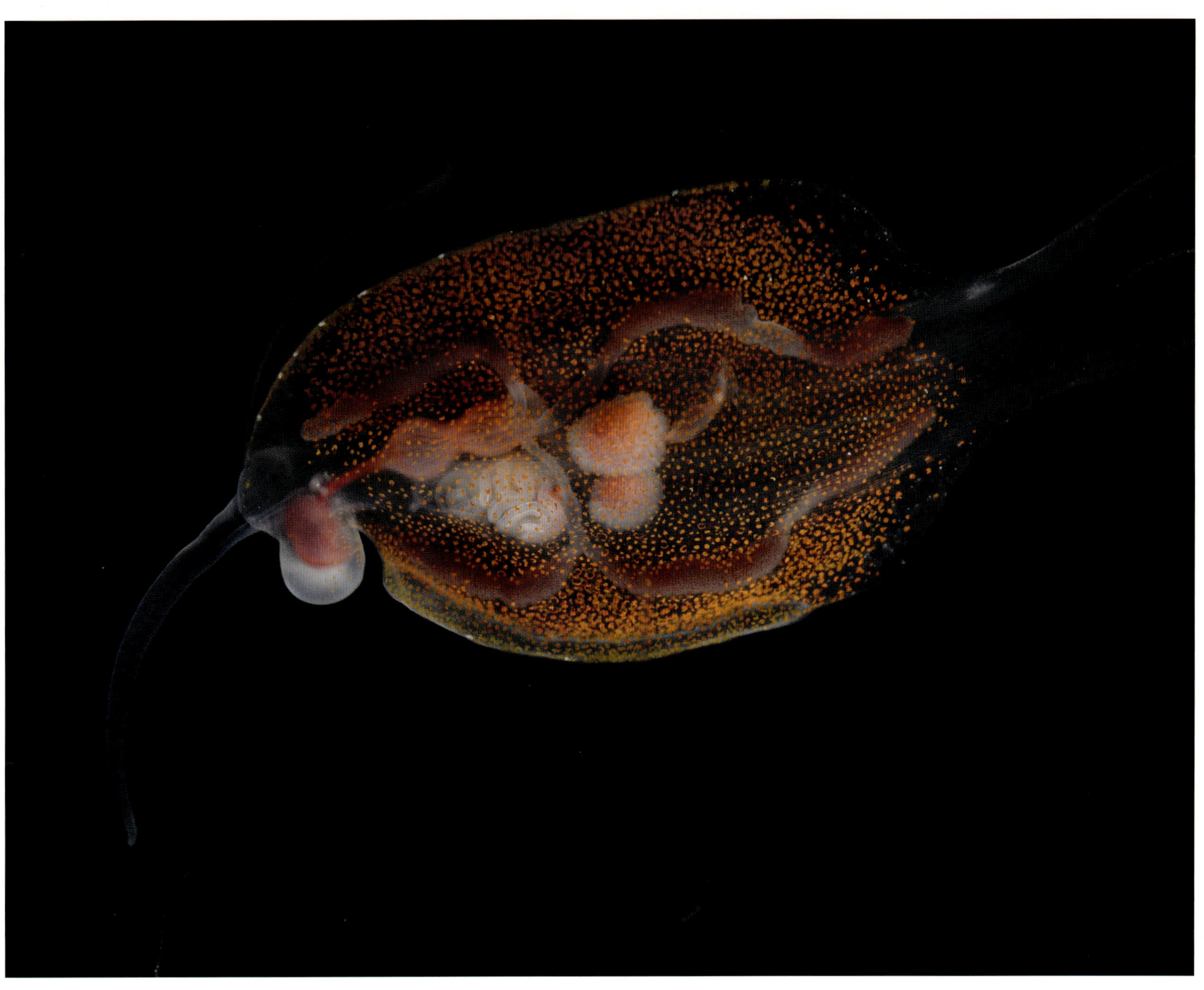

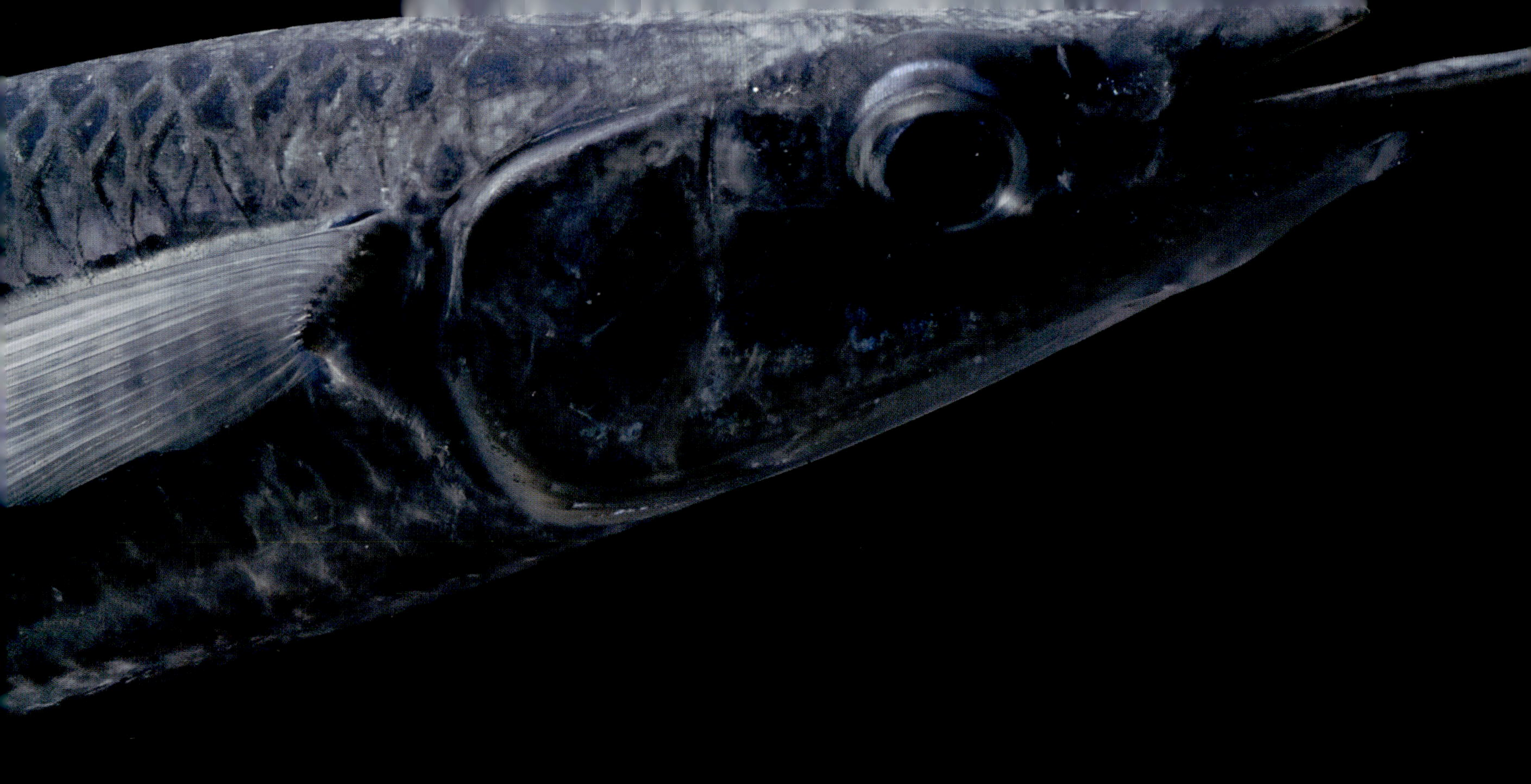

Hemiramphus saltator

The longfin halfbeak lives near the surface and can dart out of the water like a flying fish (its species name means "jumper"). When seen in white light, the long "beak" that extends from its lower jaw has a vivid orange tip, which appears to glow like a burning ember. According to our preliminary tests, the tip is neither fluorescent nor bioluminescent, so its appearance would rapidly change from bright orange at the surface to dark brown in a deeper, blue-lit environment.

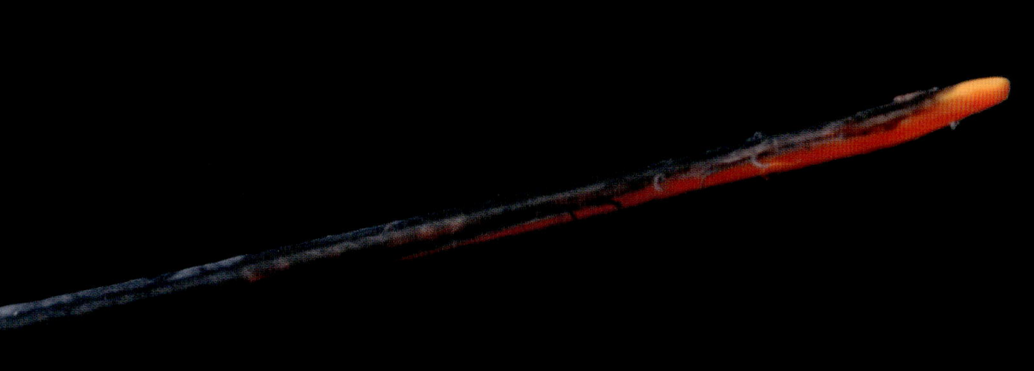

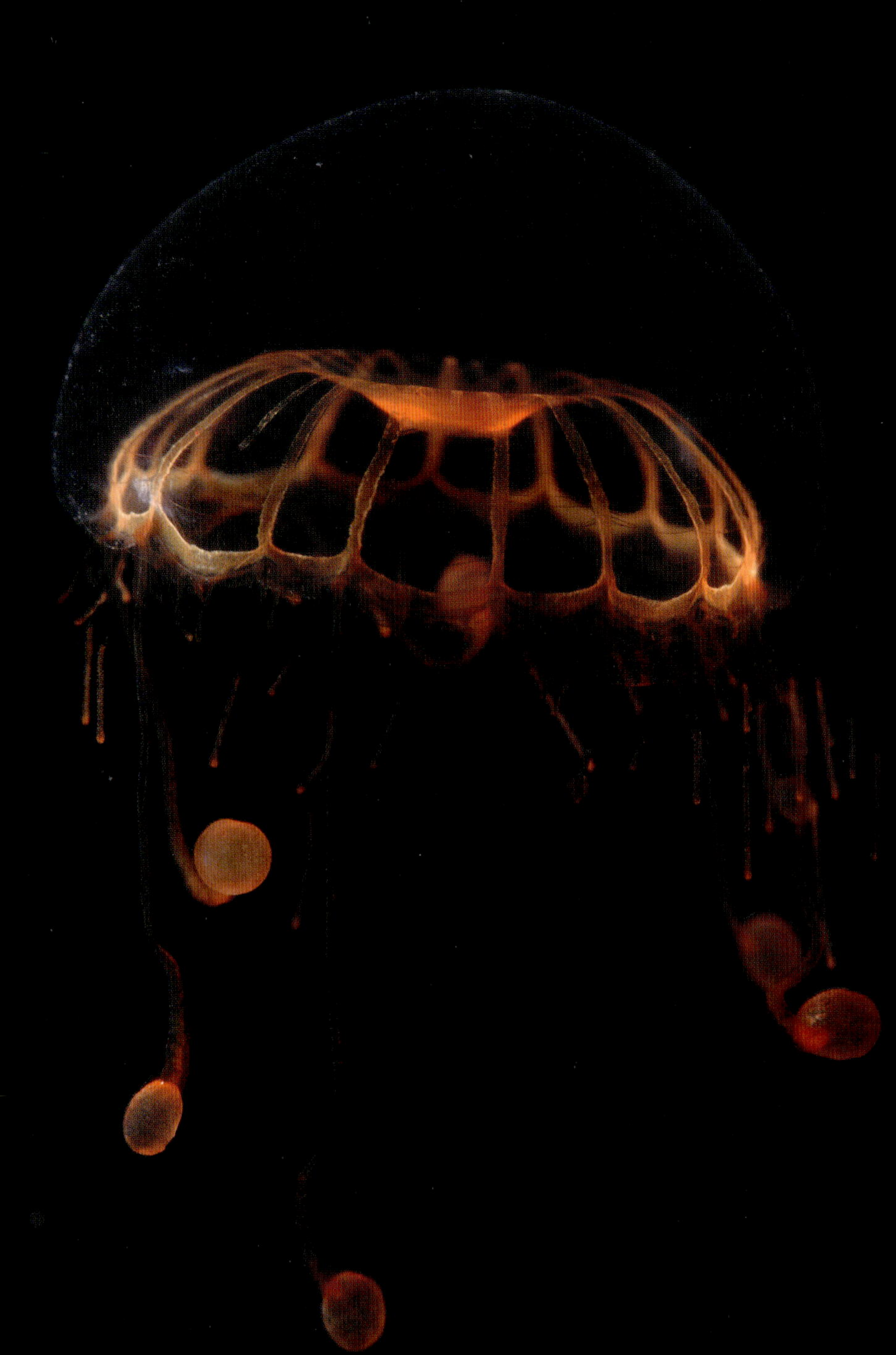

Halitrephes

Looking like a Venetian lamp, this hydromedusa has a streamlined morphology that you can enumerate with one hand: stomach, gonads, tentacles, canals ... with a finger leftover. The tips of its primary tentacles are swollen into bulbs, which may aid in the capture of certain kinds of prey. Look to the Bioluminescence section (p. 139) for an image of its distinct bioluminescent display.

Botrynema brucei

With transparent predators, it is challenging to determine whether an organism makes its own pigmentation, or instead incorporates pigments from its prey. This hydromedusa has an orange color in its stomach and peripheral canals, but the color is granular and looks like it is made up of tiny oil droplets. Therefore, we'd probably conclude this color is from remnants of its prey.

In some cases, color can be used to suggest predator–prey relationships. The rule of thumb for jellies like siphonophores is that the orange ones primarily eat crustaceans, but don't touch the black and purple ones: they get their pigments from fish prey, meaning that their stings are tailored to incapacitate vertebrates. Outside of the Portuguese man o' war (bluebottles) though, fish-eating species are rare in shallow waters.

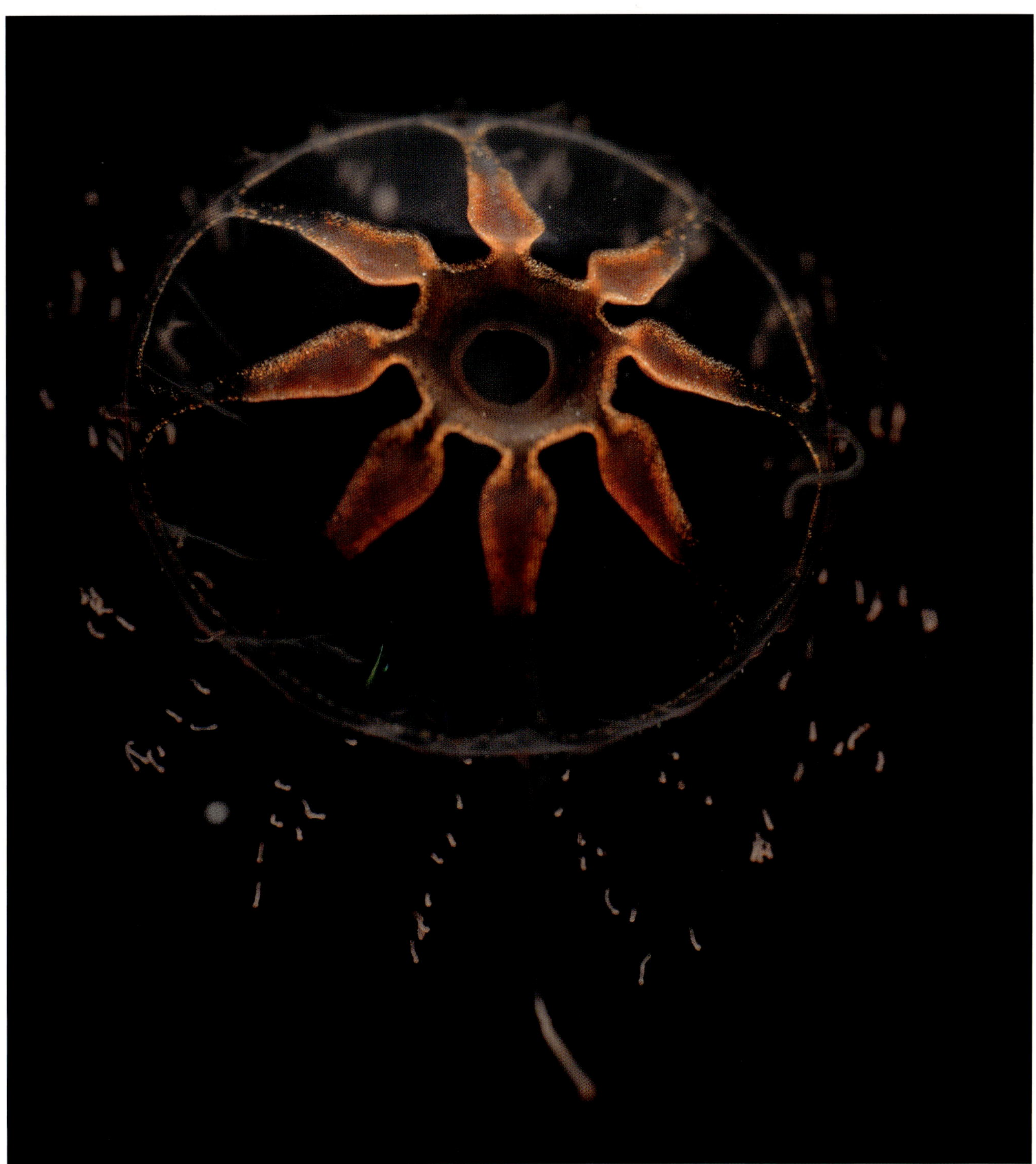

Anuropus

This pilot isopod—whose closest relatives on land are the pill bugs in your garden—is one of the remarkable natural history stories of the deep sea. There is a rare giant jellyfish called *Deepstaria*, which is beautiful in texture but has the shape and size of a kitchen trash bag. Pretty much every time we see one of those jellies, there is an *Anuropus* isopod tucked inside the bell; *Deepstaria* never has a different species of hitchhiker, and the isopod is not seen on another jelly. How these two uncommon creatures find each other in the vast deep sea is unknown, but how the isopod stays securely tucked within the bell of the medusa is easier to figure out once you see its spiky legs up close.

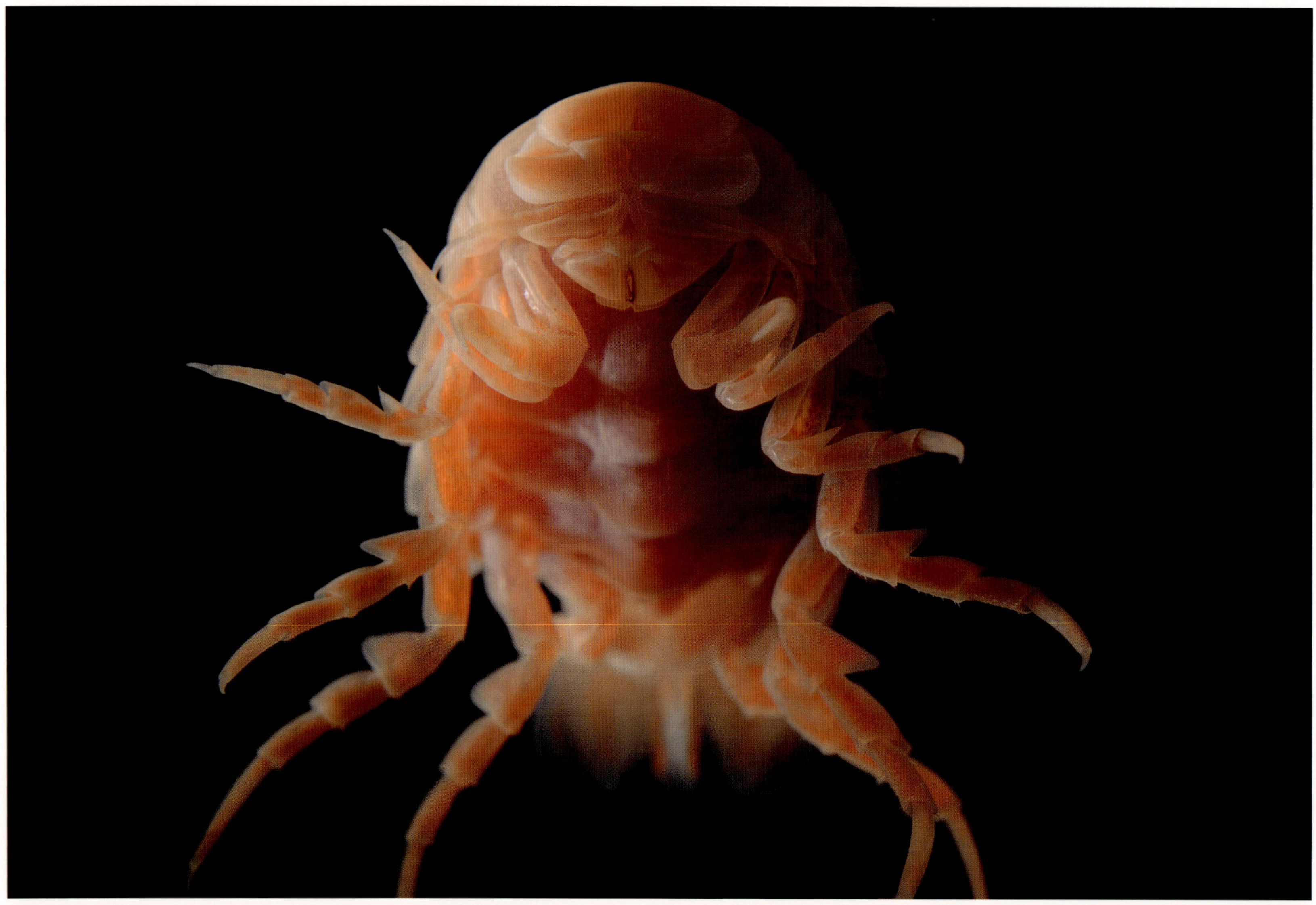

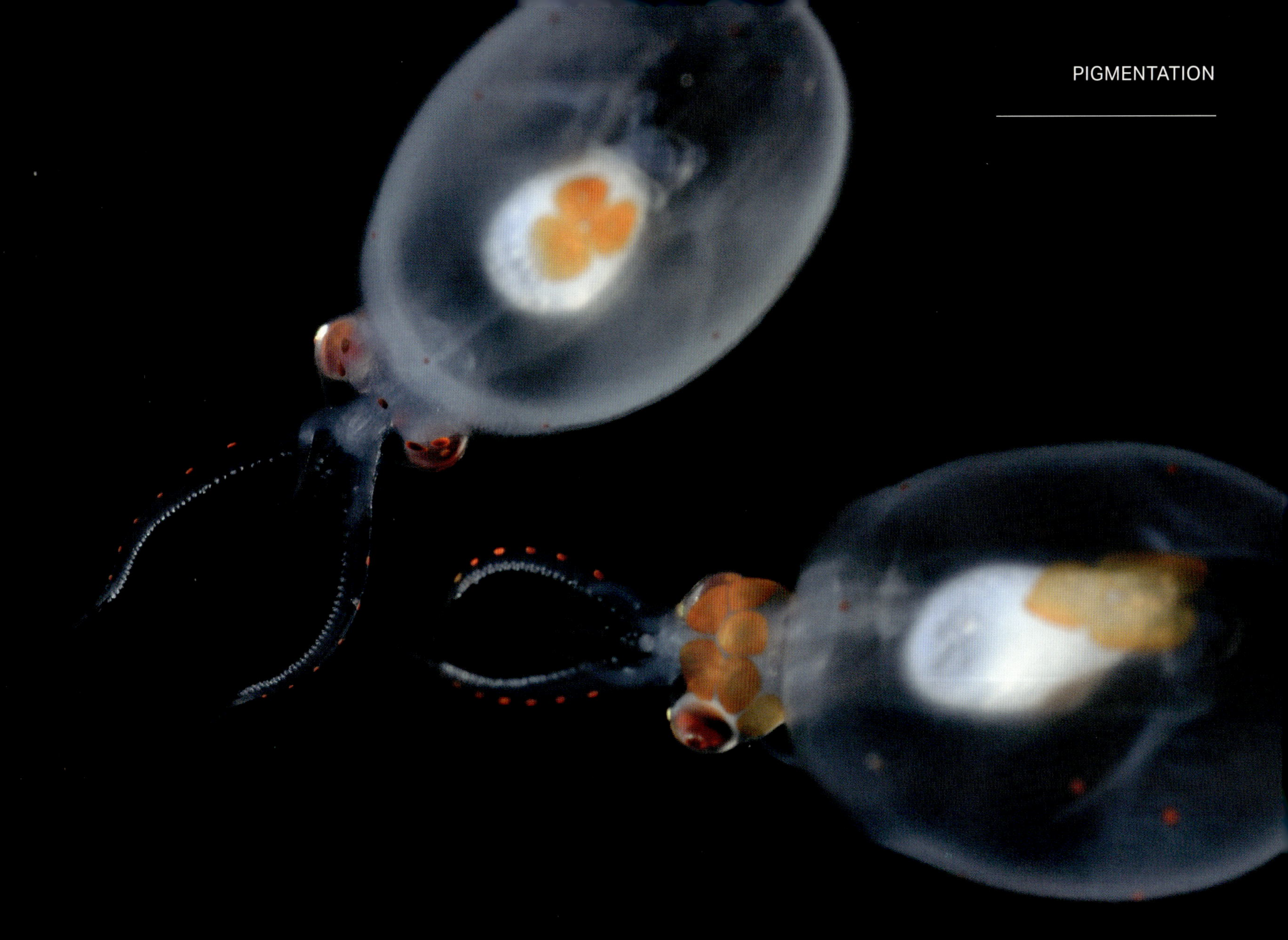

Larval squid chromatophores

Even when you're small—these just-hatched squid are only 1.5 cm long—you still need to conceal important organs with pigmentation. At this early age, these squid use three pigment patches over their internal organs, and six down each long arm. As adults, their bodies will become fully pigmented.

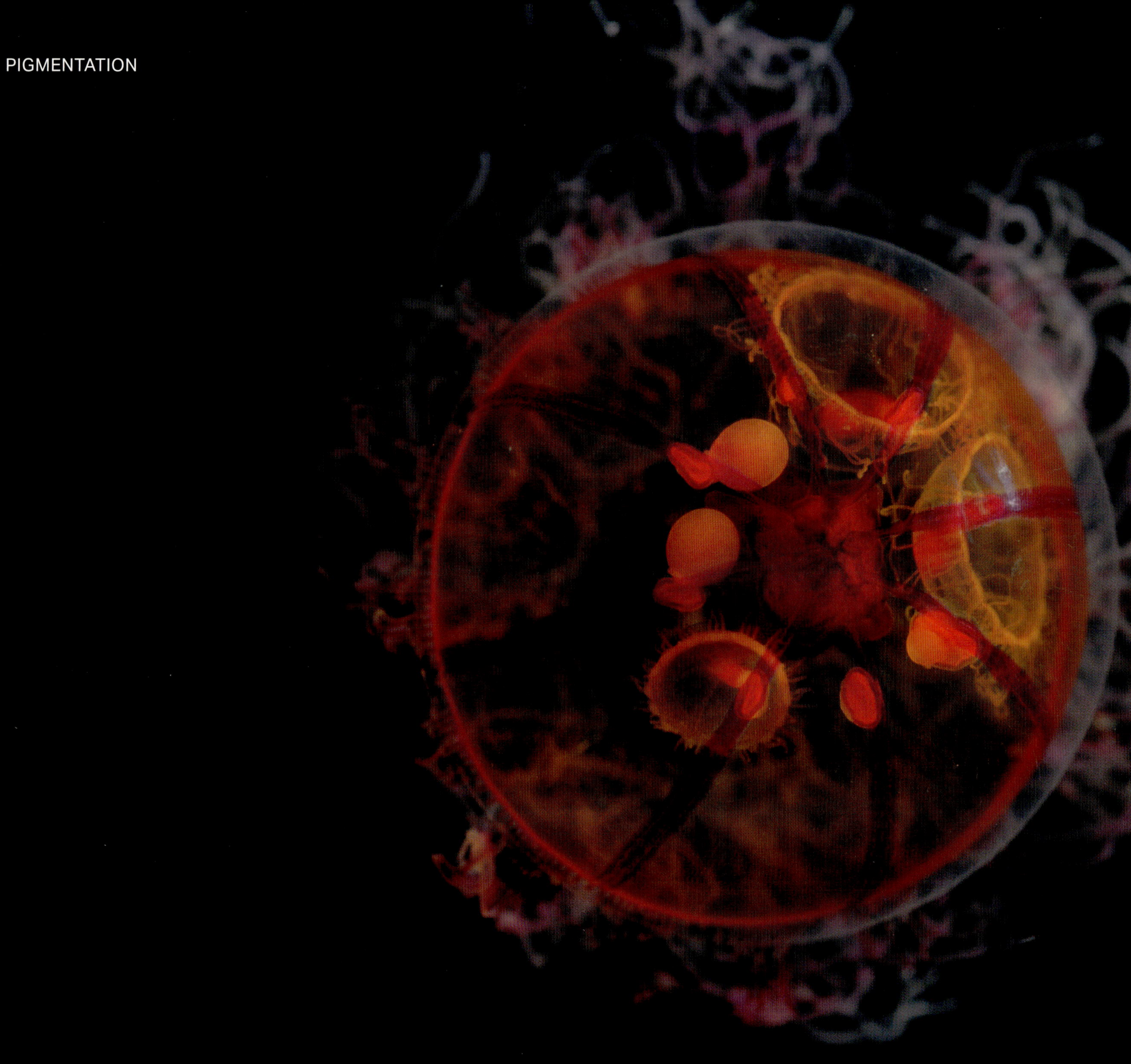

Crossota millsae

Crossota millsae, named for jellyfish expert Claudia Mills, has some of the most creative colors in its ensemble. It is appropriate that the translucent pinks and oranges look like blended watercolor paints, since Dr. Mills uses watercolors for her art. Instead of spawning into the water, this species of jellyfish and its close relatives brood their young, visible here still attached inside the bell.

Planctoteuthis

The skin of this deep-sea squid is peppered with chromatophores in varying stages of dilation. Although the amount of pigmentation is the same when dilated or contracted, the tiny dense spots are less visible than the expanded but more diffuse patches.

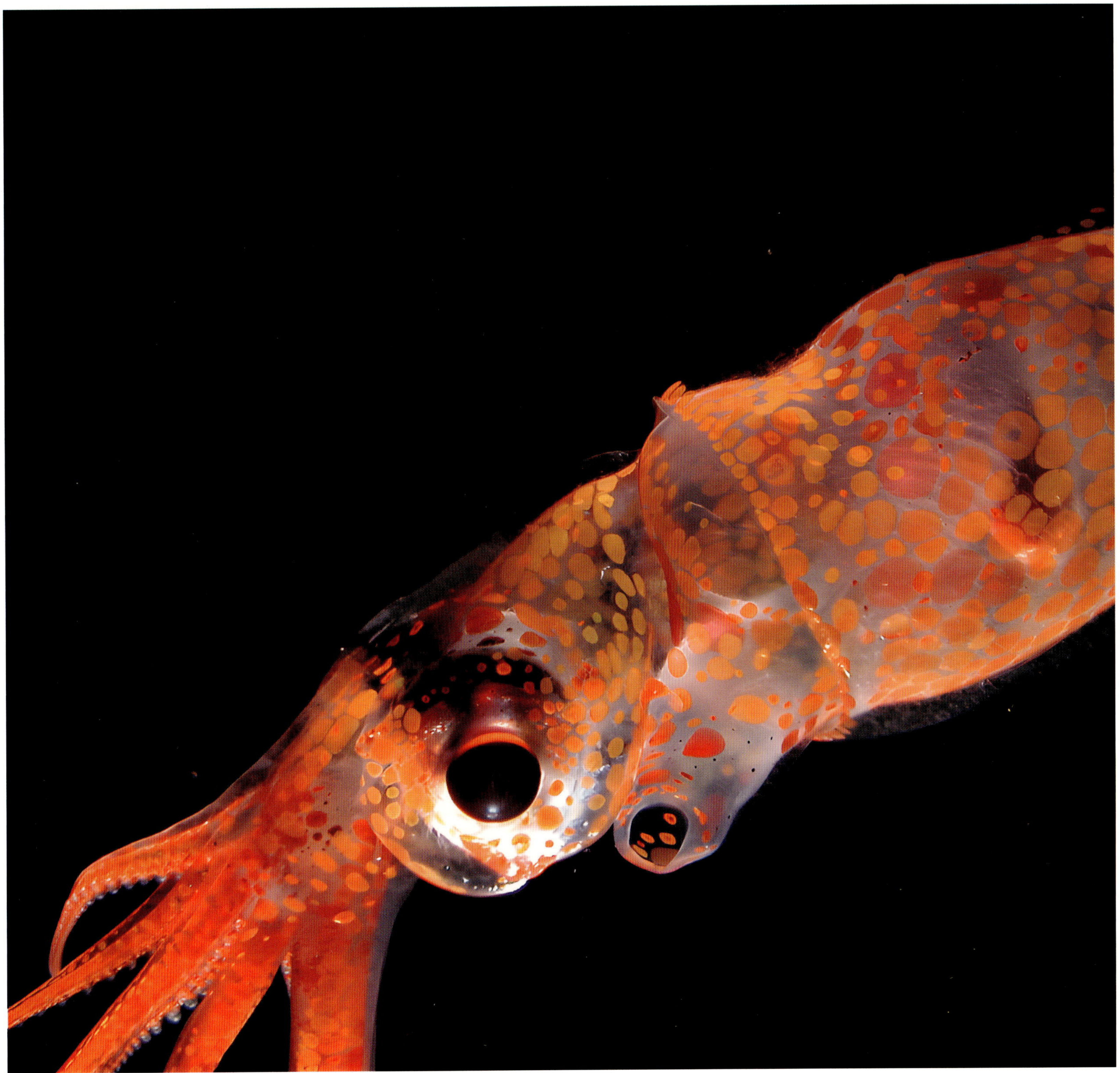

Aegina citrea

Yellow pigments are rare in the deep ocean, but this citrus jellyfish—a relative of the red-colored *Aeginura*—is named for its lemony appearance. To our knowledge, the pigment causing this yellow color is not yet known, nor do we know what wavelengths of light it absorbs.

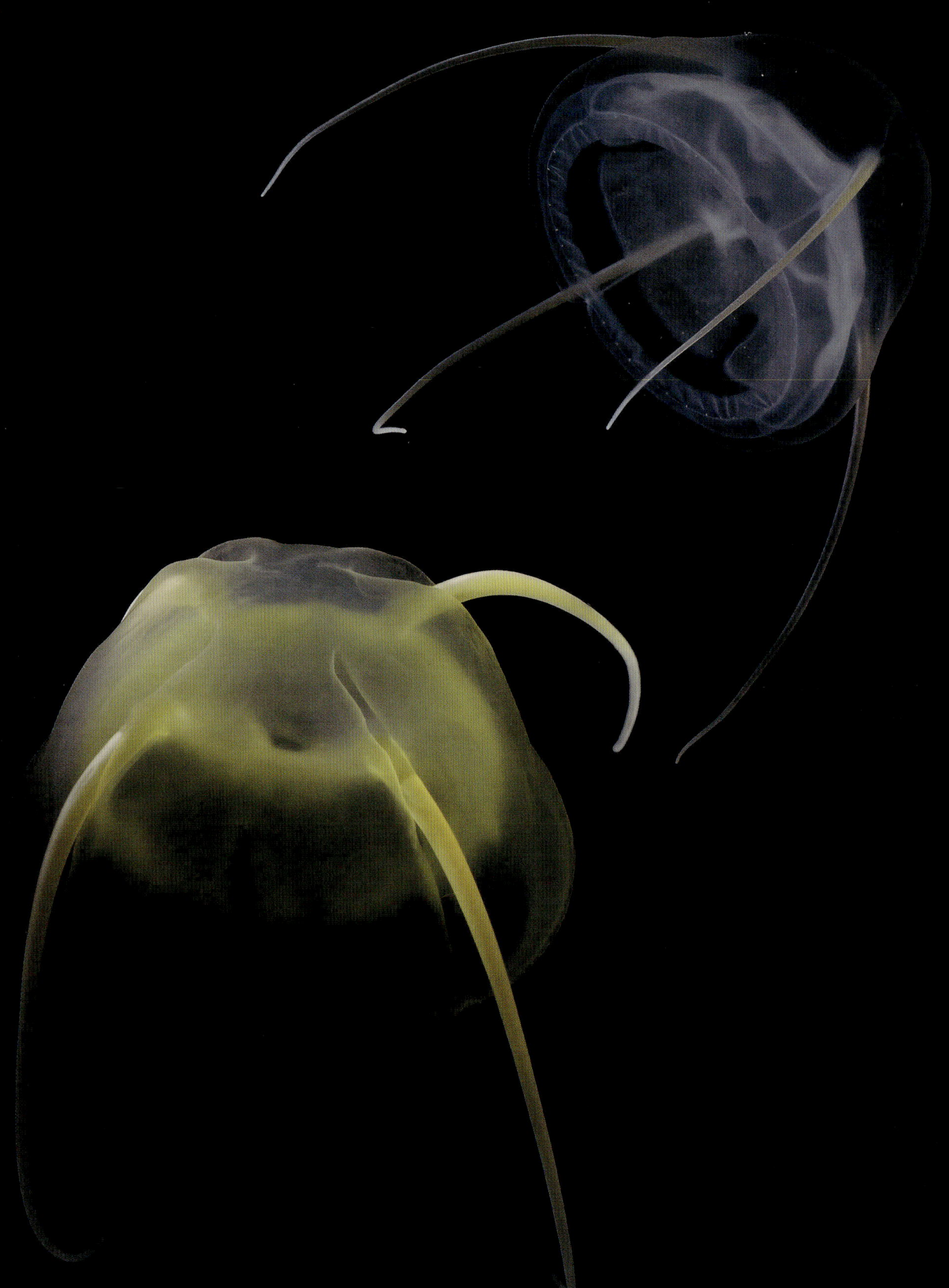

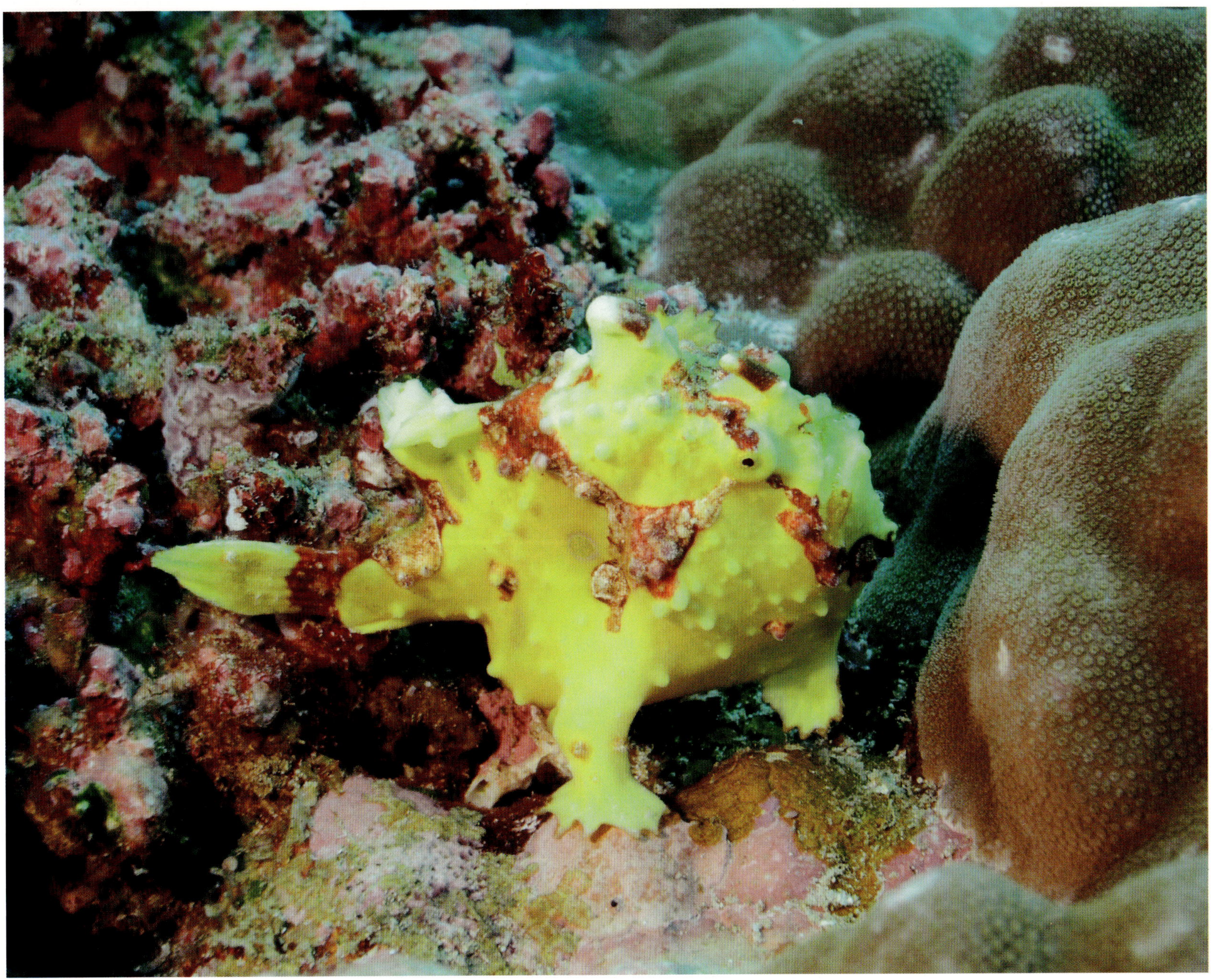

Antennarius maculatus

Taken in isolation, some bright colors would appear to be disastrous anti-predator choices. This frogfish, however, has no qualms about its lemon-yellow appearance on this brightly lit reef. In fact, frogfish are some of the most cryptic of shallow reef inhabitants, and it is easy to swim right over or past one without noticing it. The reason is that the reefs are covered with other bright animals, including yellow sponges. This is like hiding next to your kitchen sink by taking on the appearance of a blue dishwashing sponge.

Plesionika trispinus carrying eggs

We are not so sure about the ham, but green eggs certainly exist in the sea. Functional green pigment is rarely seen in the ocean—after all, why absorb everything *except* the color of light that is actually present? The color in these eggs is more likely indicative of the nutritious supplies that the mother has invested in her young. Many open-ocean animals make long journeys between egg-hood and adulthood, and the odds of survival are often slim.

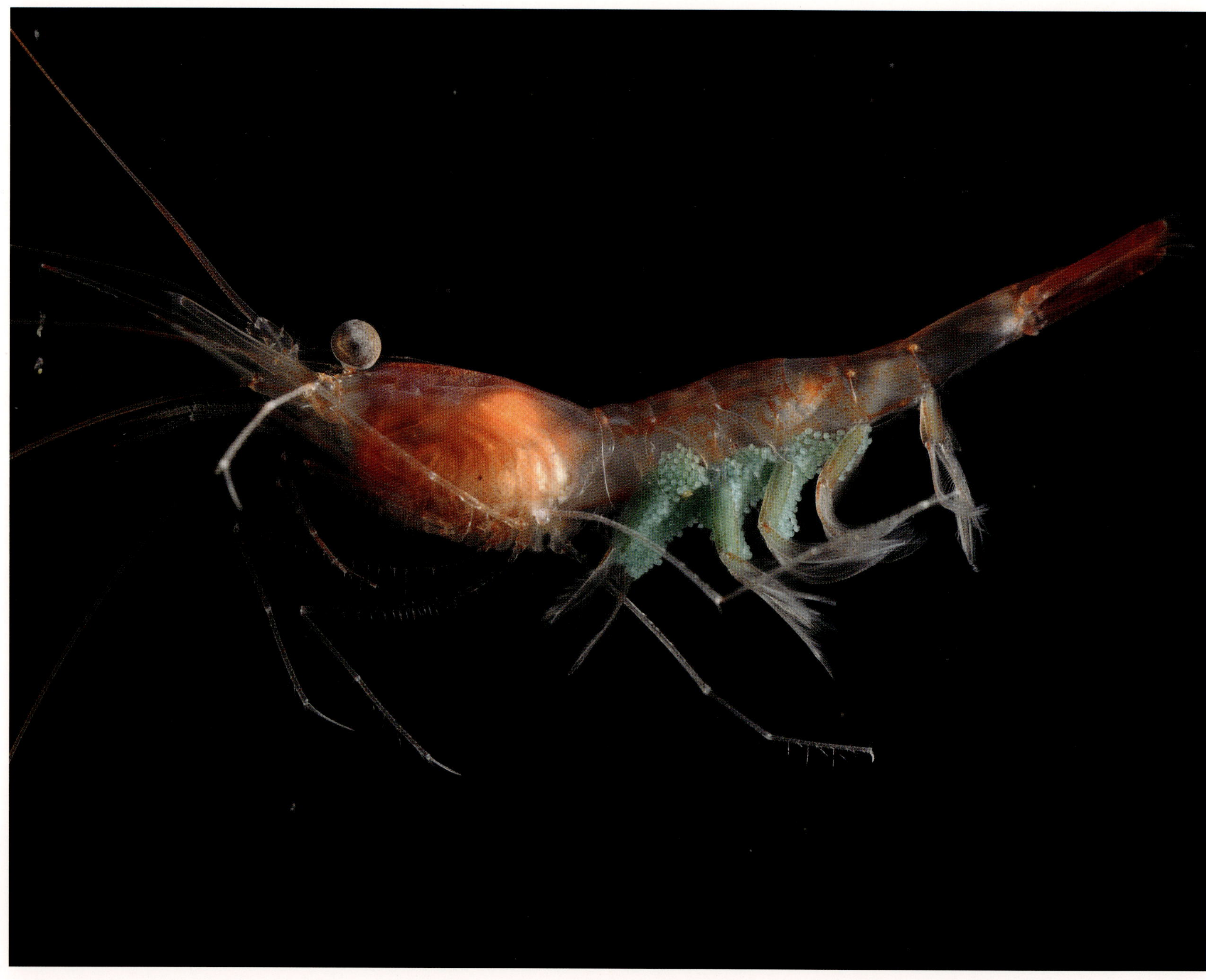

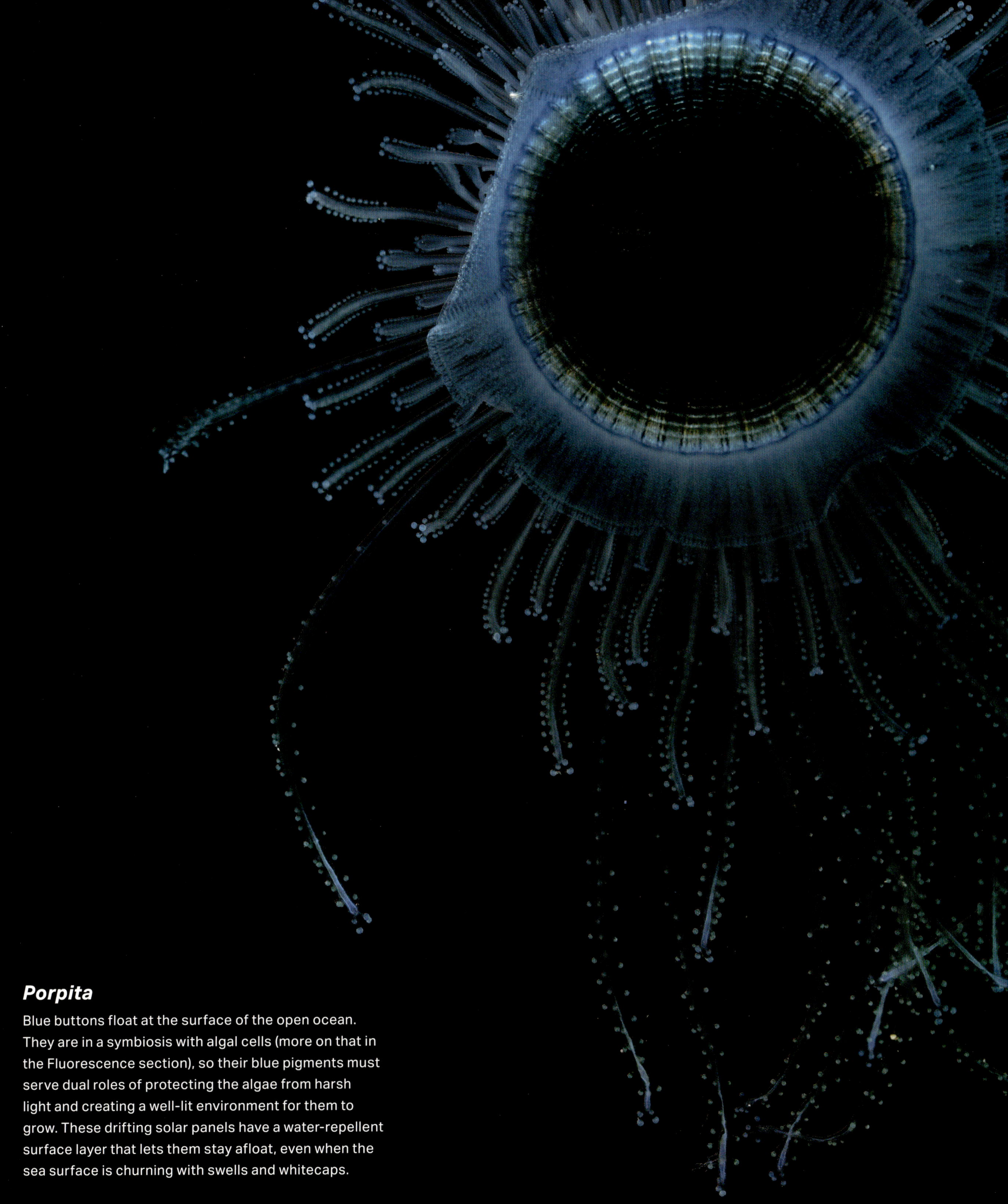

Porpita

Blue buttons float at the surface of the open ocean. They are in a symbiosis with algal cells (more on that in the Fluorescence section), so their blue pigments must serve dual roles of protecting the algae from harsh light and creating a well-lit environment for them to grow. These drifting solar panels have a water-repellent surface layer that lets them stay afloat, even when the sea surface is churning with swells and whitecaps.

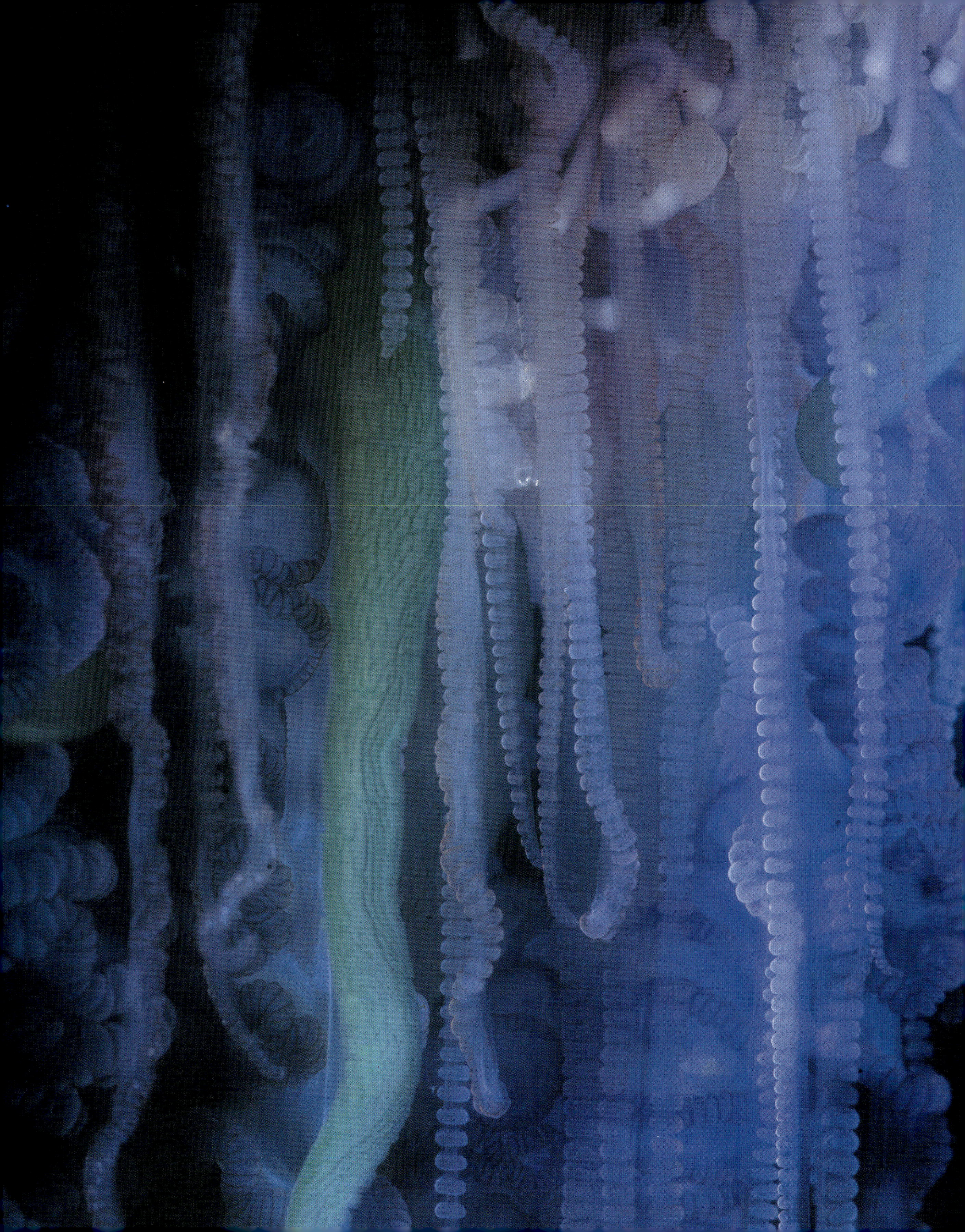

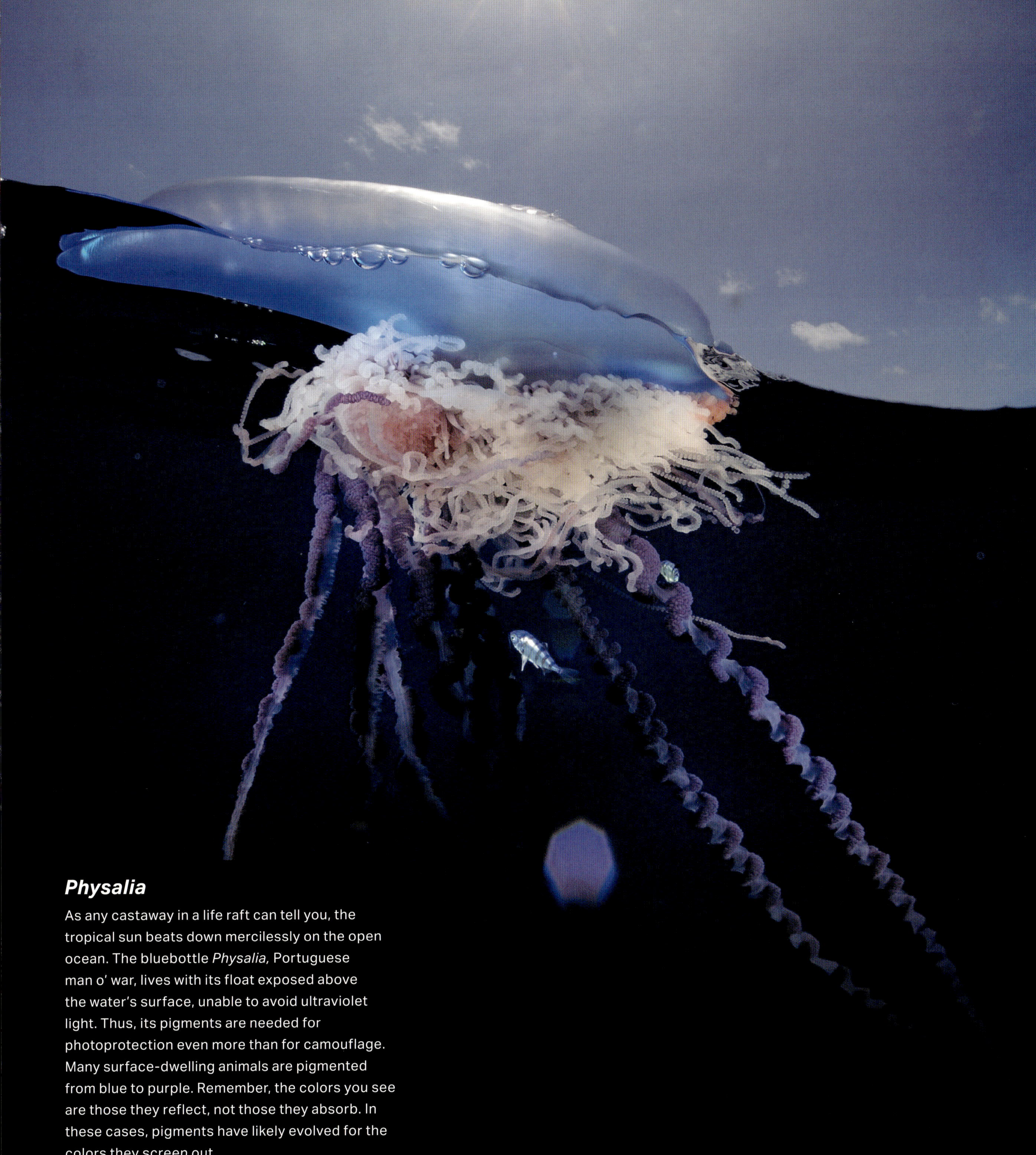

Physalia

As any castaway in a life raft can tell you, the tropical sun beats down mercilessly on the open ocean. The bluebottle *Physalia,* Portuguese man o' war, lives with its float exposed above the water's surface, unable to avoid ultraviolet light. Thus, its pigments are needed for photoprotection even more than for camouflage. Many surface-dwelling animals are pigmented from blue to purple. Remember, the colors you see are those they reflect, not those they absorb. In these cases, pigments have likely evolved for the colors they screen out.

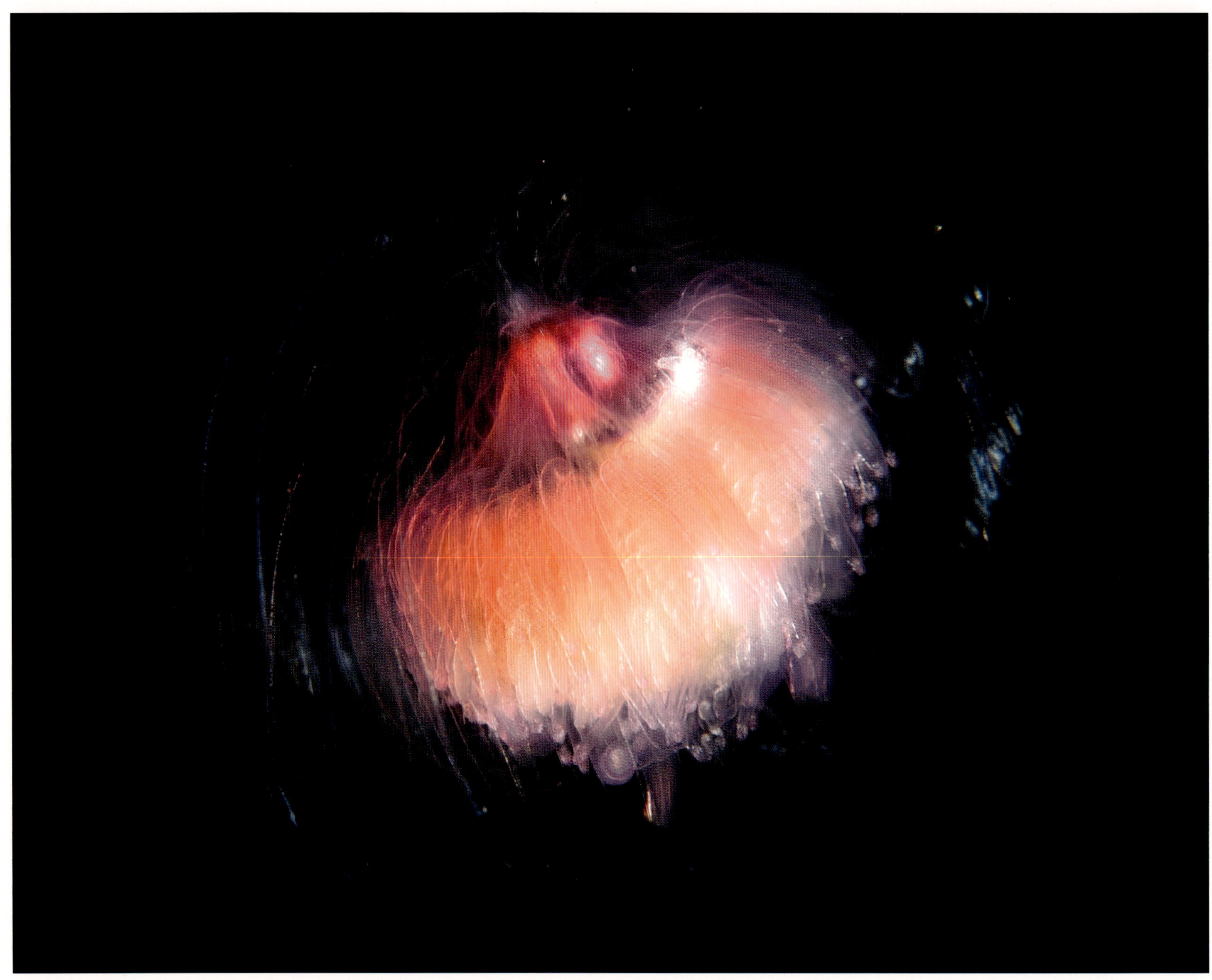

Athorybia

This pom-pom siphonophore (a relative of the hydromedusae we've discussed elsewhere) has a large gas-filled float in the center of a whorl of tentacles. Found near the surface in tropical waters, its pigmentation is a blend of pinks and oranges, which need to function in a broad-spectrum environment before light has been filtered to a more uniform blue.

Solmissus

When animals are mostly transparent in the sea, we often attribute any color variation to pigments that have been obtained from prey, then dissolved out and distributed via the plumbing of their digestive systems. In a study with Dr. Anela Choy, we found that this dinner plate jellyfish was one of the least picky eaters in the midwater food web. It was seen capturing twenty-two different kinds of prey. All that is to say that if the purple pigment in this specimen is obtained from one type of prey, then we don't yet know what that prey would be. In the meantime, we just enjoy the show.

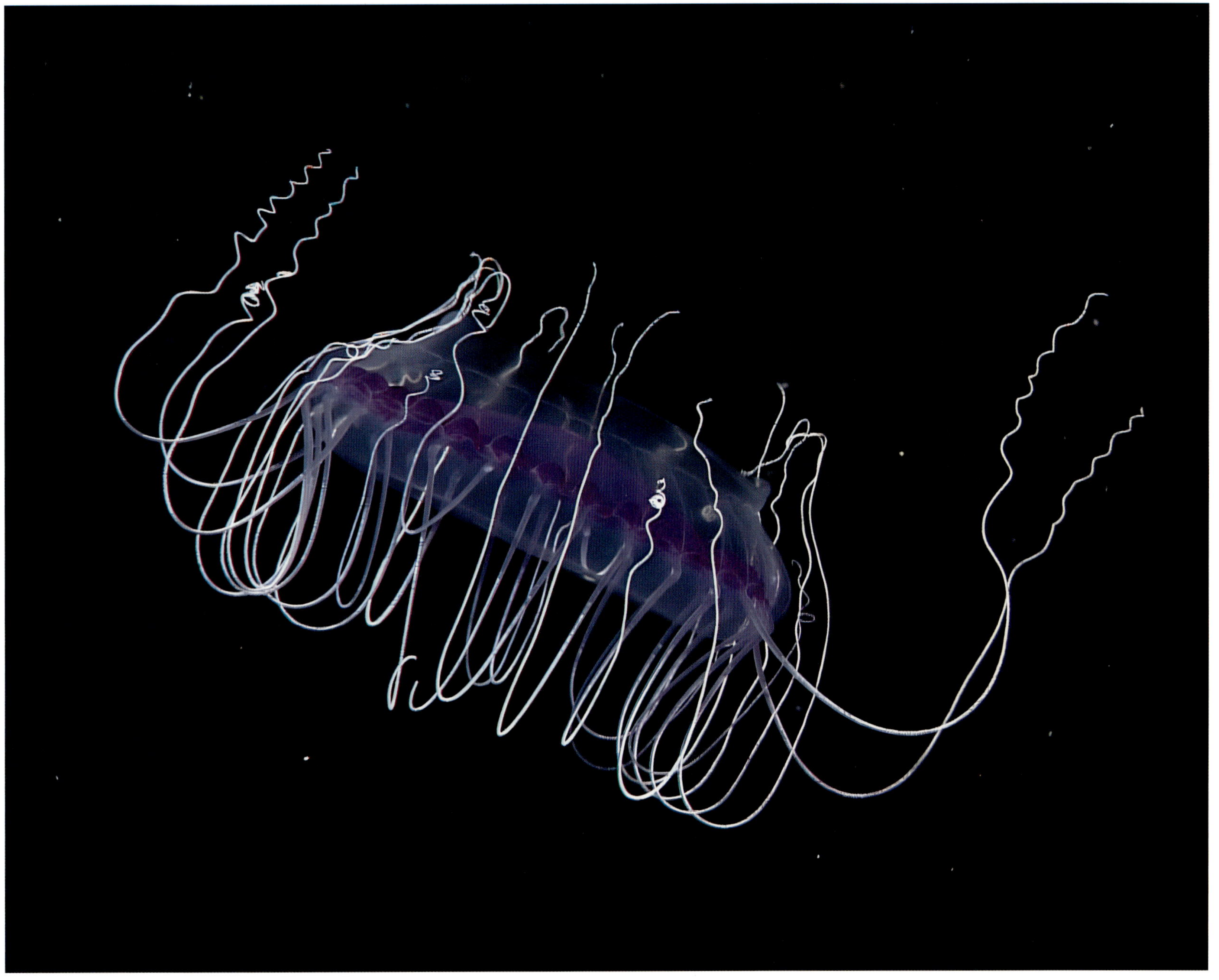

Tetrorchis hydromedusa

The pink pigmentation that is in the canals of this small jellyfish is reminiscent of the canals in *Crossota millsae* (p. 76). The deep-red gut suggests a similar function to so many others, but the pink color may be a dilute version of the red, especially as it is highlighted by the four bright white gonads attached to the inside of the swimming bell.

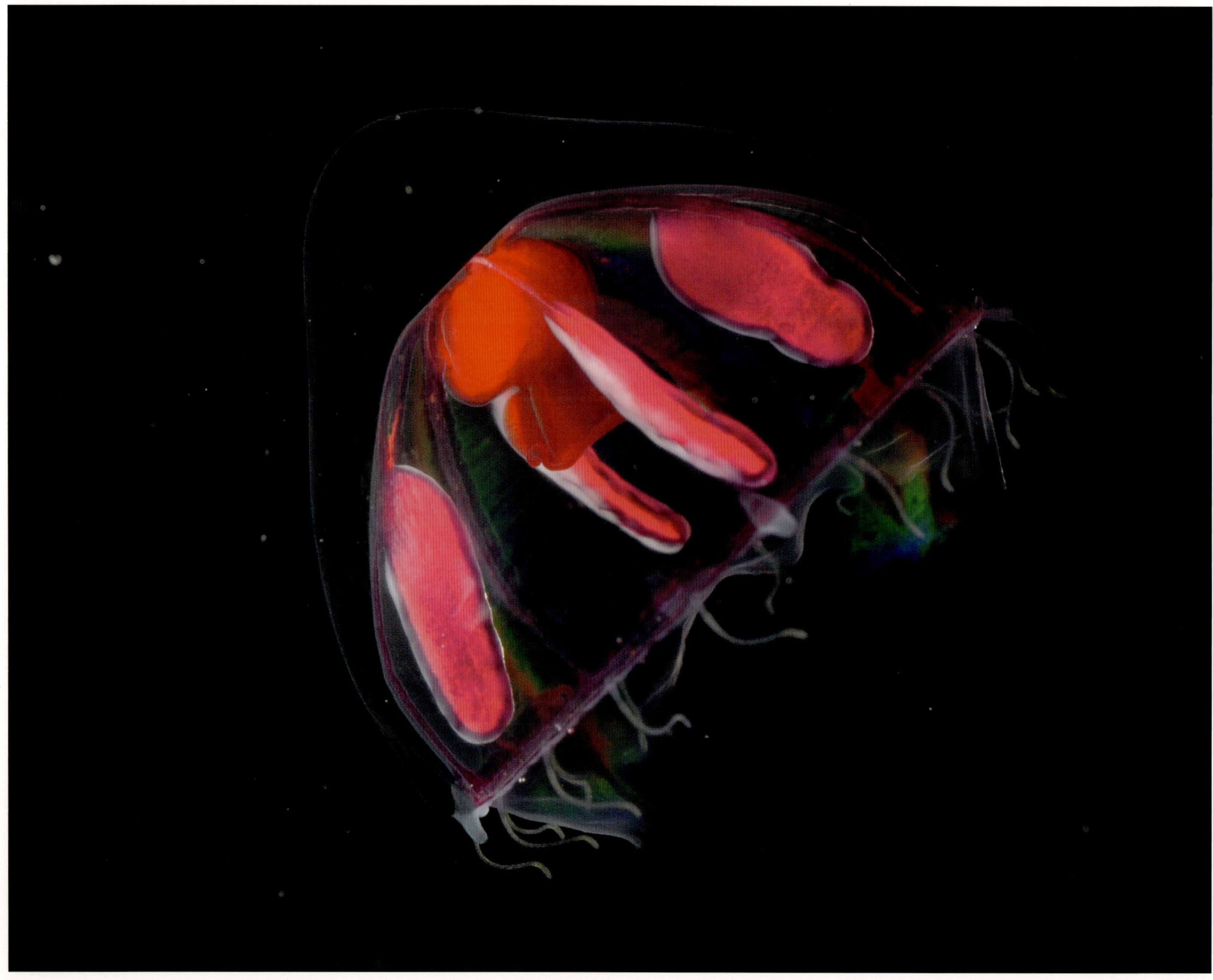

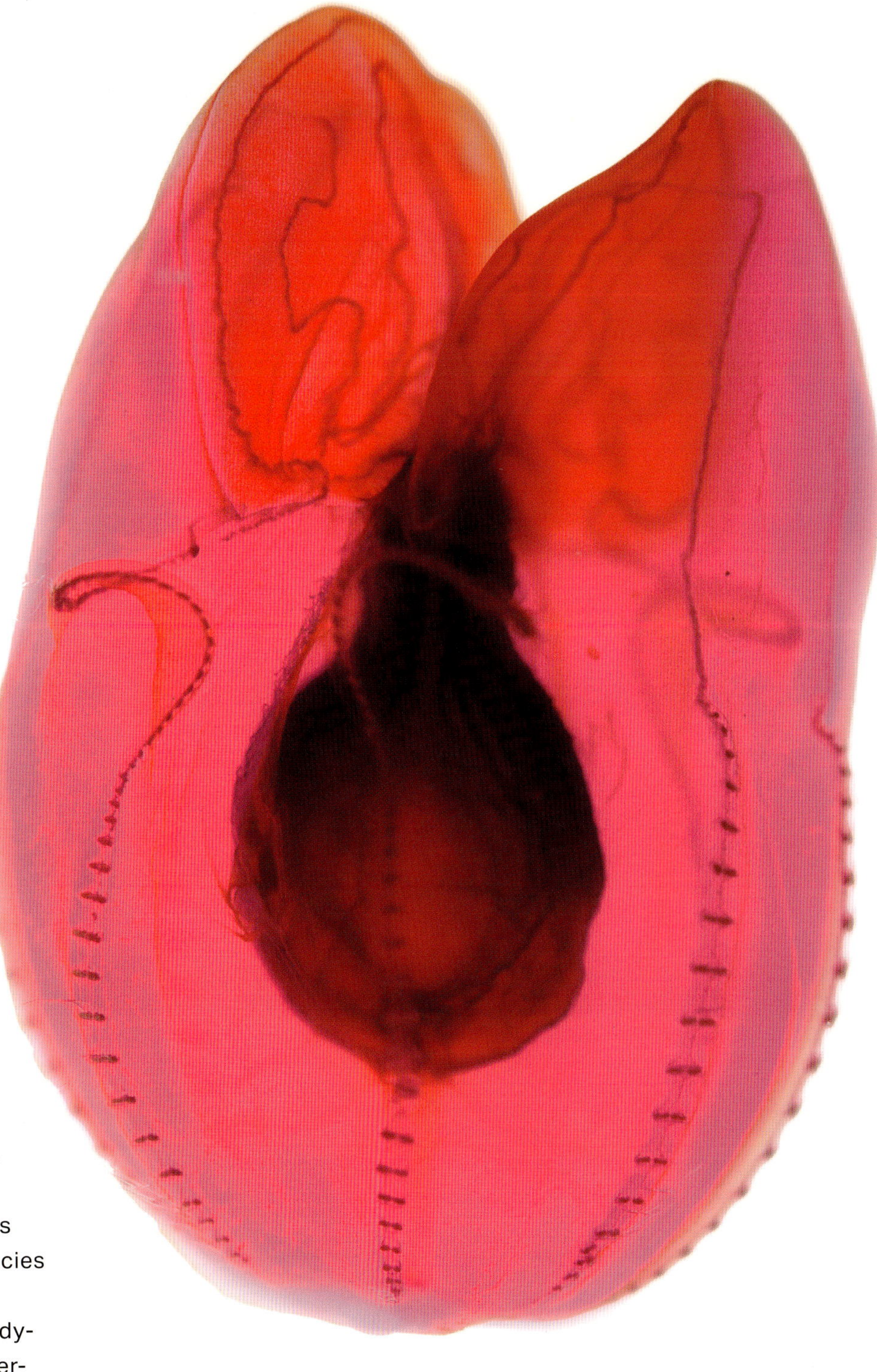

Magenta ctenophore

This unnamed species of deep-sea ctenophore has two predominant colors. The pink body registers as 100 percent magenta in any photo-editing tool, and on top of that (or rather *inside* that), the stomach has a dark-red pigment that absorbs blue light. This species is not yet formally described, but its sister species, which is red instead of magenta, is named the "bloody-belly comb jelly" (see page 2). There is even an amber-colored species as well. Although color is not always a reliable diagnostic character for identifying animals, in this case it tracks with the genetic diversity of these three relatives.

Because the function of a red pigment is to appear black in blue light, it is often difficult to distinguish the true color of a red subject on a black background. This specimen has been photographed against a white background to fully reveal its magenta/fuchsia pigmentation.

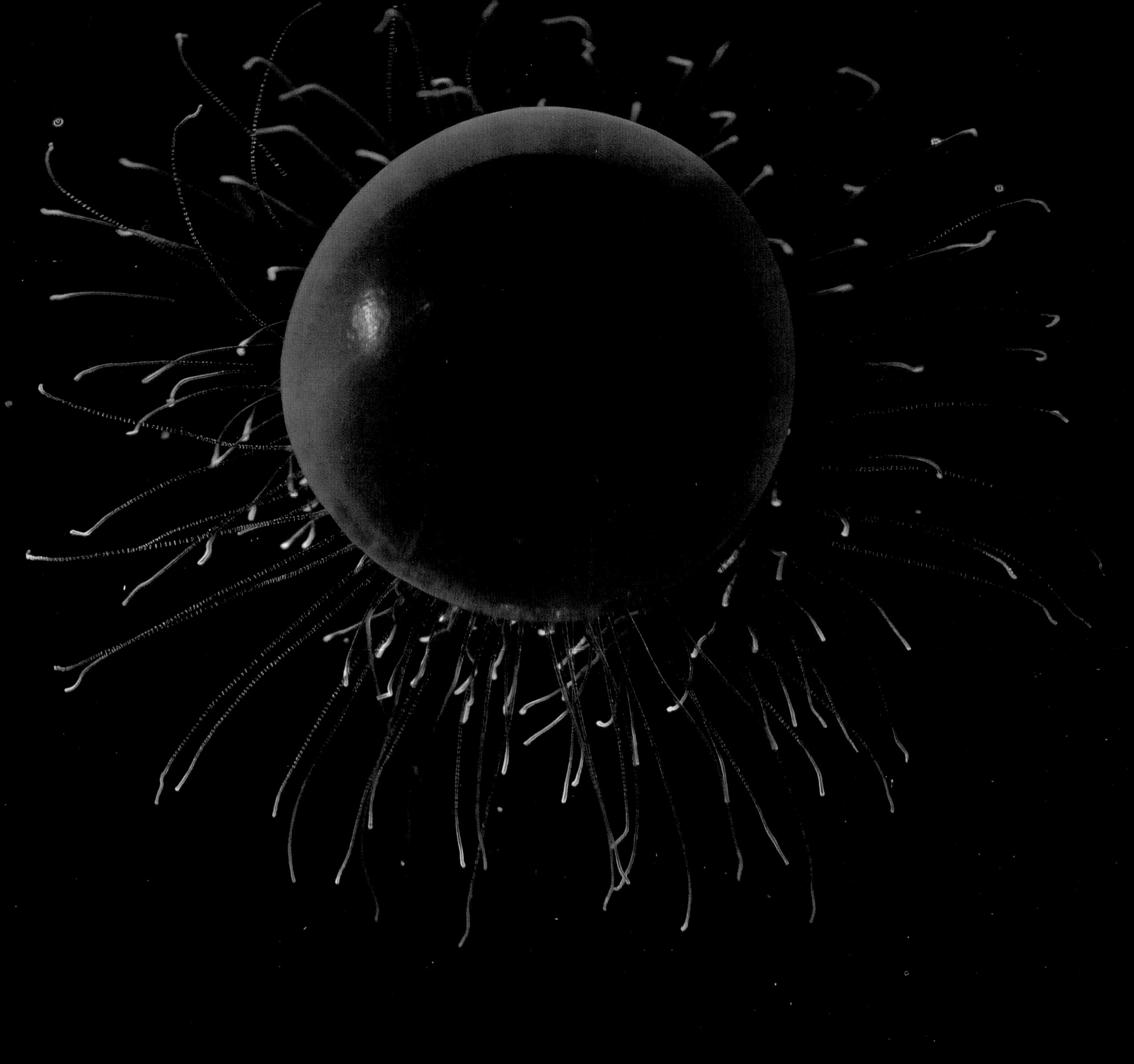

Vampyrocrossota childressi

At first glance, these medusae seem to be your typical clear jellies, photographed with a black background showing through their transparent bell. Upon closer examination, you can see that the inner surface of their bell is lined with an inky black pigment. In other organisms, like siphonophores, black or purple pigment would indicate that they'd eaten a deep-sea fish and incorporated its melanin pigment. However this species only reaches 2 to 3 cm across and is very fragile. It seems most likely that they eat small crustaceans and the pigment is self-made—a perfect research project for an aspiring biochemist.

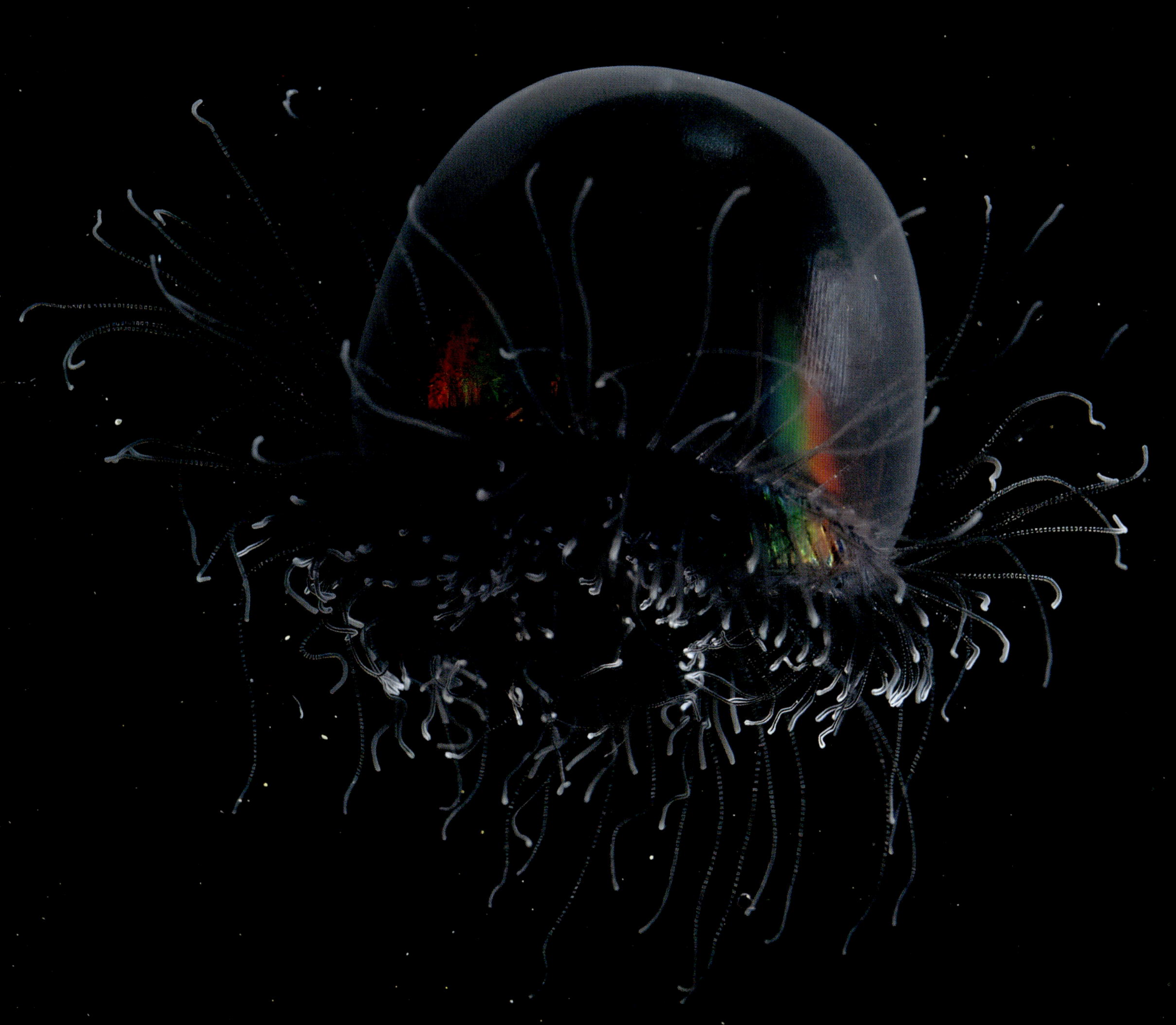

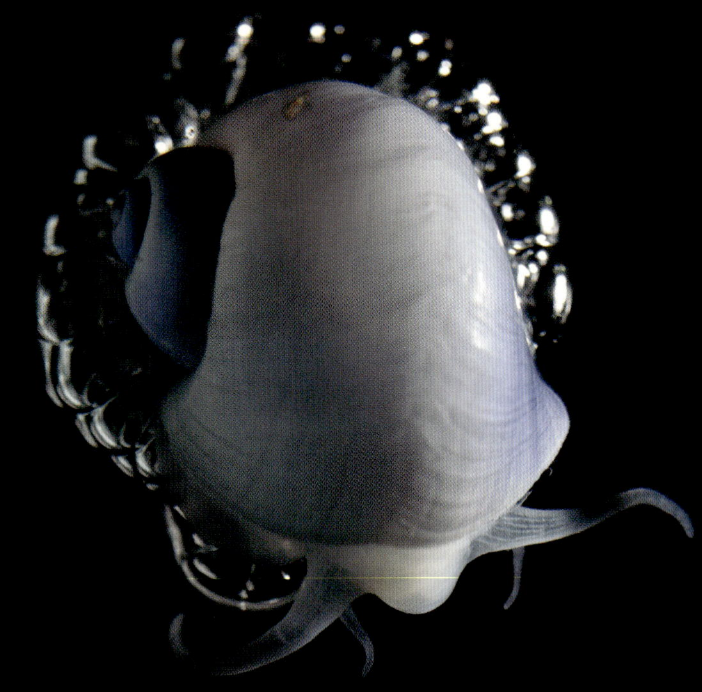

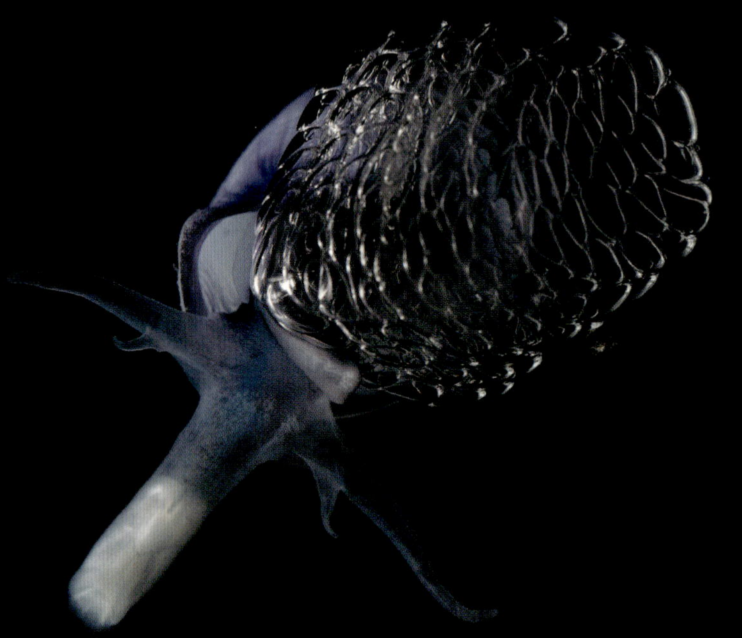

Janthina

In the Transparency section, you saw a bevy of snails that had evolved to live suspended in the water column. This raft-riding snail looks more like its terrestrial cousins, having retained a heavy lavender-colored shell. Instead of crawling across the forest floor (or through your garden), they crawl across the underside of the ocean's surface. They periodically extend their trunks to suck in a mouthful of air, blow it into a mucus bubble, and add it to their raft. This near- or above-surface environment is home to many animals that have blue or purple pigments.

Melanostigma

Eelpouts (not really eels, but definitely with a pouting expression) are shy fish that can be highly abundant in the deep sea. While some fishes will dart away from the bright lights of our submersibles, eelpouts instead curl up in a little "you can't see me" loop. Most predators probably can't see them, given their dark melanin pigmentation. The scientific name for this genus (*Melanostigma*) means "black spot."

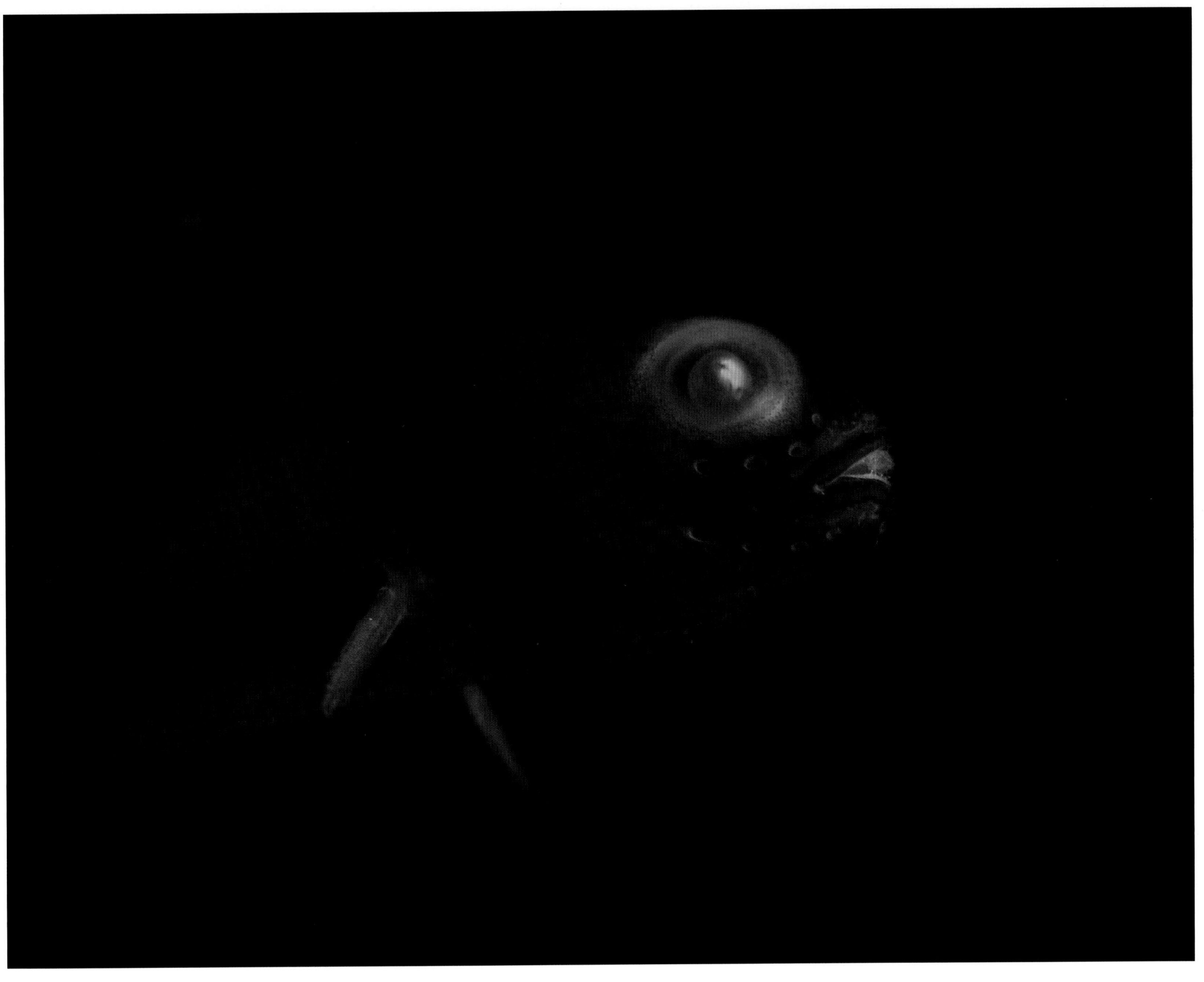

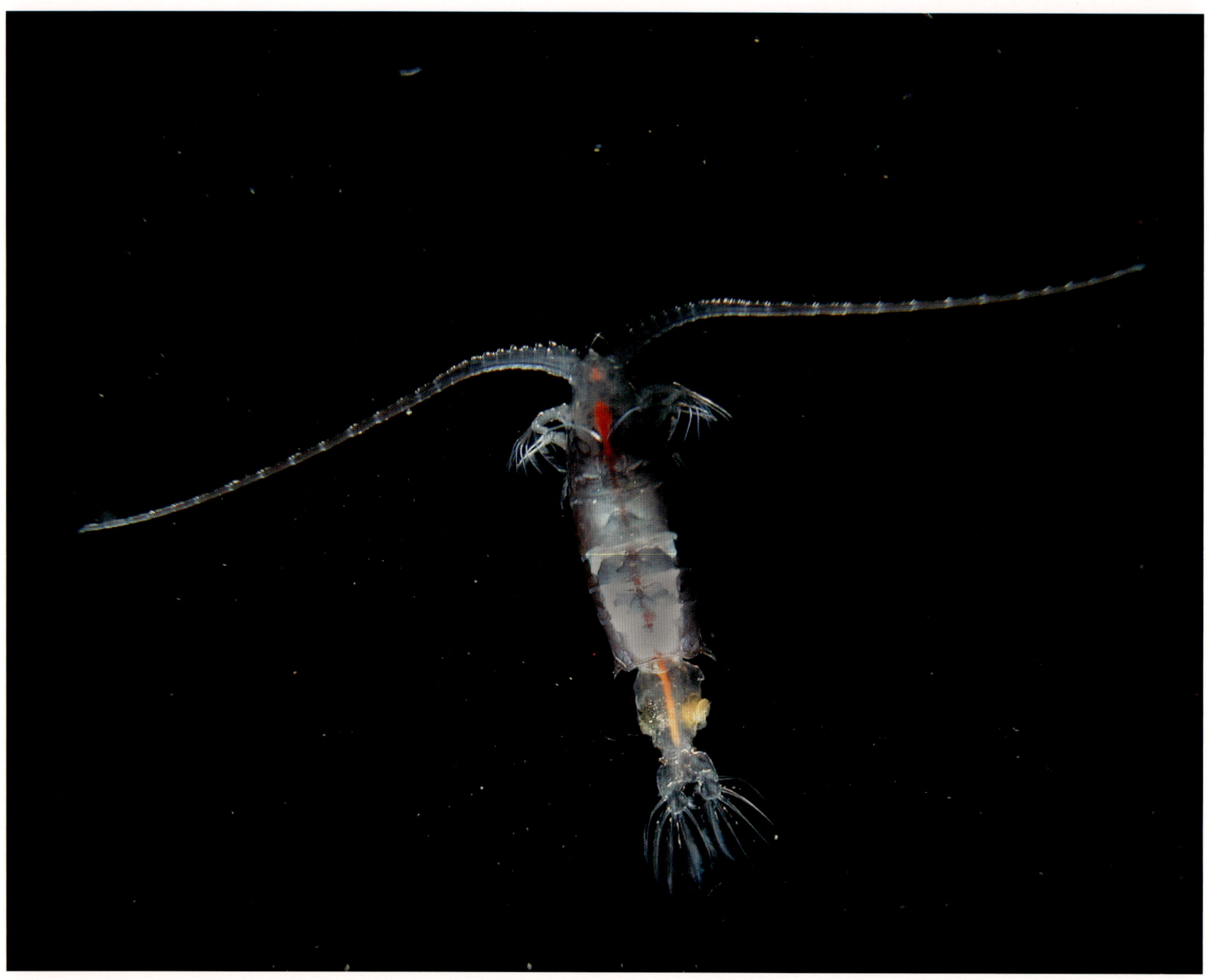

Gaussia princeps

The prince copepod is one of the largest free-living copepods in the sea, reaching the size of a large housefly. It has bright bioluminescence but also contains black melanin pigmentation—the same compound found in black fish! Since *Gaussia* don't eat fish (except maybe as scavengers), the origin and functions of its goth appearance are unknown.

Vampyroteuthis infernalis

Their name sounds intimidating, but "vampire squid" were specified for their appearance not their disposition; they are, in fact, completely docile. They get through life by having a slow metabolism and taking refuge in the lowest-oxygen waters, about 700 meters below the surface, which more oxygen-hungry predators avoid. They are said to eat "marine snow"—the fluff of particles that sink down from above—but we're not totally convinced, since we've seen them eating shrimp and other crustaceans.

The appearance of the vampire squid varies depending on their age, condition, and how they are photographed. Often the youngest look jet black (borderline purple) and the adults have a chestnut or dark-chocolate color. When photographed with the exposure or lighting cranked up, they may appear red or tan in color. In their own habitat they are about as dark as one can be without using ultra-black technology (as described in the Iridescence section).

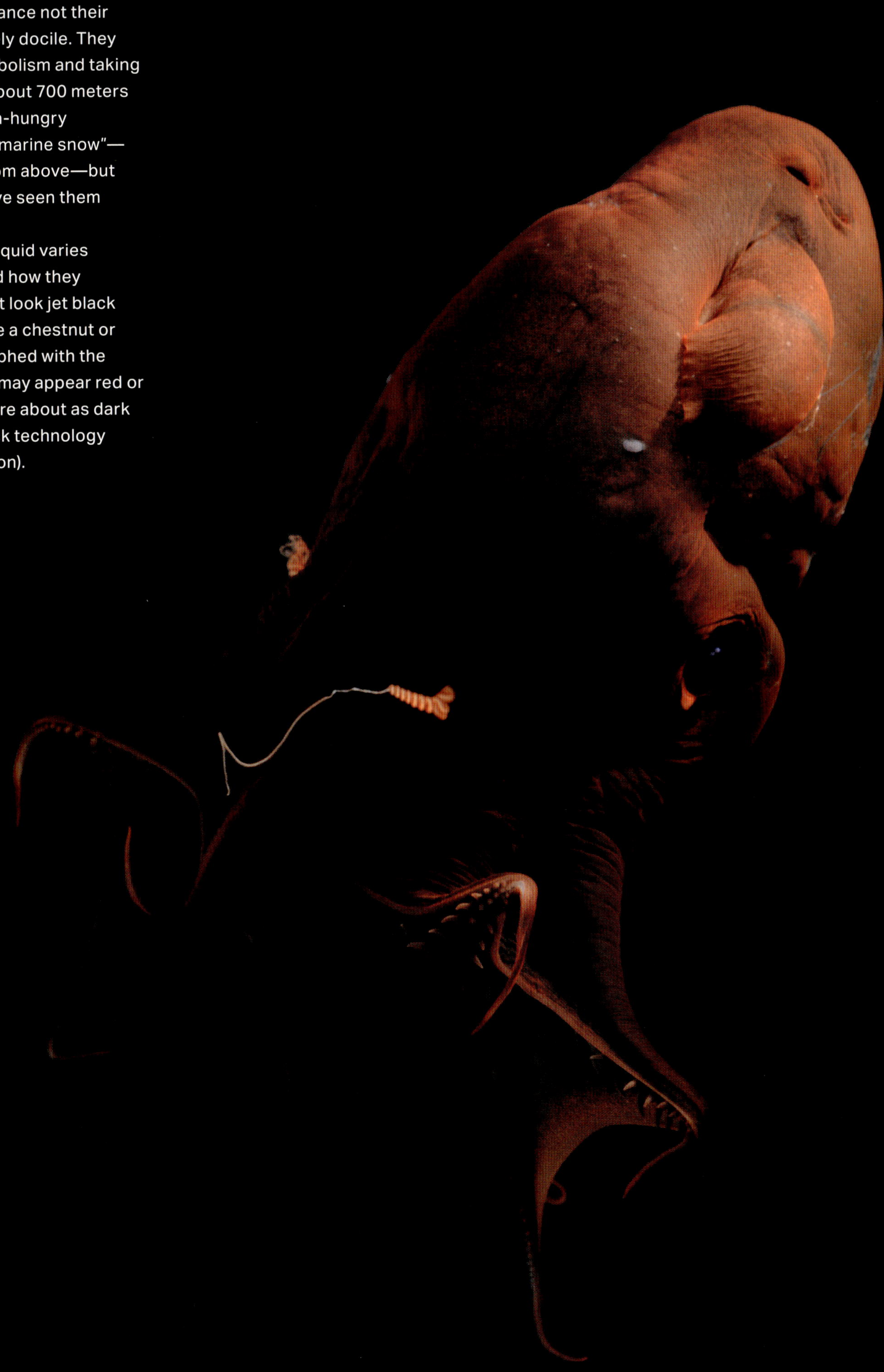

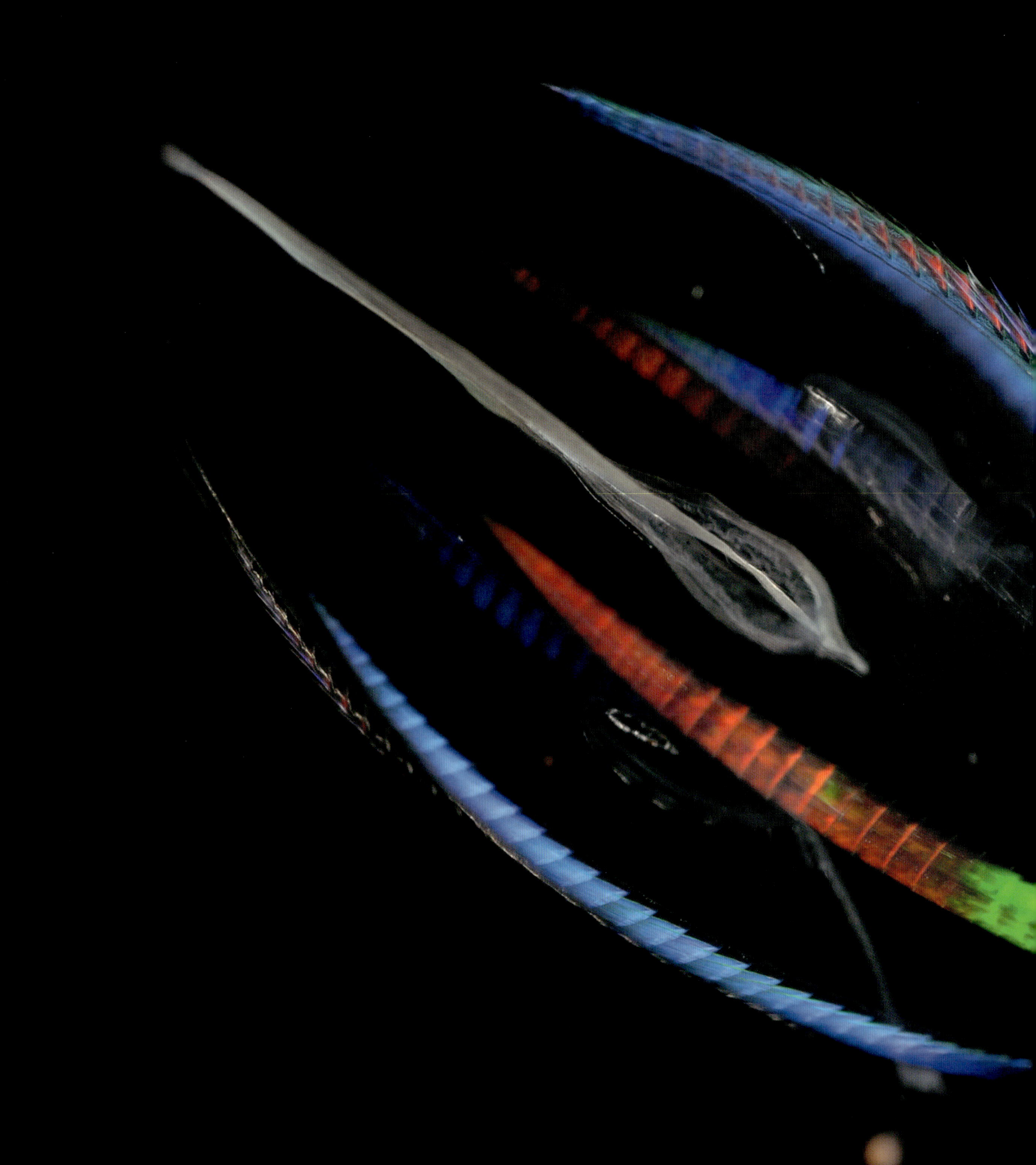

IRIDESCENCE

Color from structure

What do a rainbow, a sunset, and a soap bubble have in common?

Even seemingly transparent materials can interact with white light in ways that produce colors—many more vibrant than the colors you see in pigmented tissues. This section talks about optical effects based on the scattering of light. Absorption of light is explored in the previous section on Pigmentation.

Earlier, when we talked about transparent animals, we said that light went through them as if they weren't there. This isn't actually true. Look up from this book and through any window. The light from the scene outside does not pass through the glass. Instead, like rain dropping into a pond, the light instantly interacts with the molecules it encounters at the front surface of the glass. Each molecule emits its own light, which then interacts with more molecules. The drop of water that sets off a ring of waves in a pond does not travel to the shore itself, but it creates waves that interact with those set off by neighboring drops. Light waves interact as well—in some places piling up and in other places canceling each other out—in the same way that noise-canceling headphones can quiet a sound by adding its opposite waveform. Your window, like the pond, is full of frantic activity, but—in one of the greatest pieces of magic in our natural world—as long as the glass is uniform, the light will pop out the other side as if nothing had happened. In a way, when you think you are looking through a window, you're actually watching a movie of the outside world emitted by the glass itself.

Suppose now that the glass is not uniform but instead has some complicated geometry. Then different colors of light will interact with the surface in slightly different ways—some will go through, others will be reflected, and which color ends up where will usually depend on the viewing angle. This, in a nutshell, is structural color; vivid optical effects made not by pigments but by interactions of light caused by tiny variations in the internal geometry of transparent materials. This process is responsible for the rainbows that we see in the sky and in soap bubbles, oil slicks, older compact discs, or DVDs. It is also responsible for many other colors that we see in the natural world: the purple necks of pigeons; green beetles; blue butterflies; and the nocturnal shine from deers' eyes under headlights. On land it is especially common in insects and birds; in the ocean it is found in a diverse array of animals and even algae. The required internal structures have to be really small—less than one-thousandth of a millimeter, or about one-hundredth of the width of a human hair—but it turns out that life is especially good at making things of just this size, so we see structural colors everywhere.

In the images that follow, we showcase structural colors in the marine environment. These effects are responsible for not only iridescent colors, but also for mirrors, and—paradoxically—for making certain animals as black as possible. One must be careful, however, when pondering these pictures, because most often they are taken with a flash. Flashes produce strong, directional white light that makes iridescence particularly obvious. In contrast, light in the underwater world is nearly always diffuse and of limited colors. The striking pictures here may not represent what the animal looks like in its own world. Some structural colors, while beautiful, may have no function, although they may still tell us something about the nanoscale structure of the objects we are viewing.

COLORS FROM STRUCTURE: IRIDESCENCE

A common form of structural color is iridescence. Iridescence can arise from stacks of partially reflective materials and thin, closely spaced grooves or ridges—many conditions where wavelength-dependent scattering of light occurs. In iridescence, we usually see a bold, pure color, which is strongly affected by the angles of viewing and illumination. Iridescent materials will often shift hue and brightness as you change the position of your eyes or

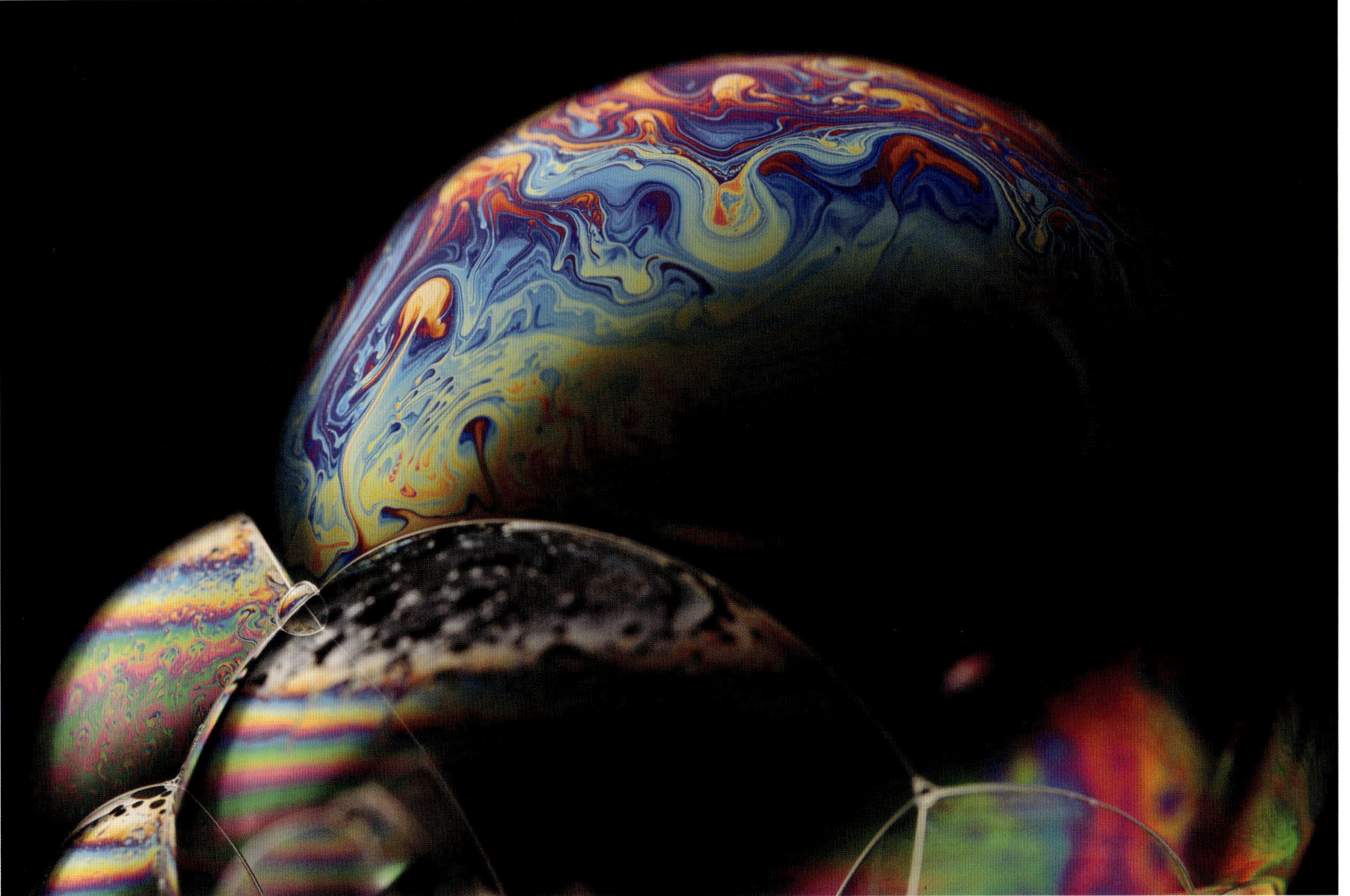

The colors in a soap bubble are a direct indication of the distance between the inner and outer surface of the liquid film. As the bubble thins, the colors progress through the rainbow, eventually reaching gold, then silver... and then the bubble pops.

as the animal moves or rotates. The vivid, changeable nature of the color makes striking displays, especially when combined with movement. As we mentioned above, iridescence is best seen under highly directional light, which is far more common in near-surface clear waters, such as tropical seas and reefs. In these habitats, you will meet sapphire blue copepods, scintillating squid, and shiny clams. Going deeper in the ocean, iridescence becomes rarer, probably because the light field becomes more diffuse, but one still finds it. A fair number of predatory dragonfish have iridescent bronze coloration on their sides, and some are mirrored like silver coins, as discussed below. We've also found a type of siphonophore that is suffused with bright, iridescent blue particles of unknown origin or function. Again, you have to be a little careful interpreting the spectacular rainbows we see in photographs because we are introducing an unnaturally bright white light into the situation.

The easiest way to see iridescent color—short of mounting an expedition—is to start blowing soap bubbles. While the true physics is more complex, the reflected colors from soap bubbles can be explained as the interference between the light wave that hits the outer surface of the soap film and the light wave that hits the inner surface. Whether light is a particle or wave is a one-hundred-year-old debate, but it does often act like a wave. In this case, the reflected color is the one with the wavelength where the reflections from the front and back surface add up. You can actually infer the thickness of the soap bubble by looking at the color it reflects. In thicker areas, multiple colors can be reflected at the same time, leading to the pale greens and pinks we often see.

MIRRORS FROM STRUCTURE

As anyone who has opened a can of sardines or anchovies knows, many fish are silvery. This makes them obvious on land, and probably makes a school of fish a confusing sight in near-surface waters, but the primary purpose for silvery sides in fish is camouflage. On land, to make an invisibility cloak, you have to project the light field hitting your back so it comes out of the front. In the ocean, the ambient light coming from the sides is similar, so a vertically oriented mirror reflects light that looks just like the light you'd see if you were able to look through it. We know that fish, via evolution, are paying attention to this fact because the mirrors inside their scales are all mounted vertically, even if their bodies are curved. In other words, even many fairly cylindrical fish have evolved to act optically like a vertical mirror. This is common, especially in open water, and is found from large predatory fish such as jacks and tarpon to small bait fish such as sardines; even down to deep-sea hatchetfish and lanternfish. In this chapter, you'll also see two fish with amazing mirrors: the lookdown and pompano.

Marine mirrors aren't only found on the sides of fish. As we discussed in the Transparency section, they're also used to hide the food in stomachs

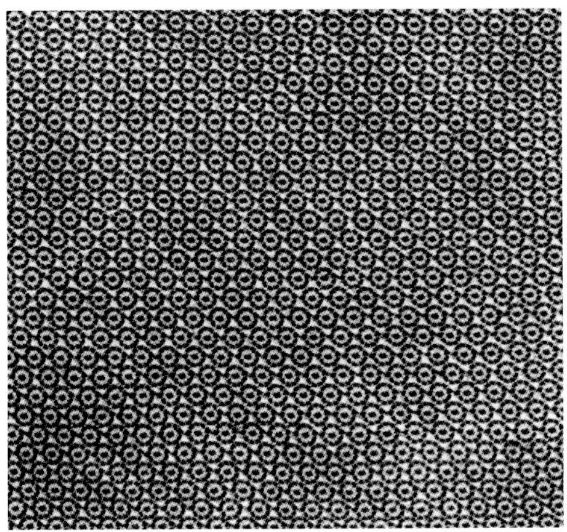

Using an electron microscope, you can visualize a cross-section through a cluster of cilia from a ctenophore. Each circular flower-like shape shows the filaments of one long tube, less than 200 nanometers in diameter—that's less than half the wavelength of visible light. The whole cluster shown is less than 5 microns across—15 times narrower than the width of a human hair. This uniformly packed array makes an ideal structure for interacting with light, reflecting shorter wavelengths of light toward more perpendicular viewing angles, and longer wavelengths (more red) when viewed at a shallow angle. Although the most prominent arrays of cilia are found in the long rows used for propulsion, this micrograph taken by ctenophore biologist Sid Tamm shows a ciliary "tooth" from the mouth of the ctenophore, *Beroe*.

and the black retinas in eyes. Some marine animals even use mirrors to gather more light for vision, much like modern telescopes. Surprisingly, silver mirrors are extremely rare on land, although there is a truly stunning silver scarab beetle in the rainforests of Central America.

Mirrors—at least most of them—don't have color, but they work in the same fashion as other structural colors. Within the mirror are a number of stacked plates, but the difference between these stacks and the ones that produce an iridescent color is that the distances between the plates vary all over the place. Each plate–gap–plate combo thus is best at reflecting a different color of light, so the silvery color we see is actually the sum of a wide variety of colors. You can see this if you look at a silvery fish under a microscope. Up close, you don't see silver but a mosaic of all of the colors of the rainbow.

There are a few colored mirrors in the ocean. For example, some deep-sea fish look like they are made of bronze. What is going on here remains a mystery, but the answer may lie in the fact that the light hitting the fish when it is in the water comes from the top, not the side. Therefore, the reflected color of the mirror may not actually be bronze.

ULTRA-BLACK AND ULTRA-WHITE FROM STRUCTURE

In a seeming paradox, one of the most amazing uses of structural color results in an animal that is completely black. As we've discussed before, the open sea is a dangerous place, full of predators coming at you from all directions. In the deep sea, this is made worse by the fact that many predators are swimming around with their own flashlights. More on these in the Bioluminescence section, but for now we need to discuss what this means for the animals "in the headlights." Because water itself is mostly empty and reflects almost no light, any little bit of reflection from these flashlights will give the prey away. Even being transparent is not good enough, because transparent animals—like windows illuminated at night—reflect some light. In addition, fish that use bioluminescent lures need to make sure that their lures don't light themselves up and spoil the trick. No one feels comfortable venturing forth for a snack and being greeted by a leering mouth full of long teeth!

As a result, some fish have evolved to reflect as little light as possible. This can be achieved using black or brown pigments, but the most remarkable ones use structure to give them a boost. These are one hundred times blacker than the usual objects we think of as quite black (black paper, tires, and so on). In fact, they are as black as Vantablack, the blackest known human-made substance. This ultra-black coloration makes them almost invisible to animals with flashlights, and also lets anglerfish use their lures without being given away.

You might suppose that making something extremely black is simply a matter of adding more and more pigment. In the case of fish, the pigment would be melanin, which we discuss in the Pigmentation section. However, this by itself doesn't actually work because even the most pigment-laden material still reflects at least a few percent of light from its surface, sometimes much more. So the material—as is true of the other materials we discuss in this section—needs internal structure. In this case, the structure is there to make the light bounce around and around, as if in a tiny pinball machine, until it has bounced against so many pigmented internal structures that it is all absorbed.

In ultra-black fish, we found that this is achieved by having skin that is filled with tiny particles of melanin called melanosomes. In addition, these particles are not spheres, but are stretched a little so that they resemble Tic Tac candies. It turns out that the size and shape of these melanosomes are just right, so that nearly all the light that enters the skin is trapped inside it and ends up being absorbed. This clever trick is now being used by engineers who want to make ultra-black materials of their own for cameras, solar cells, and other devices.

Finally, a number of animals use similar structures without the melanin to make ultra-white materials. These are often used by marine animals to advertise themselves. Cleaner shrimps use tiny spheres embedded in their antennae to make them ultra-white, thus a good signal for telling fish to please not eat them while they are cleaning them! The most remarkable ultra-white animal, though, is the disco clam (*Ctenoides ales*, p. 123), which uses thousands of tiny glass spheres to make it look like lightning is flashing inside its body.

In sum, animals can dramatically change the way they look by altering their microscopic structure. Animals are particularly effective at manipulating structures of the sizes that interact best with light. Human engineers are not nearly so good at doing this, but the animals have provided remarkable inspiration for what can be done.

Sapphirina

A classic example of iridescence is the sapphire copepod, which is found near the surface in blue tropical seas. Copepods, which are distantly related to shrimp, might be called the breakfast cereal of the sea—common and important but not exciting. They're typically only a few millimeters long and are nondescript (at least to our eyes). *Sapphirina* can be 10 mm long and males have minute crystals in their cells that are arranged in layers. The distance between the layers of crystal and the angle of viewing determine which colors are reflected. As you might expect from their name, some of the species of *Sapphirina* have strikingly blue males, while others are more golden. At other times, and when the light strikes from a different angle, the layers are spaced so that the animal is totally transparent. When in their "formal attire," the males are on display. While scuba diving, we've seen them stacked in a vertical column, each orienting to its downstairs neighbor like a squadron of jets. Other times, you can see them falling through the sea like tumbling jewels. The reflected light is so bright and is such a pure blue that it is often mistaken as a bioluminescent flash, and has even been reported in literature as such. This animal doesn't produce blue light, though; it only *reflects* blue. In its shallow, well-lit habitat, that is more than sufficient.

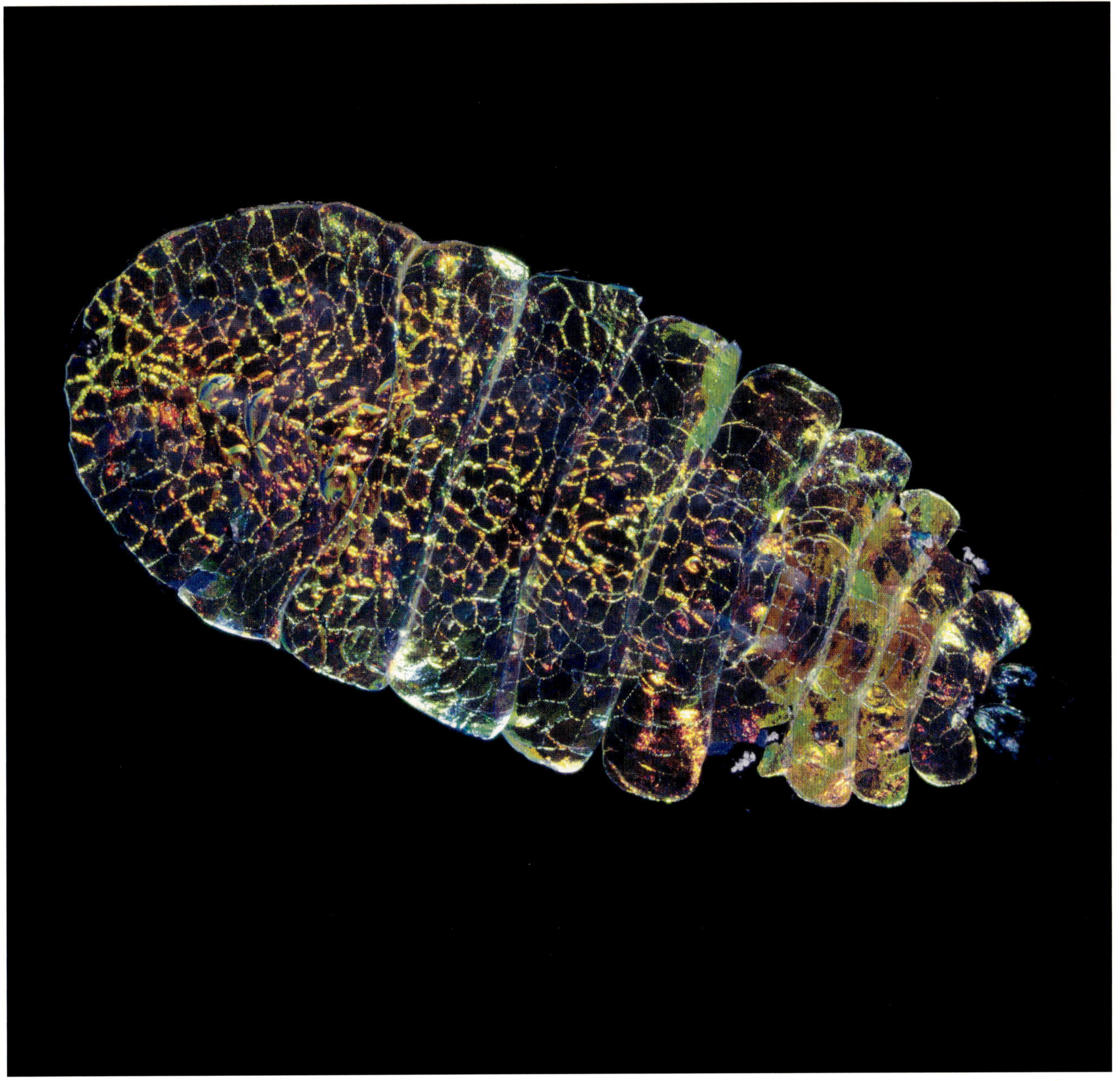

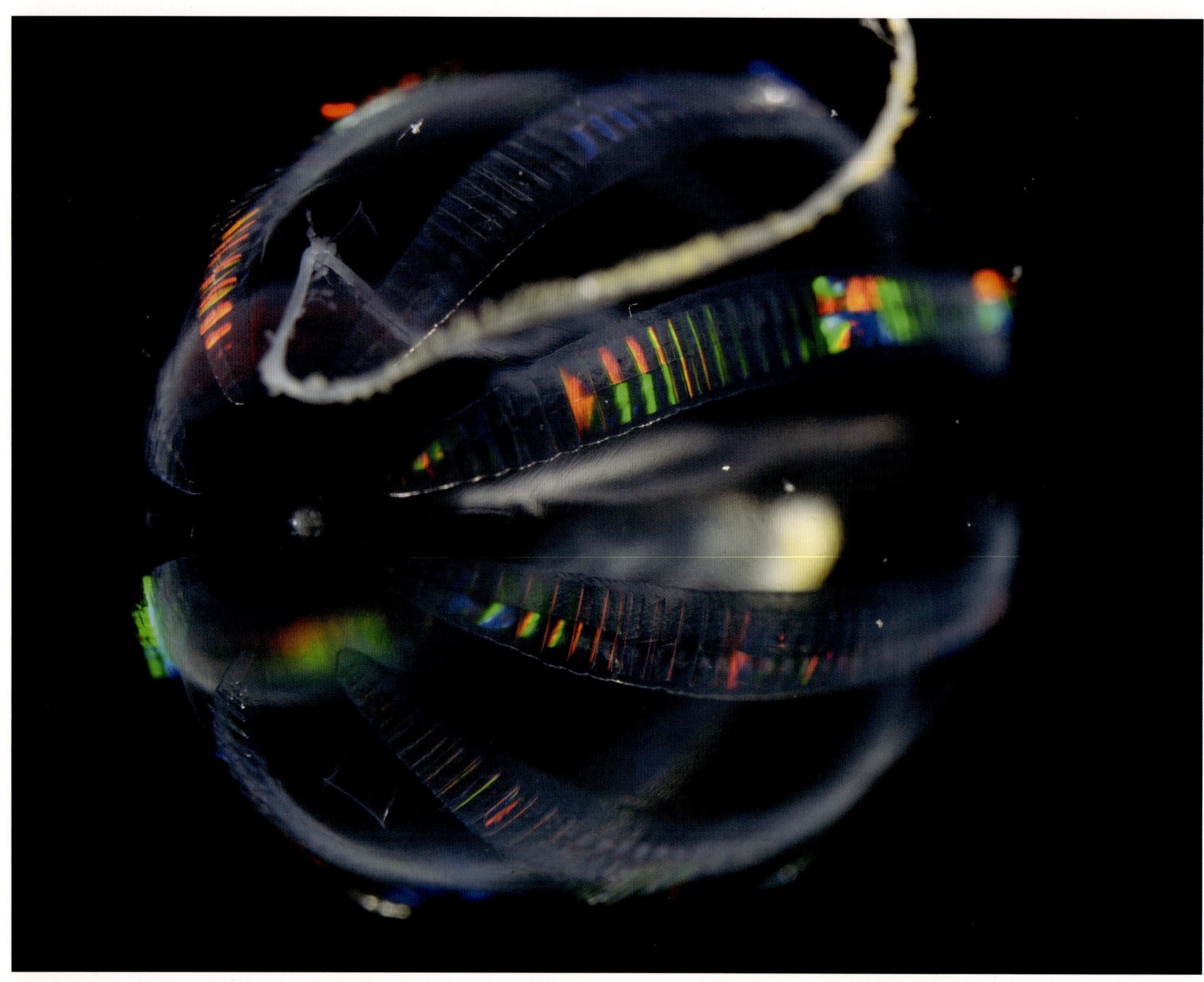

Ctenophore comb rows

Ctenophores, commonly called comb jellies, are named for the eight rows of paddles, or combs, which they beat to move through the water. One can think of it as a small thousand-person canoe. Each of these paddles in turn is made from a large number of tiny filaments (known as cilia), almost like the bristles of a microscopic paintbrush. (Imagine an ant painter, not a mouse artist.) A side effect of all this structure is that the plates of cilia act very much like the grooves in compact discs, and are highly iridescent. You might read that they produce beautiful rainbow "bioluminescence," but their bioluminescence is purely blue-green. The iridescence is beautiful to see, yet it only occurs when the animal is illuminated by a bright light from the side, such as a camera flash. In the more diffuse light of the sea it's much less noticeable. Sometimes on a scuba dive you can see a flicker of brighter light, but not the beautiful multicolored iridescence. Sadly, this beauty likely has no ecological purpose; however, if it stirs someone to care about ocean inhabitants then it has done its job.

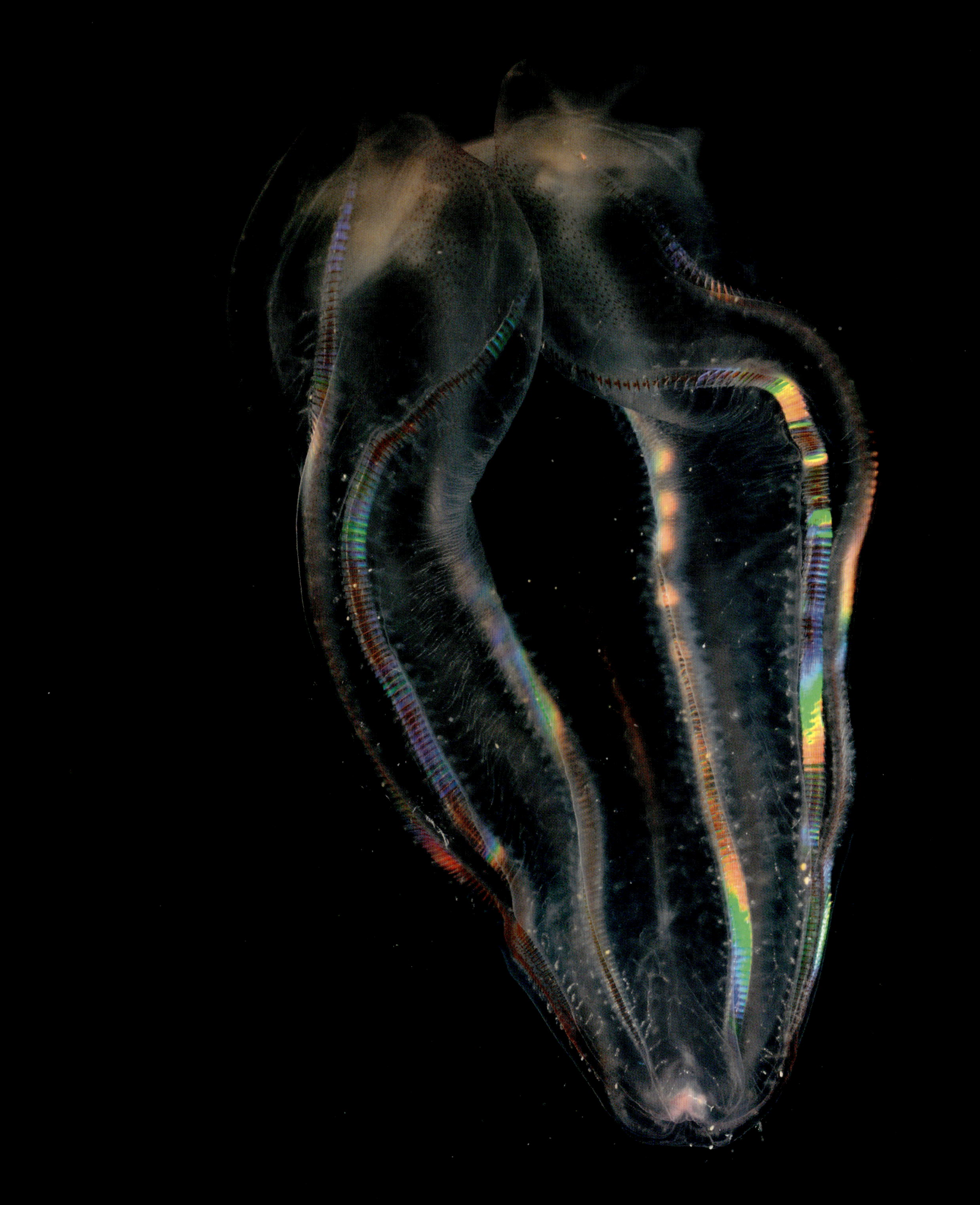

Hapalochlaena

On land, brightly colored animals are often toxic or taste bad. The color serves as a warning that it's in both of our best interests if you don't try to eat me—a phenomenon biologists call aposematism. The same adaptation occurs in the ocean, and in at least one of the cases, the display is enhanced by the color-bursting innovation of iridescence. The diminutive blue-ringed octopus *Hapalochlaena* is highly venomous: a 15-cm animal packs enough tetrodotoxin that it can kill a human if provoked. Given human behavior, this is one case where the beautiful bright pattern may encourage more interactions than evolution intended.

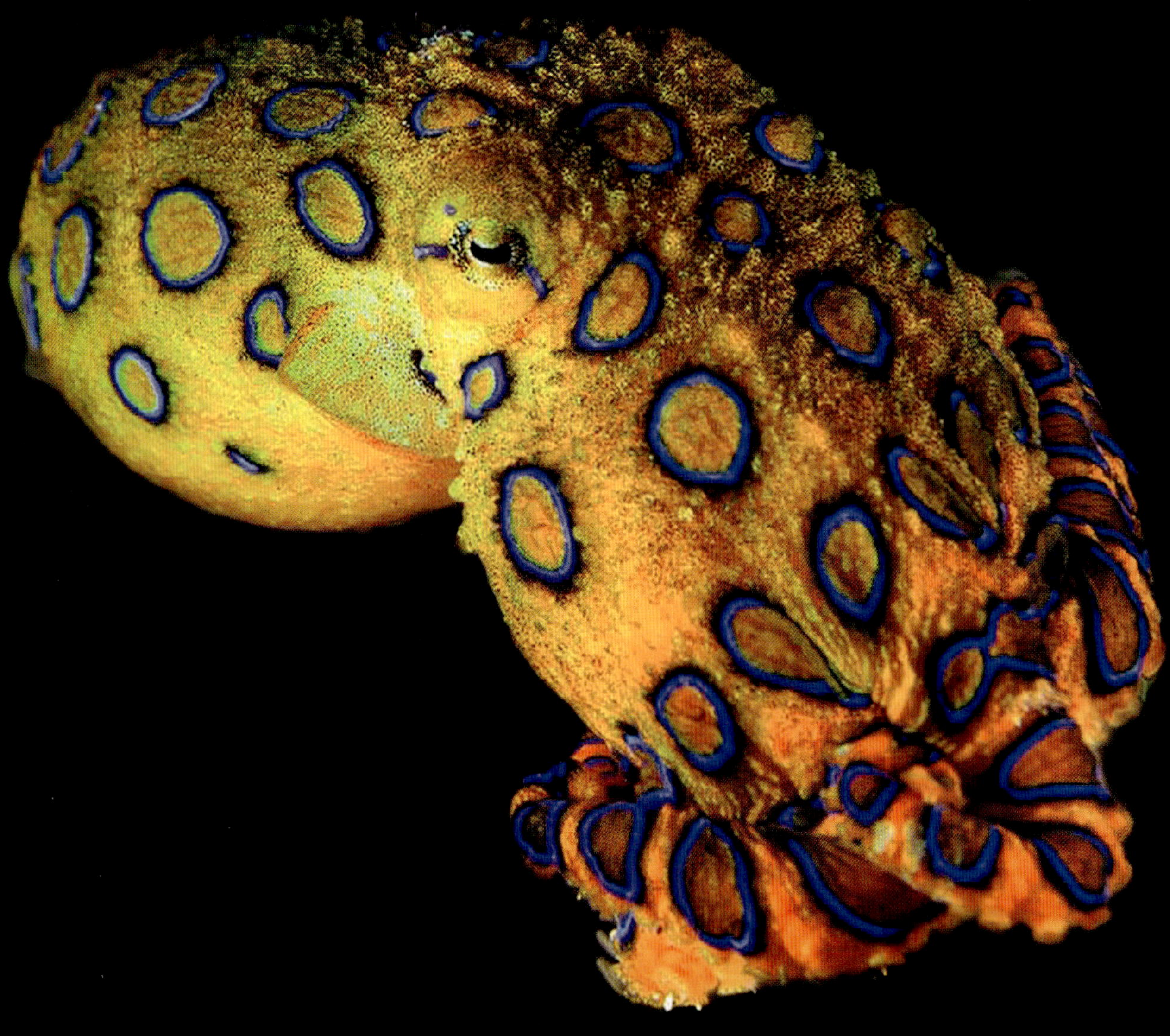

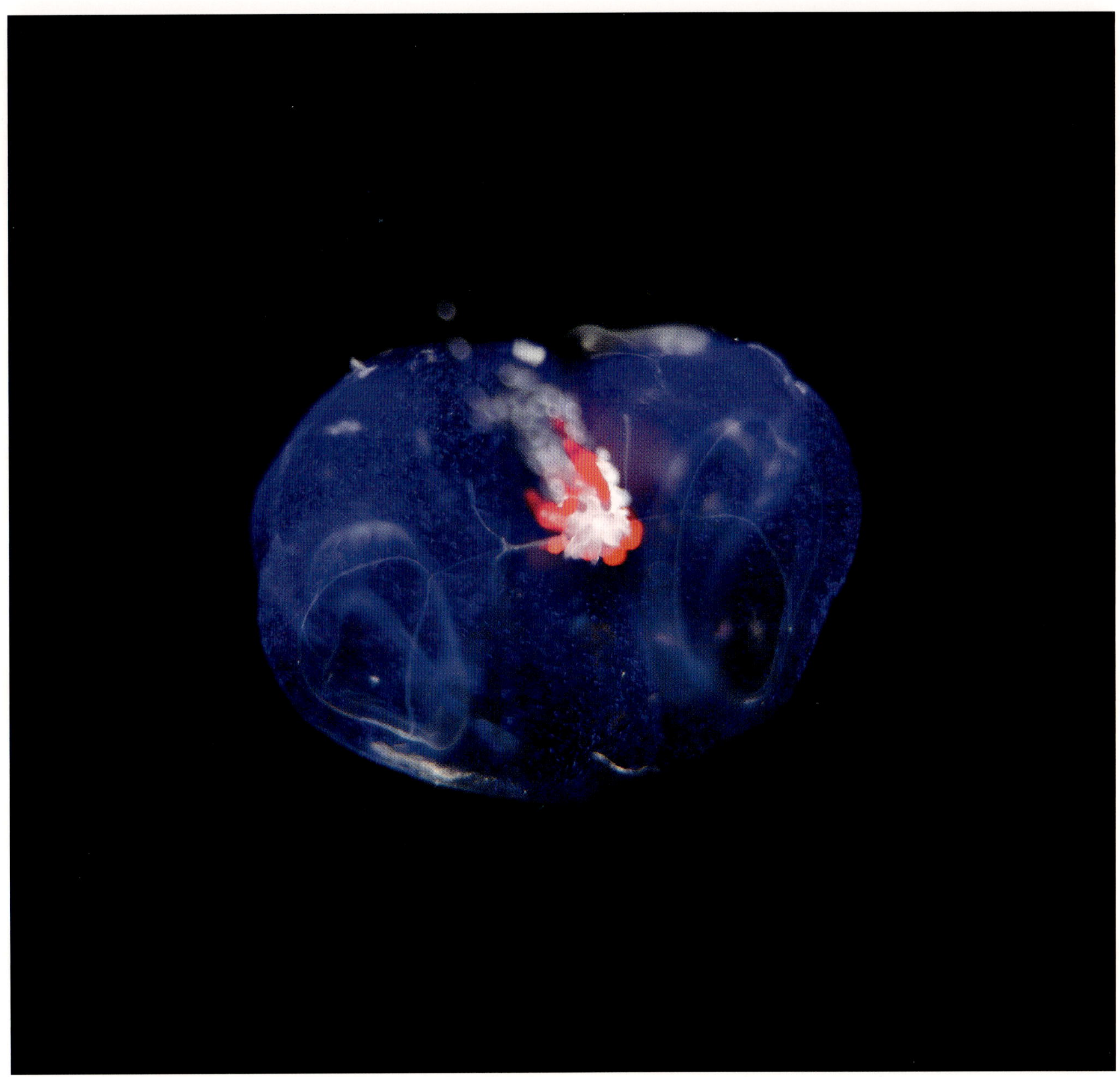

Siphonophores

Steve: Siphonophores are hard to explain. They are relatives of hydromedusae, but their growth occurs along the length of a central tube. Different body parts serve specialized functions; some can only catch and consume prey, some can only pulse for propulsion, and others are used for reproduction. This division of labor leads them to be categorized as colonies, but they are still individuals whose parts bud off, rather than coming together, and who can't function when disassociated. Using scuba and submersibles, we find dozens of siphonophore species that haven't yet been given names. We described this blue species, *Gymnopraia lapislazula,* in the same paper where we described a vividly green species of siphonophore. The blue color is produced by tiny spherical particles embedded in the gel of its two swimming bells. The blue is most vivid when the animal is illuminated from about 45 degrees behind the shoulder of the observer—similar to how rainbows appear in the sky at 42 degrees offset from the sunlight. We haven't yet figured out the optics of this blue color, but it seems likely that it is a wavelength-selective internal reflection—the same mechanism found in so-called "quantum dot" technology. These human-synthesized molecules have precise geometric arrangements, which give rise to unique optical effects.

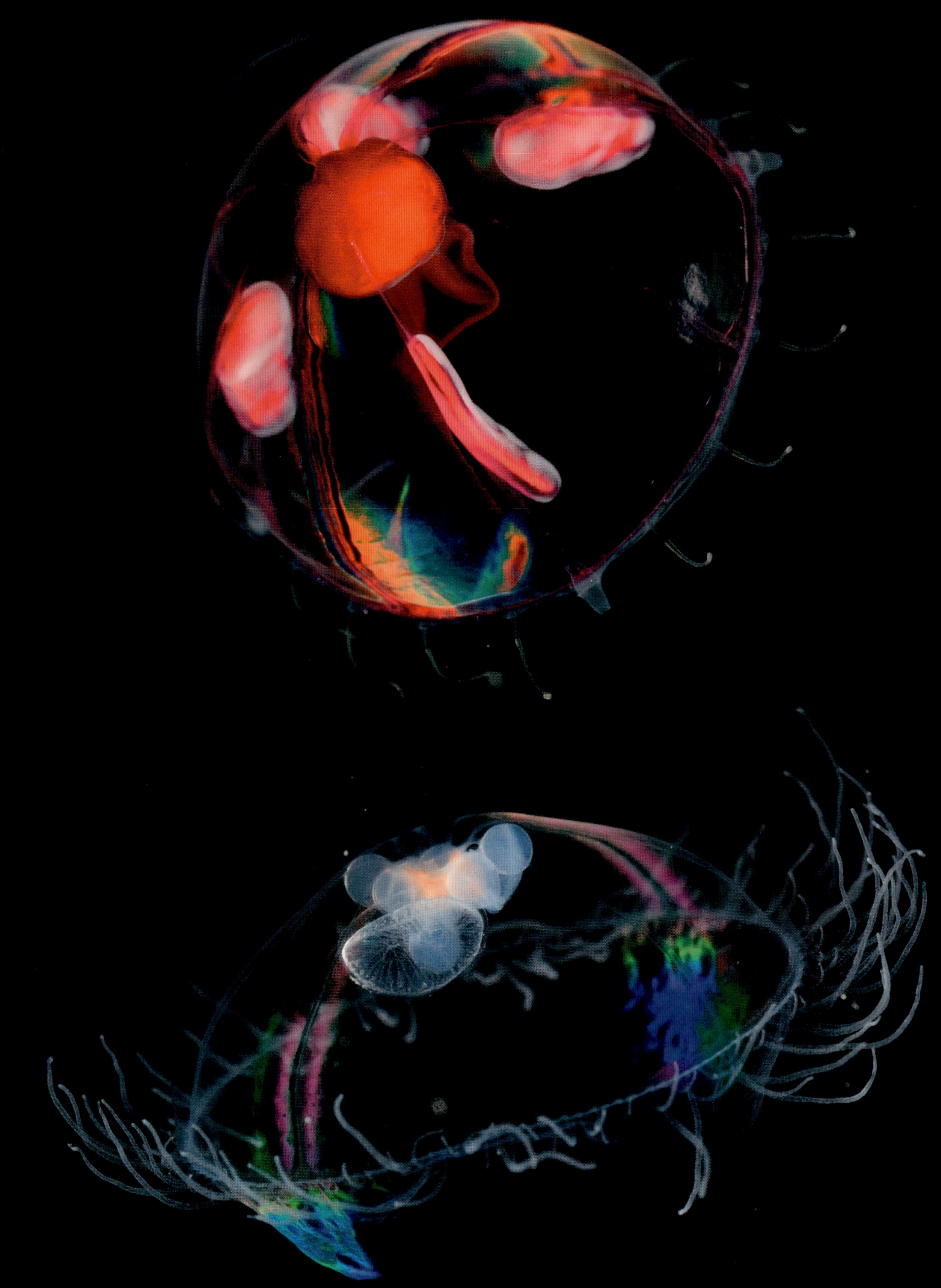

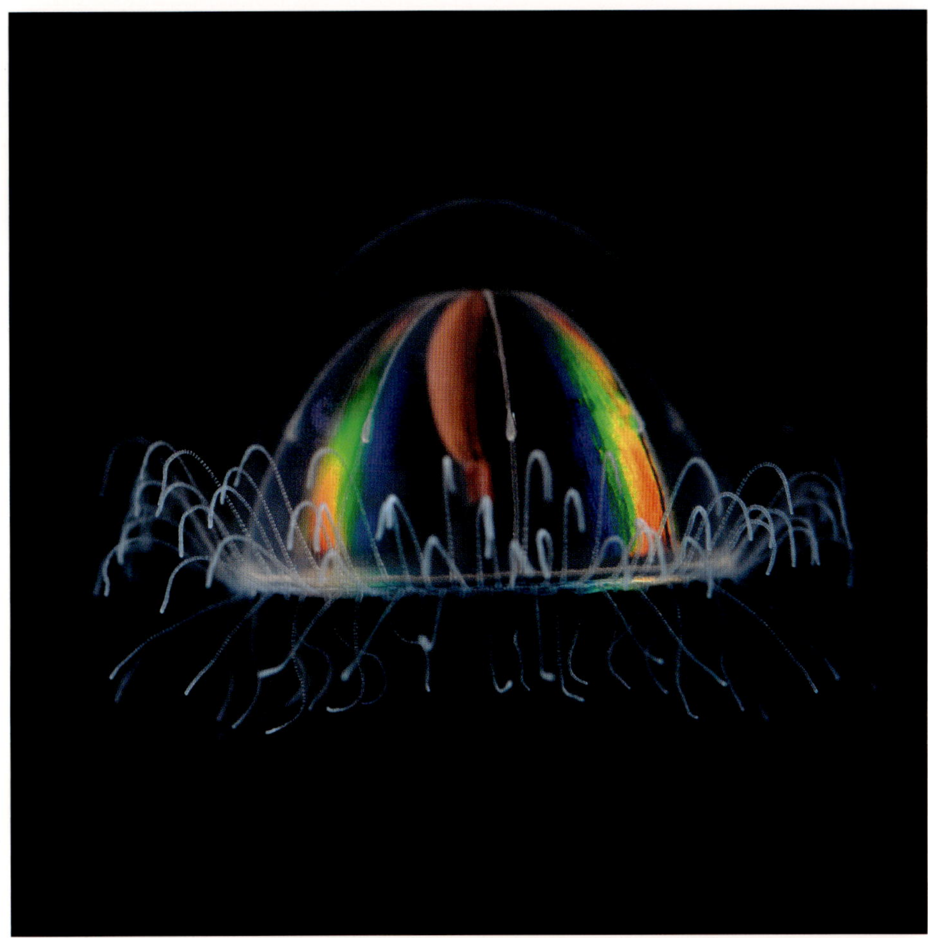

Fragile hydromedusae

Unlike their larger cousins (sea nettles and moon jellies) hydromedusae are typically smaller (only a couple of centimeters across) and more diaphanous. Their simple anatomy includes a dome-like bell, which jets out a burst of water when contracted. The bell and muscle sheet are just a couple of layers of tissue separated by a thin gap. In many species, the thickness of this gap is the correct dimension to create vivid iridescence when illuminated with white light. In our now familiar refrain, these colors are not produced under normal deep-sea conditions, and they likely serve only to make these jellies even more beautiful when photographed.

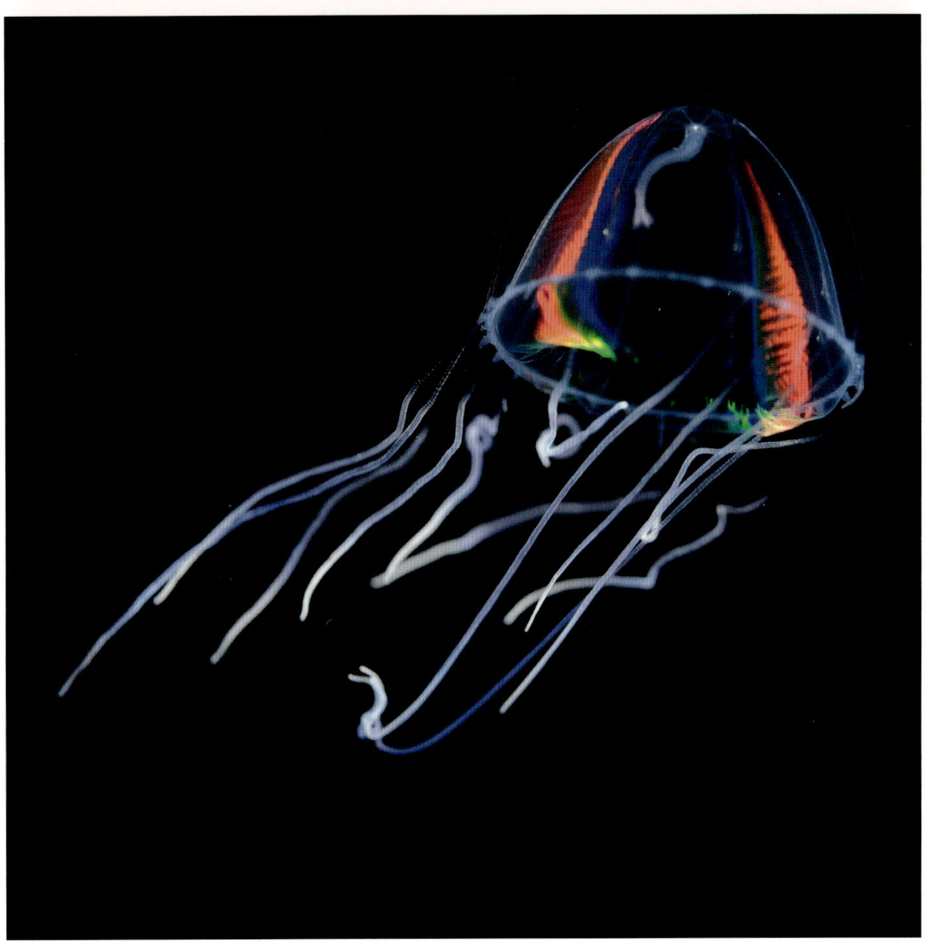

Japetella

Not only does the deep-sea pelagic octopus *Japetella* exhibit transparency and pigmentation—and wear bioluminescent "lipstick"—but its tissue is embedded with reflective iridescent structures with angle-dependent iridescent color. It is possible in some species that iridescence gives them a low-cost way to break up their silhouette.

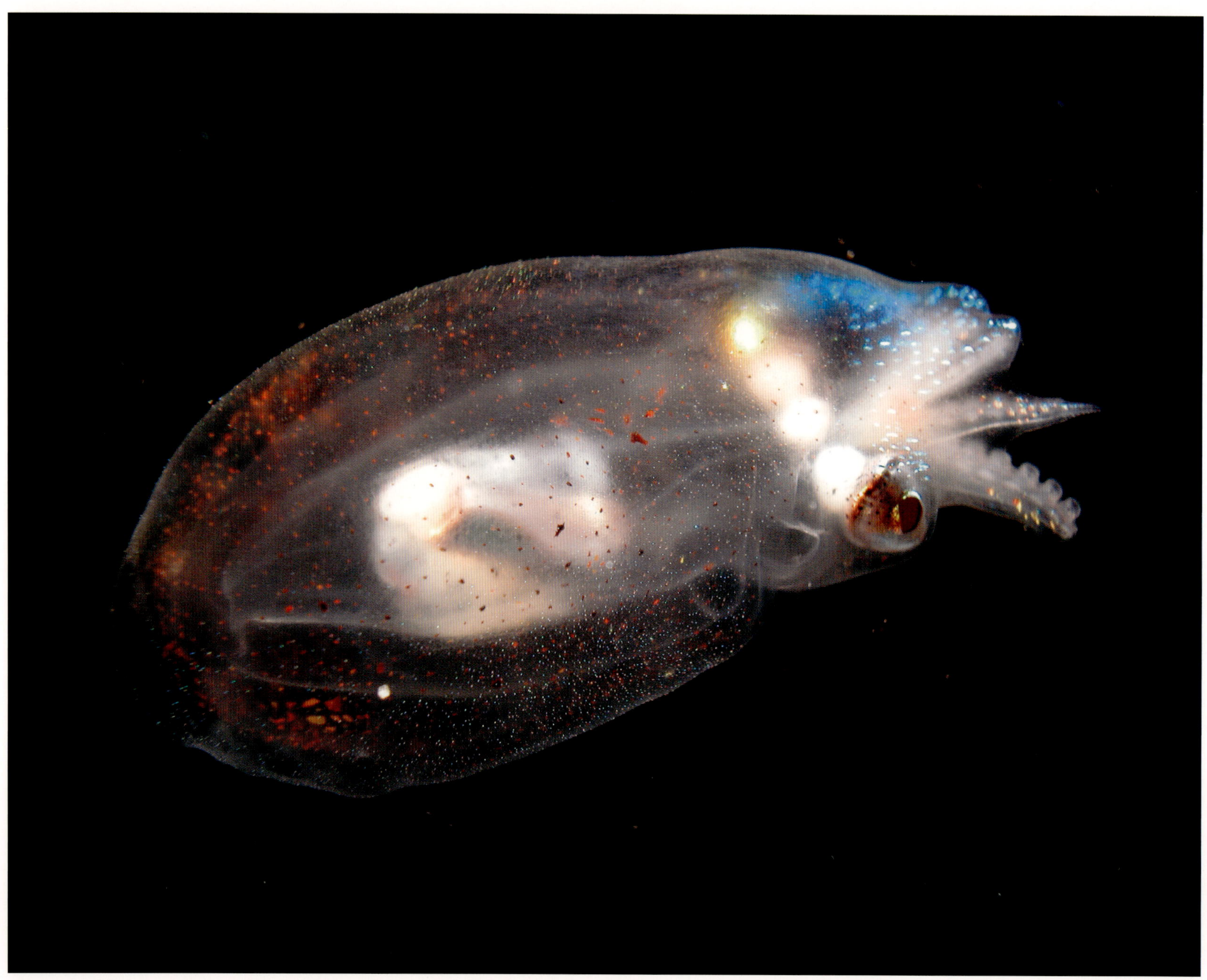

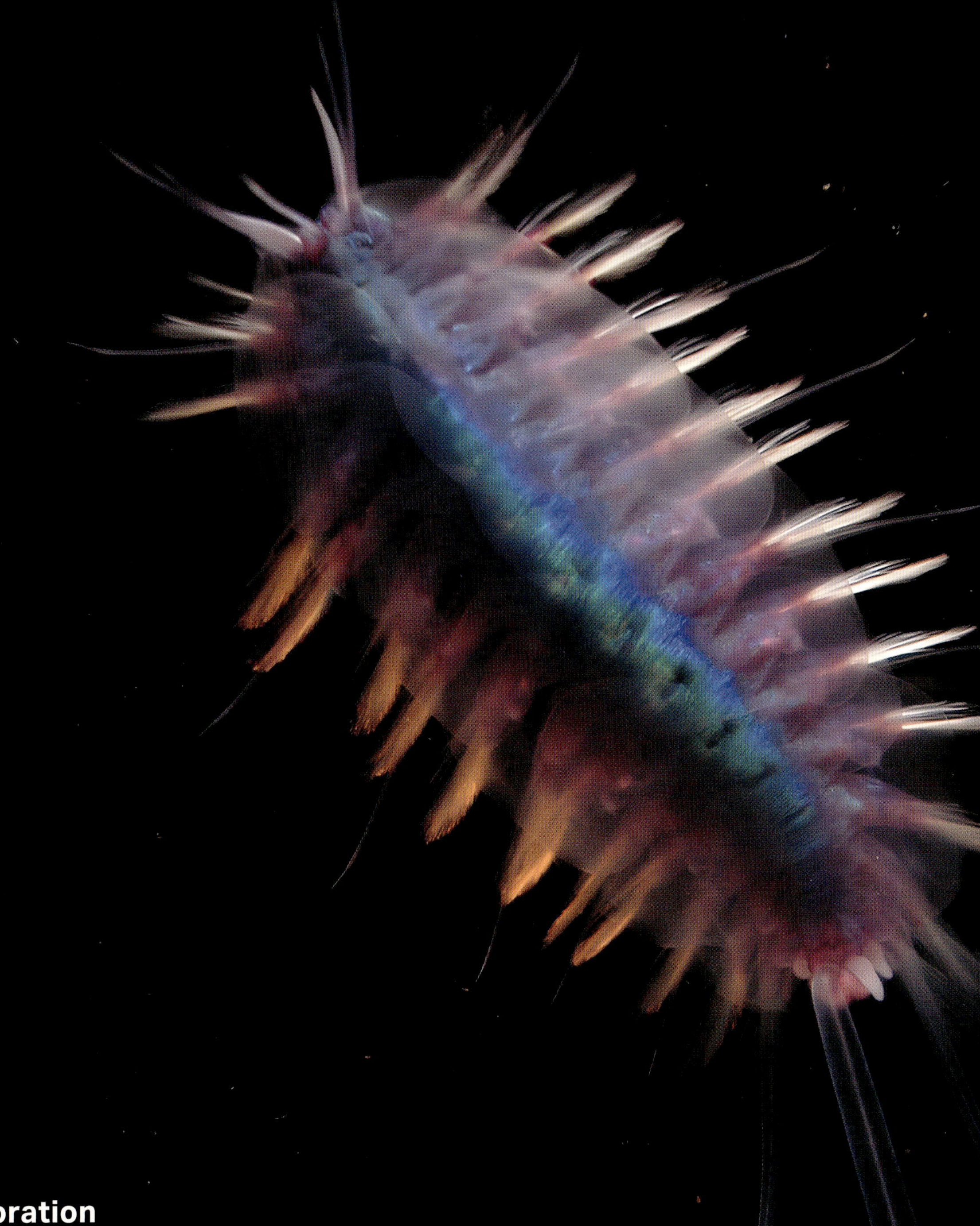

Scale worm coloration

Scale worms, which typically live on the deep seafloor, are distant relatives of earthworms. Their scales—like the plates along the spine of a stegosaurus—serve a protective function. Some species (of worms, not dinosaurs) drop their scales when disturbed, and the scales produce rapid bioluminescent flashes to distract the would-be predator. In other species, such as the polynoid *Aphrodita*, the microstructure of the scales themselves produces strong blue iridescence under white illumination.

In contrast, the iridophores can reflect greens and blues. Iridophores are made of stacks of crystal plates that are spaced at particular distances. Incredibly, the animal can control the spacing of the crystals and thus change the reflected color at will. Changing the iridophores' reflectance happens more slowly than the chromatophores—over a few minutes instead of a few seconds—because it requires them to pump water into and out of the iridophores.

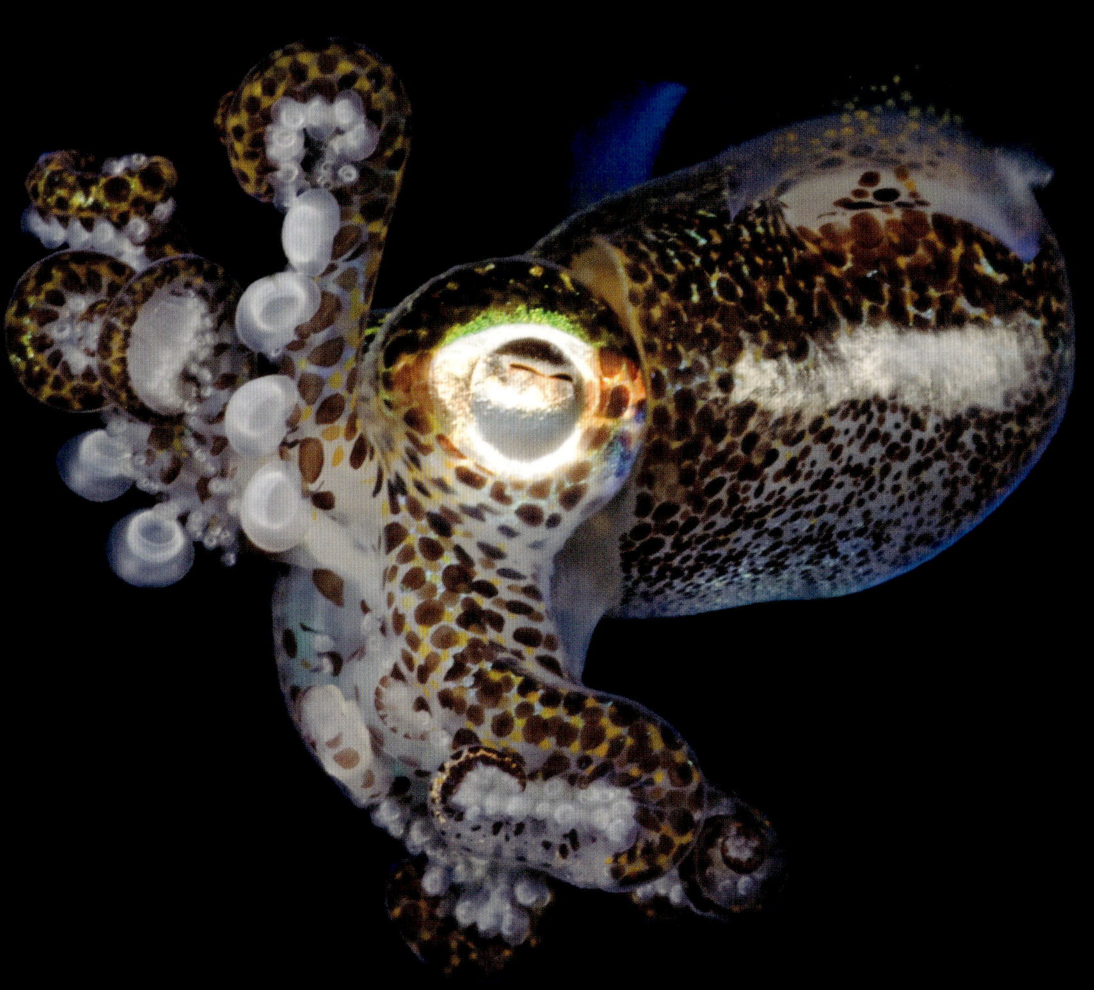

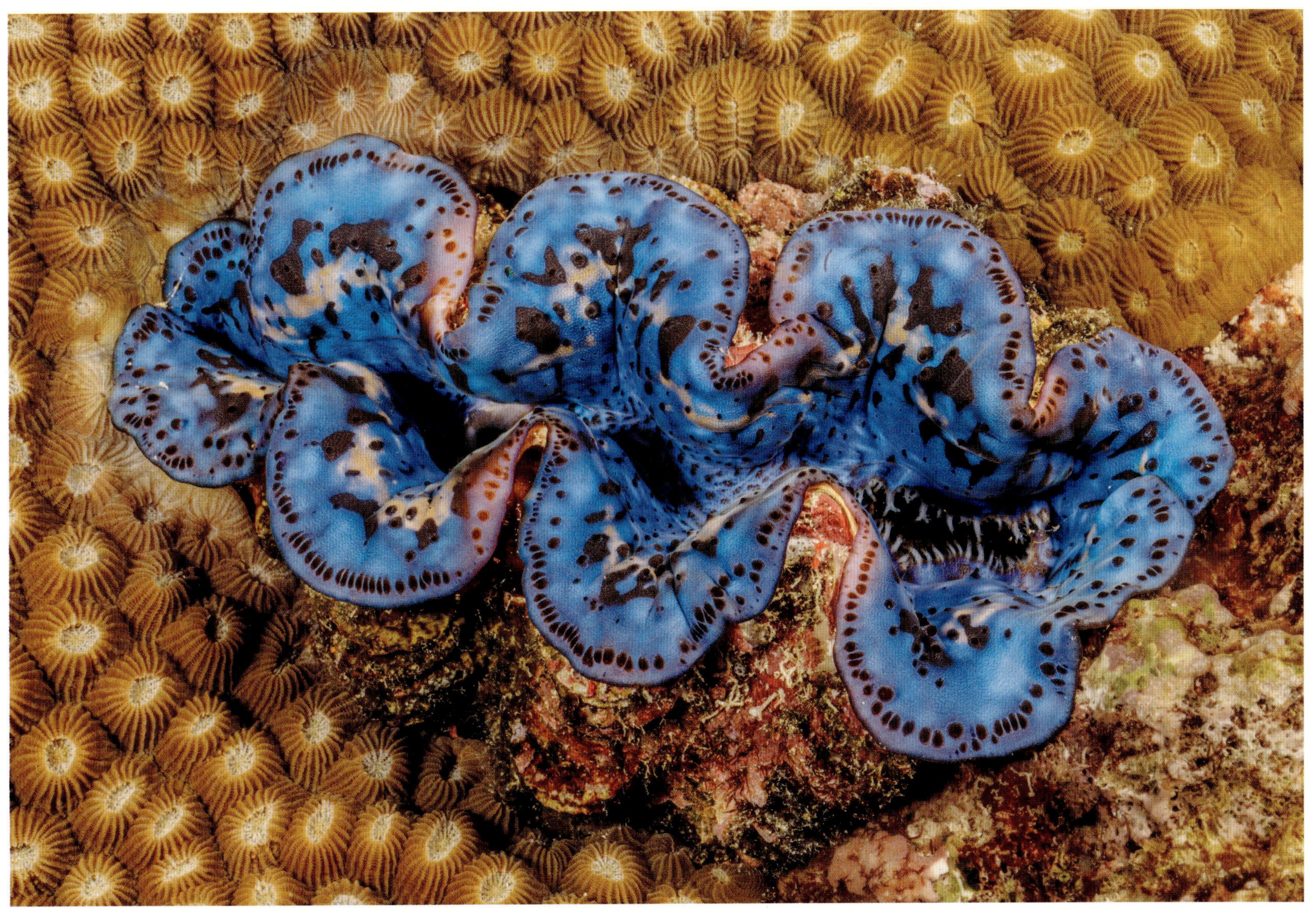

Tridacna

Corals may be the most famous animals to live in symbiosis with algae, but they are far from the only ones. A number of other marine animals also harbor and protect algae, which in turn make food from sunlight. One of the most striking is the giant clam. It's usually found in shallow water with its two shells slightly separated, like the jaws of a napping giant. Within this "mouth" is a special tissue that not only harbors algae but also appears to create just the right optical environment for them, dotted with iridescent spots and streaks. The internal structures of the clam have evolved to maximize the passage of light of the colors that the algae needs, while at the same time reflecting harmful ultraviolet radiation back out of the body.

Bathophilus flemingi

While most deep-sea fish are reliably black, a number are surprisingly bronze in coloration, being both brown and shiny. The shiny bronze coloration is so striking in captured specimens that it is tempting to assume that the tint must have some purpose beyond the mirror-like properties. It will be interesting in the future to test the wavelength-specific reflectance of their skin and see if blue light has some special kind of interaction, since metallic colors can be difficult to characterize intuitively. Note that the *Bathophilus* also makes special appearances in the Transparency and Fluorescence sections (p. 41 and 210), and is also bioluminescent.

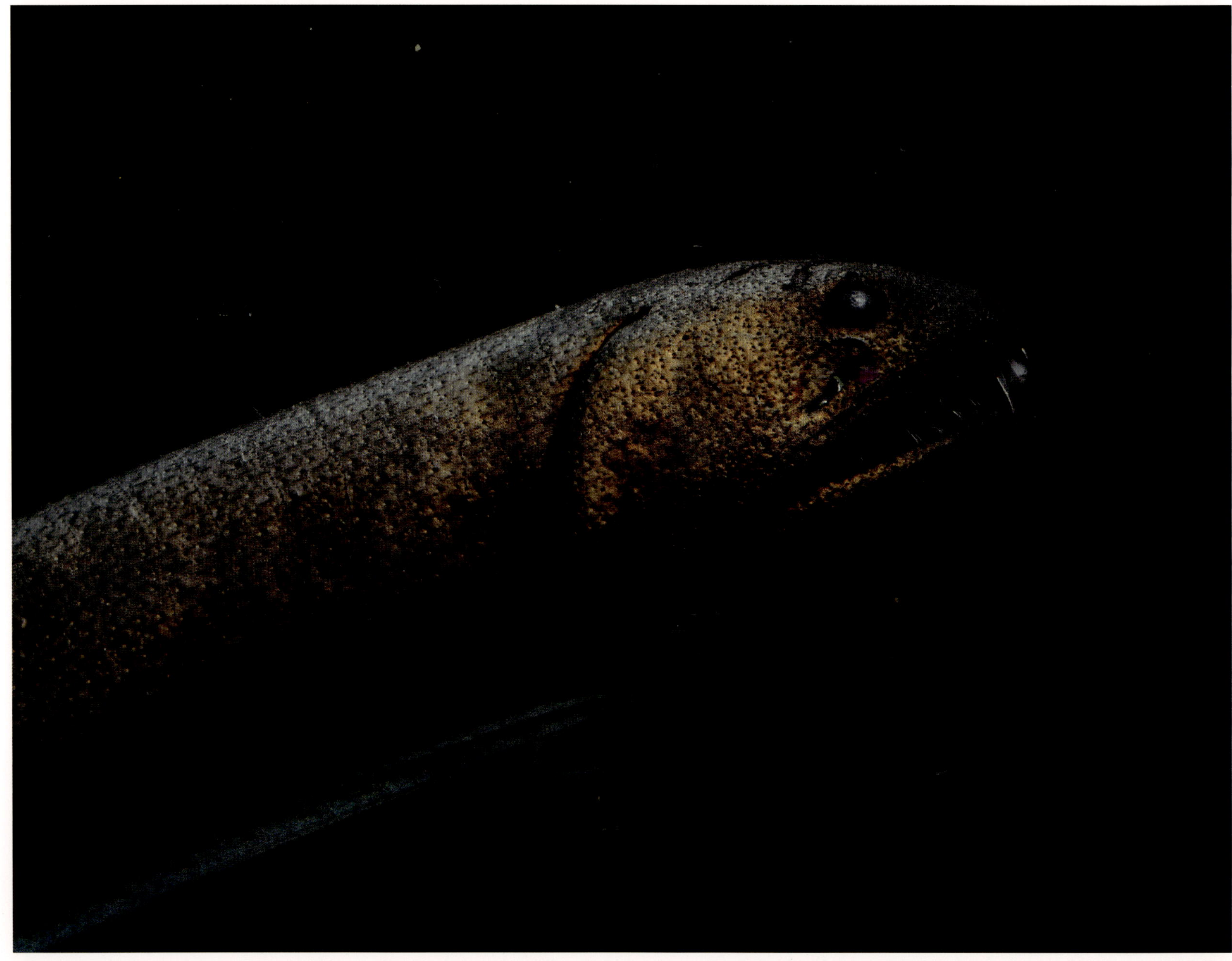

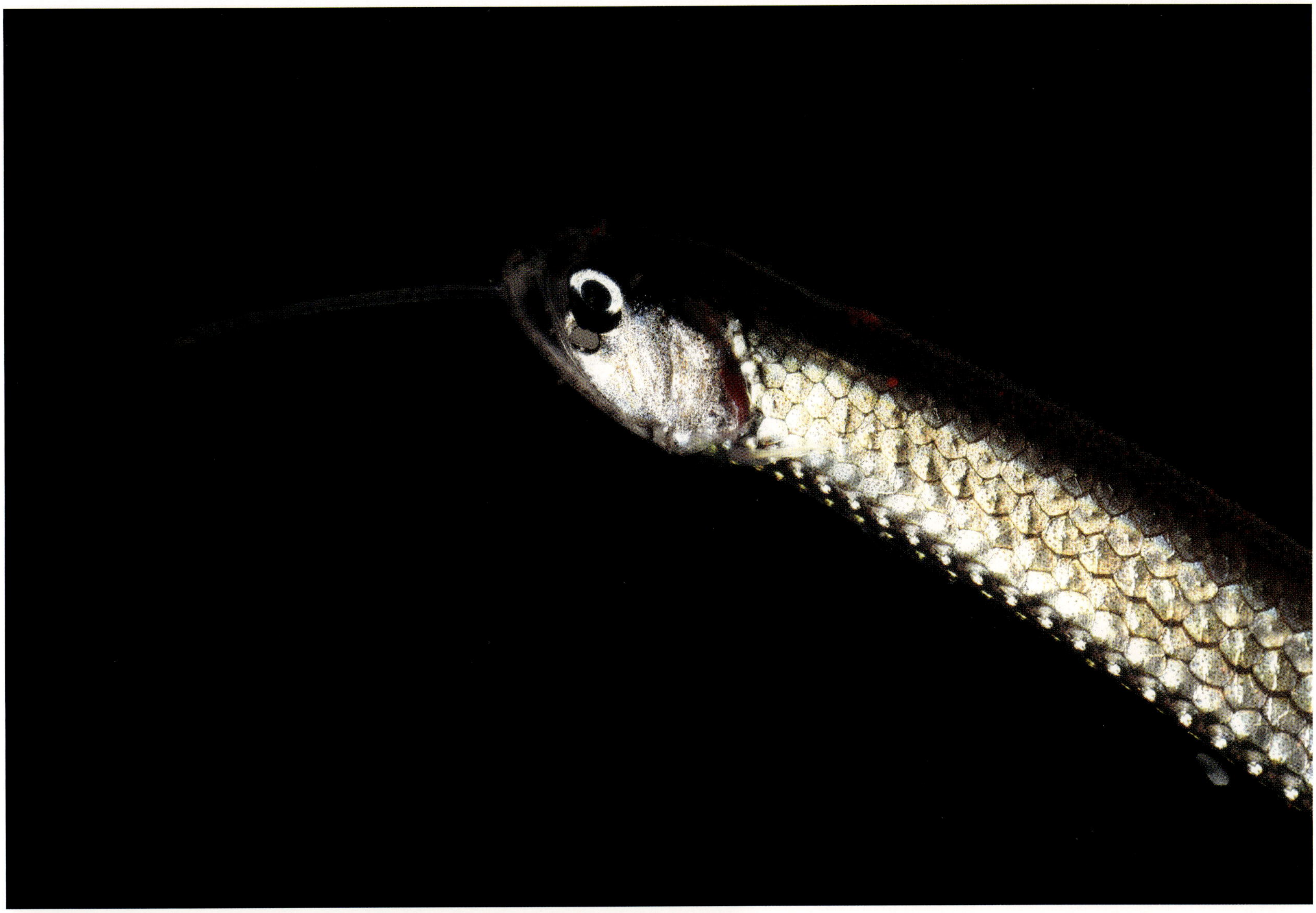

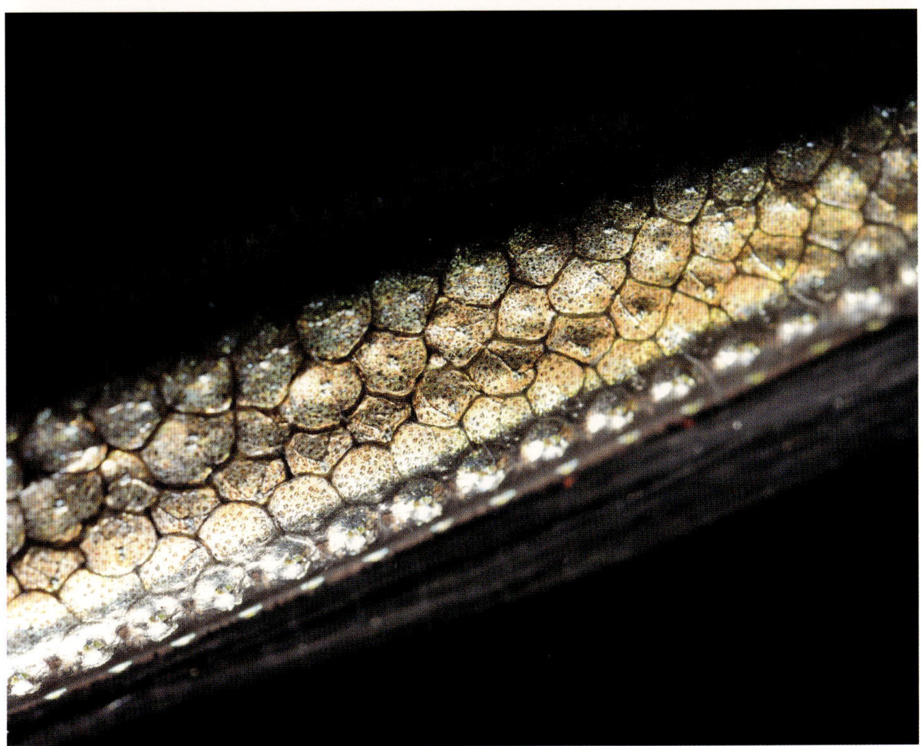

Stomias

The scaly dragonfish is covered with hexagonal plates of reflective tissue, but it seems undecided whether it should be a black or a shiny fish. Brown pigment spots covering the plates give it a gold-to-coppery hue under certain angles of illumination. One fun aspect of deep-sea diversity is that if you make a matrix of specializations, you can find species embodying almost every mix and match of those adaptations.

Although usually depicted from damaged specimens with their jaws gaping open, in life *Stomias* are actually sleek and agile. Despite their modest size—under 20 cm—they are one of the more fearsome predators patrolling the depths.

Selene vomer

Many oceanic fish are quite silvery. Near the surface, this can make schools of them flashy and confusing to predators, but further down, the primary function of all this mirroring is to hide. For this camouflage to work, the mirror has to be vertical, otherwise it will reflect brighter light from above or the darker depths. Because fish bodies are usually curved, the mirrors all have to be embedded at different angles relative to the surface of the body. For some fish, such as these lookdowns, this arrangement is easier to achieve as they are remarkably flat. Given its frankly comical appearance, it seems likely that this shiniest of fish got its common name from the condescending expression on its face.

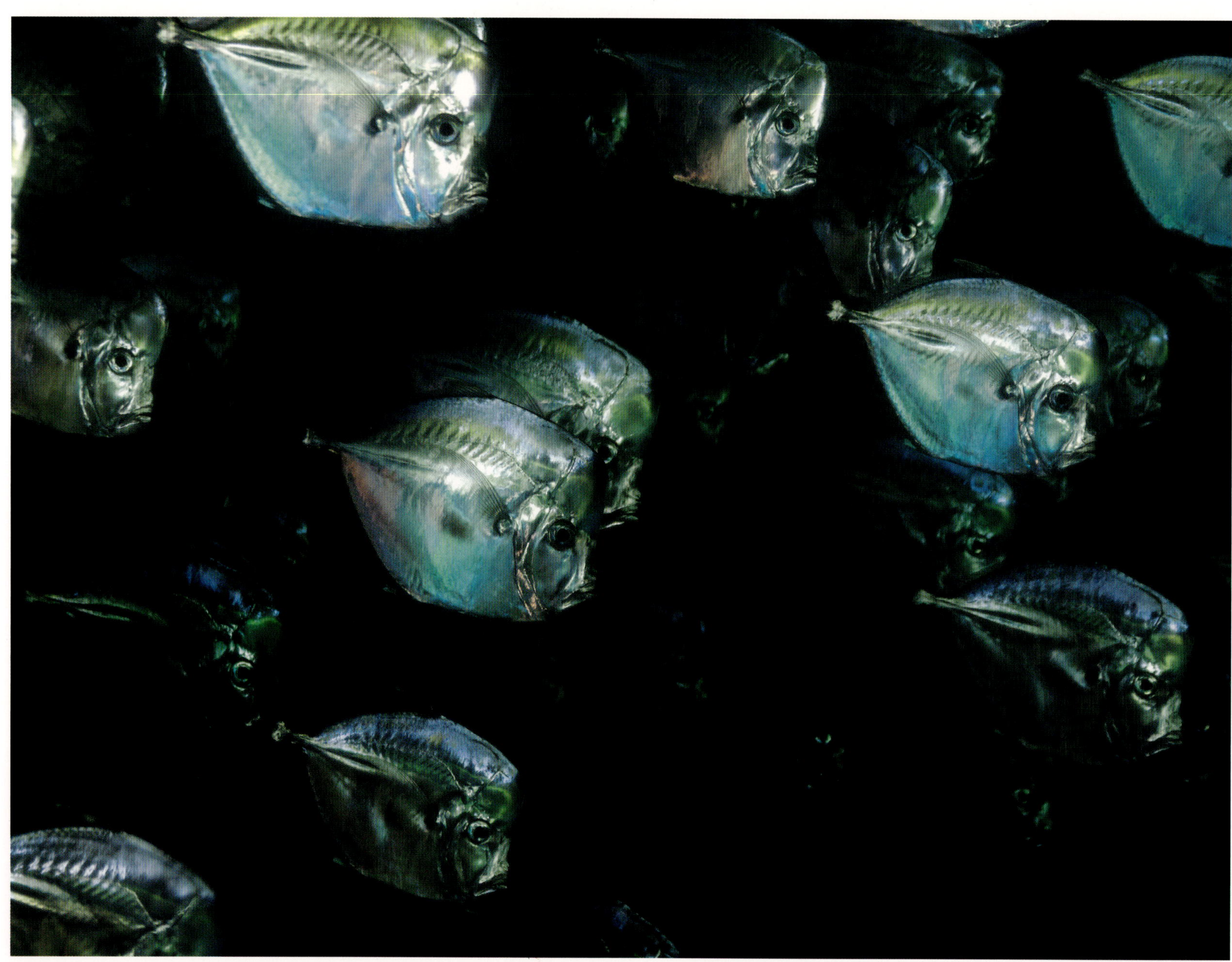

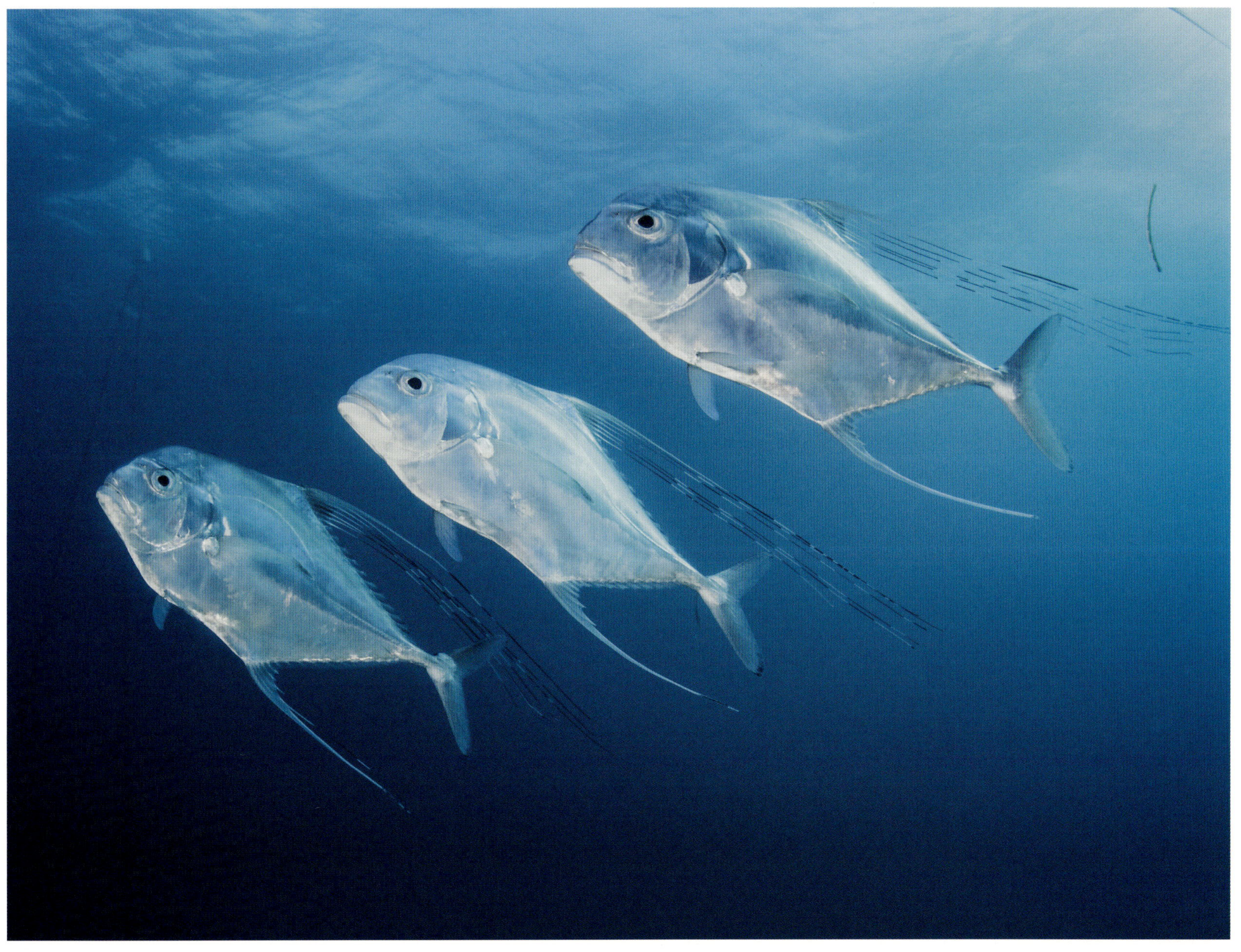

Alectis ciliaris

African pompano fish, like many other species in the upper water column, are laterally flattened with mirrored sides. These biological mirrors are made out of thin plates of randomly spaced guanine crystals. They act like a collection of soap films of different widths and therefore don't reflect just one color but all colors of the rainbow. This makes them silvery white. You can't see your face in them, but they do reflect the ocean very well. So when you look at these fish, their mirrored sides reflect light from behind your head that looks almost like the background water you would see if you could look through the fish. This trick allows these large, shiny animals to be surprisingly invisible underwater.

Sternoptyx

Apologies for the anthropomorphism, but we think their wide-eyed look of surprise and Muppet-like mouths make hatchetfish one of the cutest out there. Their reflective sides give them the appearance of a silver coin. To ensure the integrity of their camouflage, hatchetfish use transparent light-guides to direct their bioluminescence downward, not to the side. Like most ocean animals, they really do need to pay attention to the details, because they are one of the main prey items for toothy predators like lancetfish and snake mackerel.

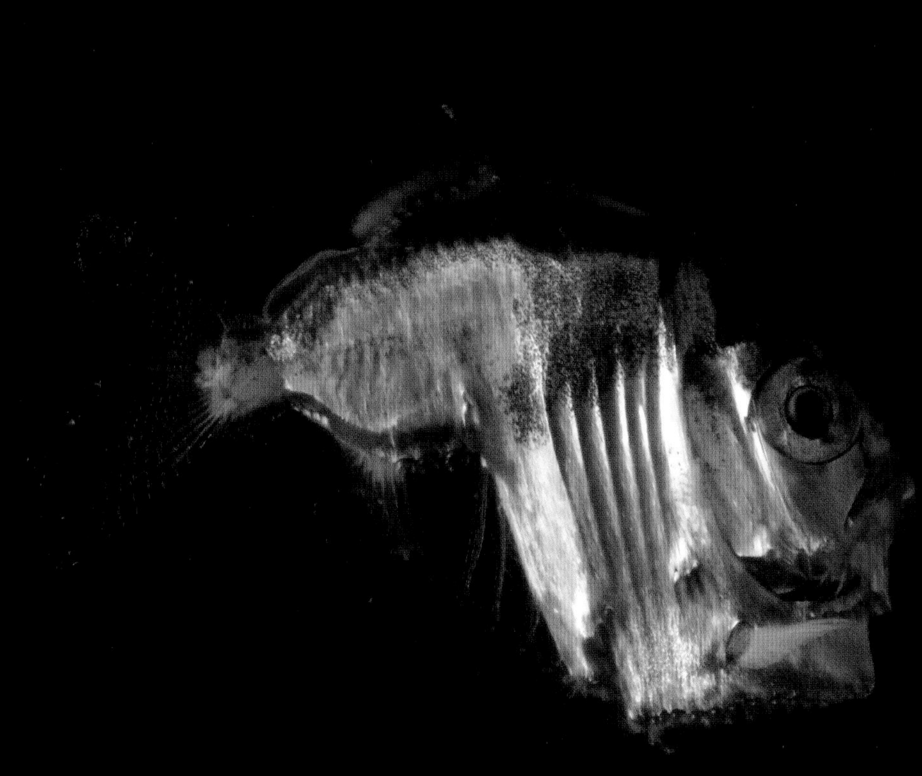

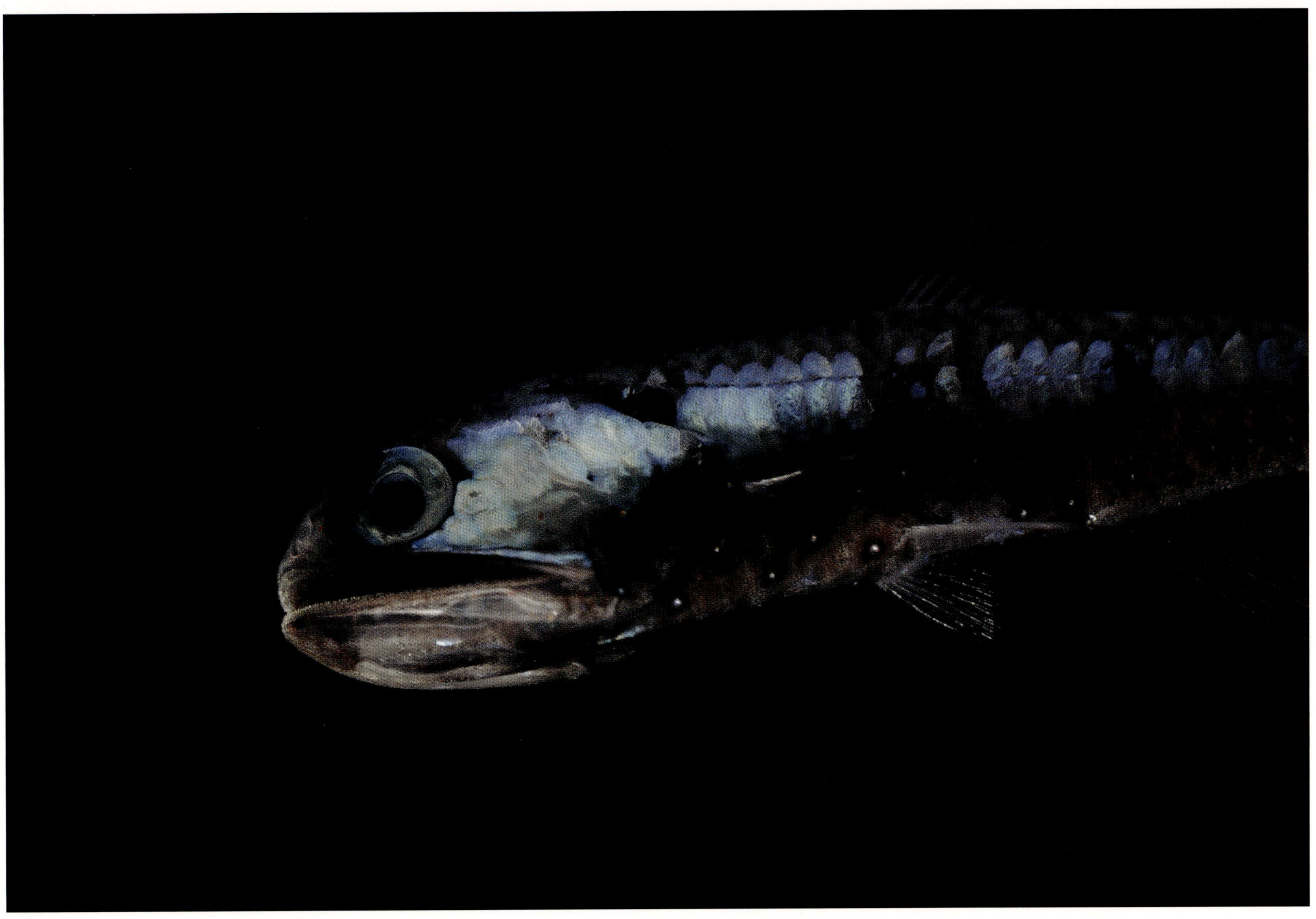

Myctophid fish

Lanternfish are one of the most important and abundant of the deep-sea fishes. They serve as prey for countless animals, including humans, since they are part of a seafood dinner in some countries. Many species of myctophids migrate right up to the surface at night. On a night dive, you may even have one of these bump into you, having swum so far, only to be disoriented by your flashlight.

Perhaps because they inhabit such a span of habitats, myctophids exhibit a wide range of adaptations, including bioluminescence, dark pigmentation, and mirroring. In this specimen, you can see the reflective racing stripes along its otherwise brown body, with the smattering of bioluminescent dots along the lower surface. The reflectance of these fish is very angle dependent. In photos taken before and after this, the fish sometimes appears uniformly brown. Perhaps this adaptation is more sophisticated than we thought, and functions only to reflect light when seen from the side, not from below.

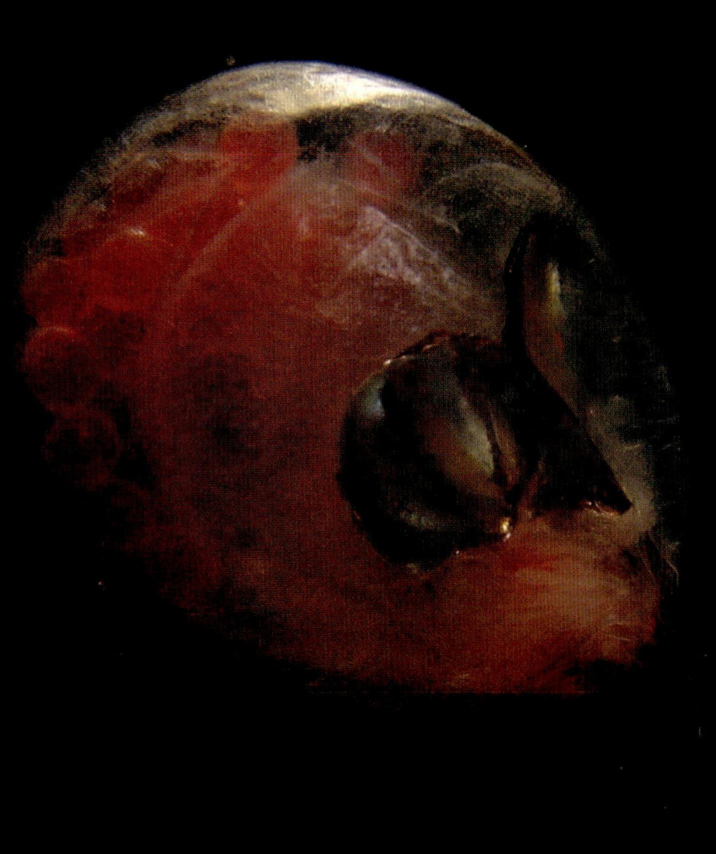

Gigantocypris

Reflectors are not just for camouflage. Animals like this giant (2 cm-plus) ostracod (a kind of crustacean) can use parabolic reflectors (curved mirrors) to direct more light to their eyes. Incredibly, *Gigantocypris* continuously flexes these mirrors, like a camera's autofocus that cannot quite lock on to its subject. It is thought that this behavior makes light from distant objects (predators) fade in and out of focus, while keeping close objects (prey) in focus.

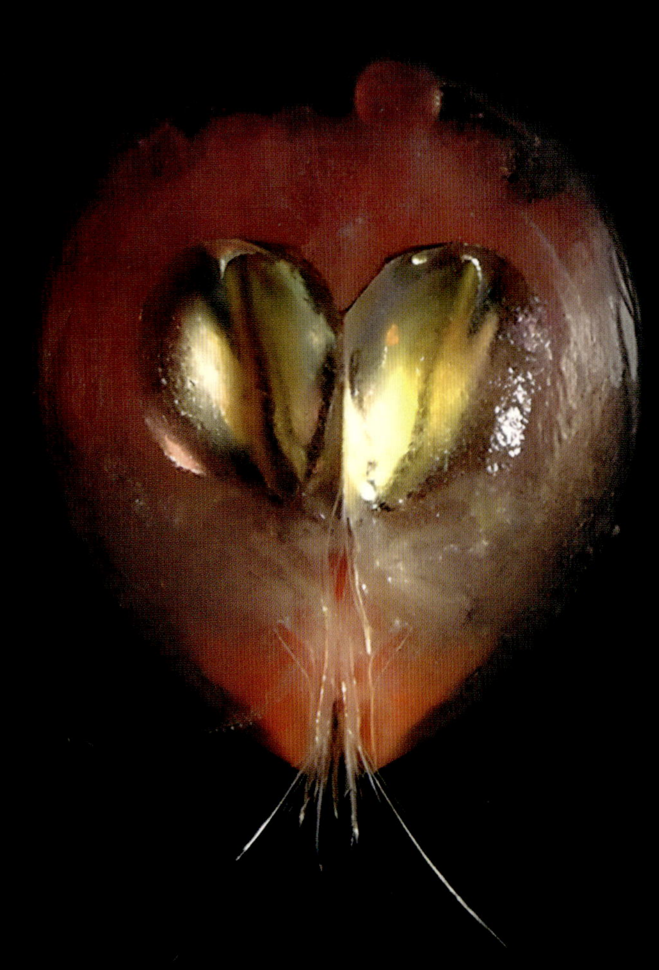

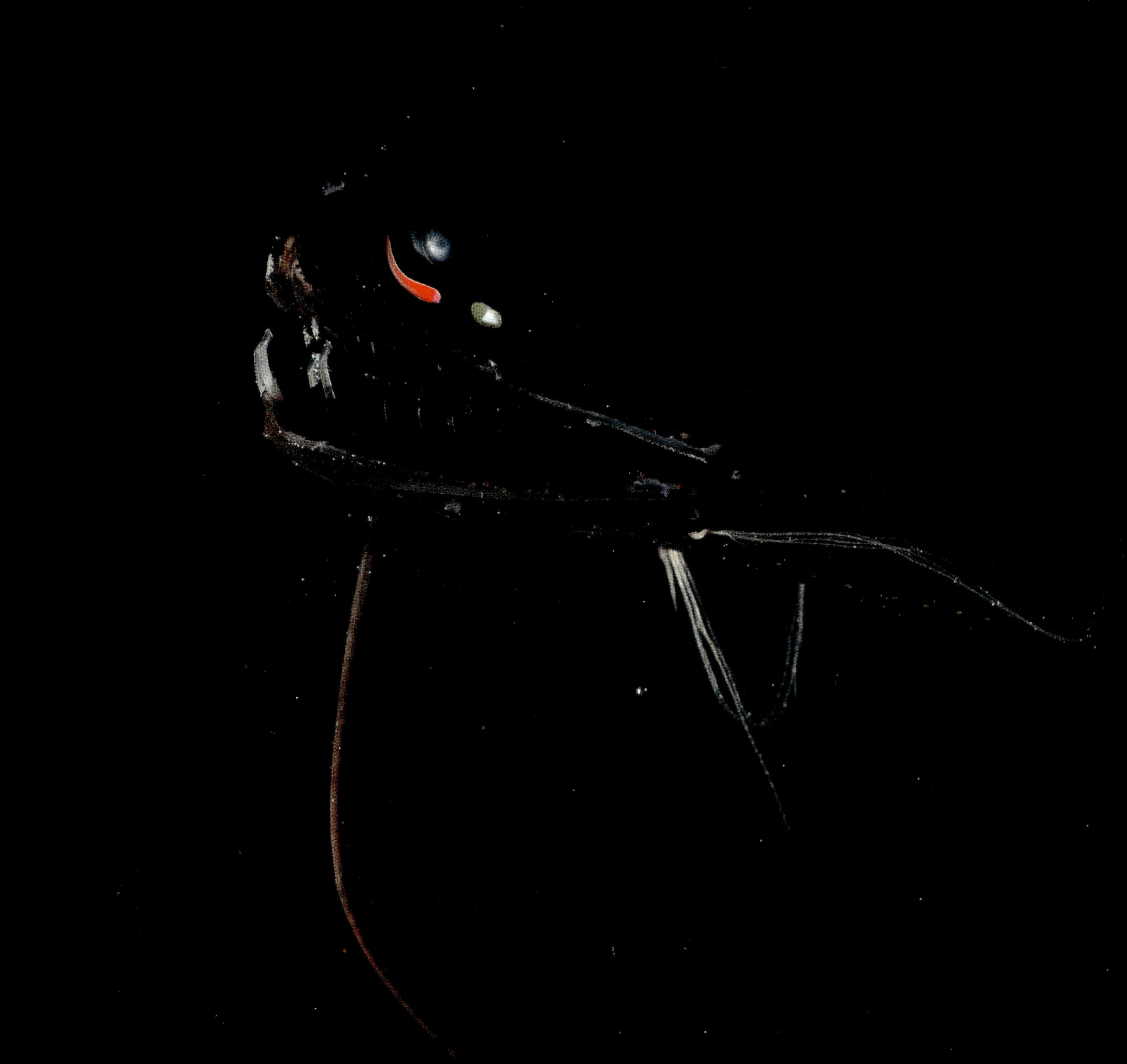

Aristostomias

Aristostomias—one of the black dragonfishes—is an absolute favorite for people studying bioluminescence. It doesn't just have standard green photophores like other fish, it also has a pair of red photophores beneath its eyes. We will talk more about this in the Bioluminescence section, but in short, it uses these red lights to hunt prey without being detected. It is also very hard to photograph properly because the bright reflections of its photophores (and admittedly its gleaming teeth) tend to be overexposed if you try to properly illuminate its ultra-black body.

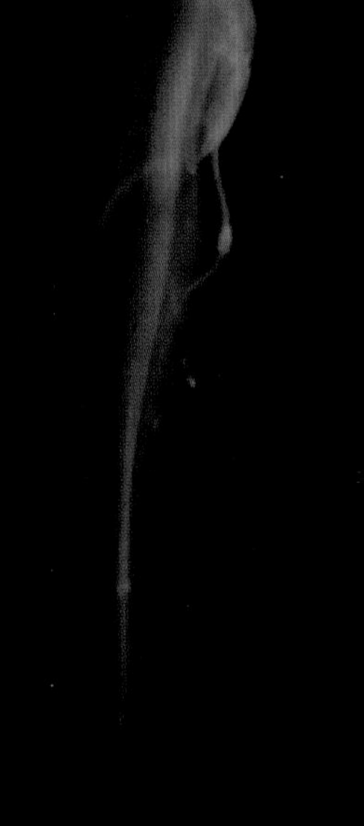

Idiacanthus antrostomus

Since there are many deep-sea fish, shrimp, and squid swimming about with bioluminescent flashlights next to their eyes, it pays to reflect as little light as possible. The black dragonfish is extremely good at this and is about fifty times blacker than the blackest objects in our day-to-day lives. As discussed previously, it does this both by having black pigment and by packaging it in a way that maximizes how much light the pigment can absorb. These tricks are now being co-opted by engineers interested in making their own materials with very low reflectance.

Chaenophryne

Sönke: The only animal we found that is blacker than a black dragonfish is an anglerfish that is about twice as black, and as black as the darkest known substances. They may need this low reflectance to hide from predators with bioluminescent "flashlights," but it's also possible that they need to keep their own bioluminescent lure from lighting up their faces. We recently painted a deep-sea camera system black so that its illuminated lure (which looks a lot like one of those blinking restaurant pagers) wouldn't light up the whole apparatus and scare animals away. This is yet another time when what we have learned from the animals gives us better tools to study them.

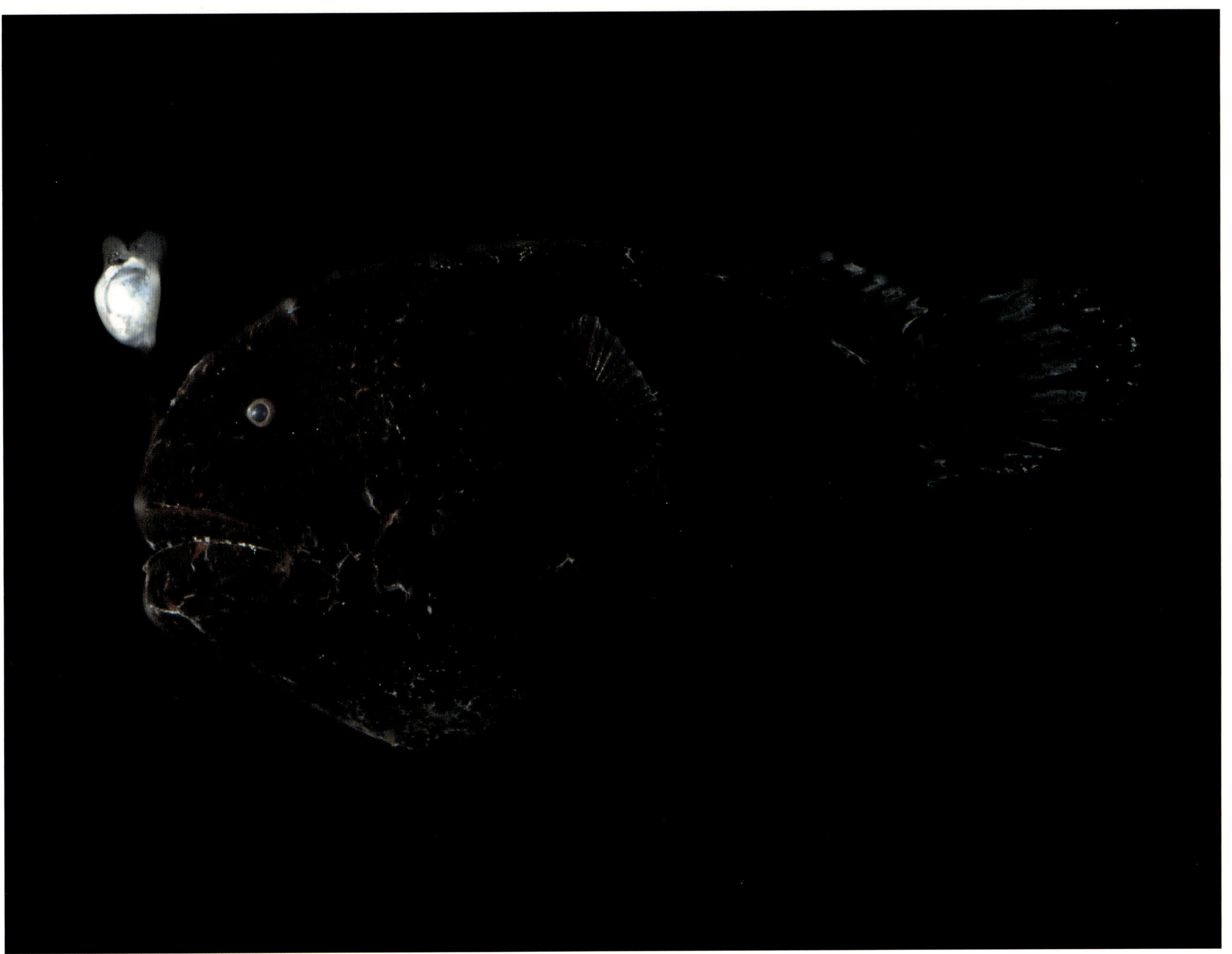

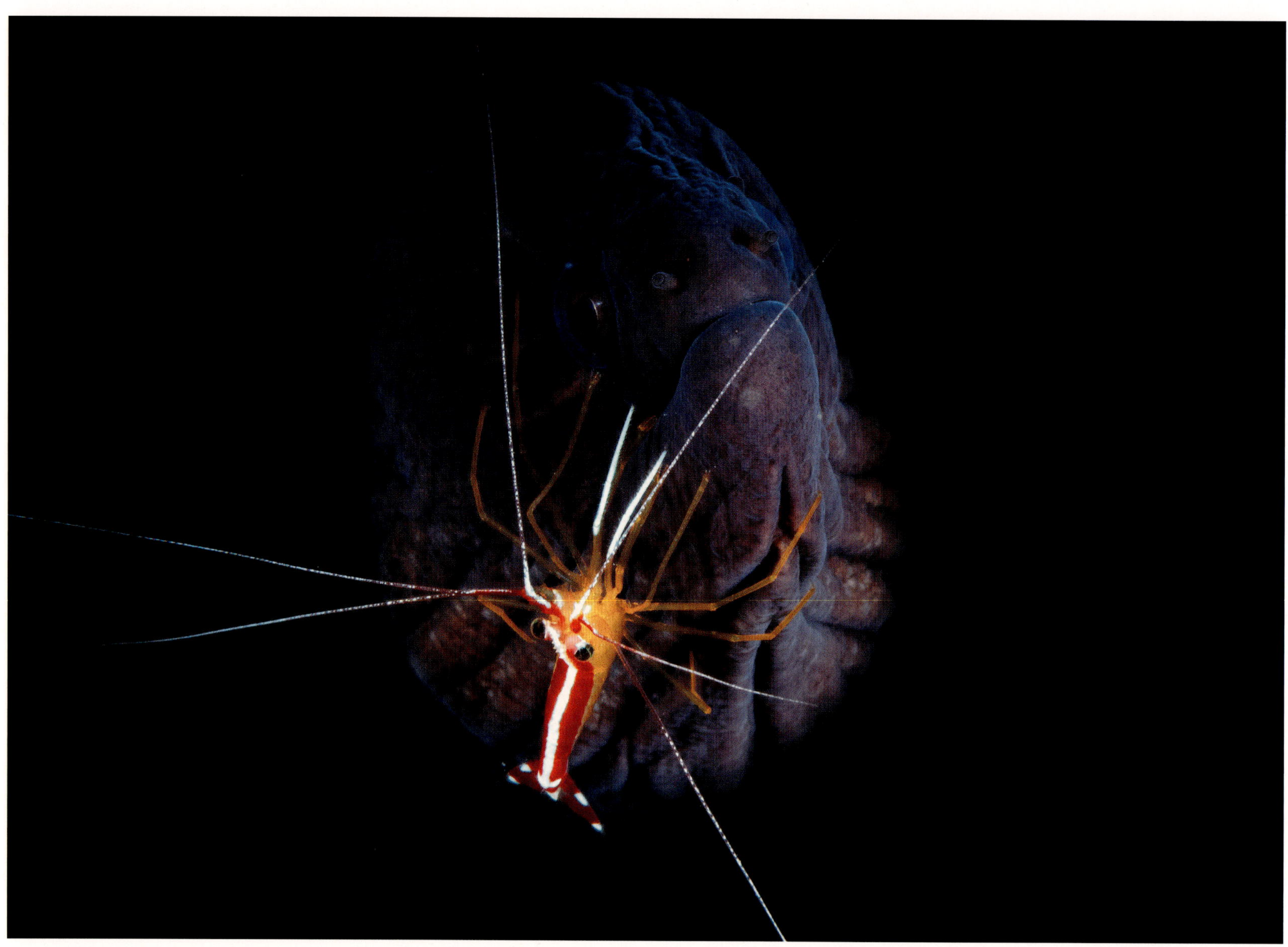

Lysmata amboinensis

Sönke: Being a cleaner is for the most part a good deal. You hover by an anemone and wait for fish to show up, sit down, and want to be cleaned. The only problem is reminding the fish that you are there to clean them of parasites, not to be eaten yourself! Cleaner shrimp solve this problem by signaling the fish, usually by waving their antennae or clapping their long, thin claws. This signal needs to be seen, which can be hard in the ocean where colors are affected by depth and water clarity. Nearly all cleaner shrimp seem to use white to get attention, which looks as bright as possible under all conditions. We learned that they have evolved nanostructures inside their antennae (and maybe their claws) to make them uniformly white across the visible spectrum.

are just the right size to reflect a great deal of light of all colors. The animal can curl and uncurl this white edge rapidly, so when the shell is open, it looks like a small thunderstorm is happening inside. The effect is made all the more dramatic because the rest of the mantle is a deep orange-red. No photo can truly do justice to this effect, so we suggest you also watch a video. At the moment, we again do not know why the disco clam does this. The presence of the spheres, their size, and other factors suggest that this has evolved for a purpose, but it's difficult to know what this is.

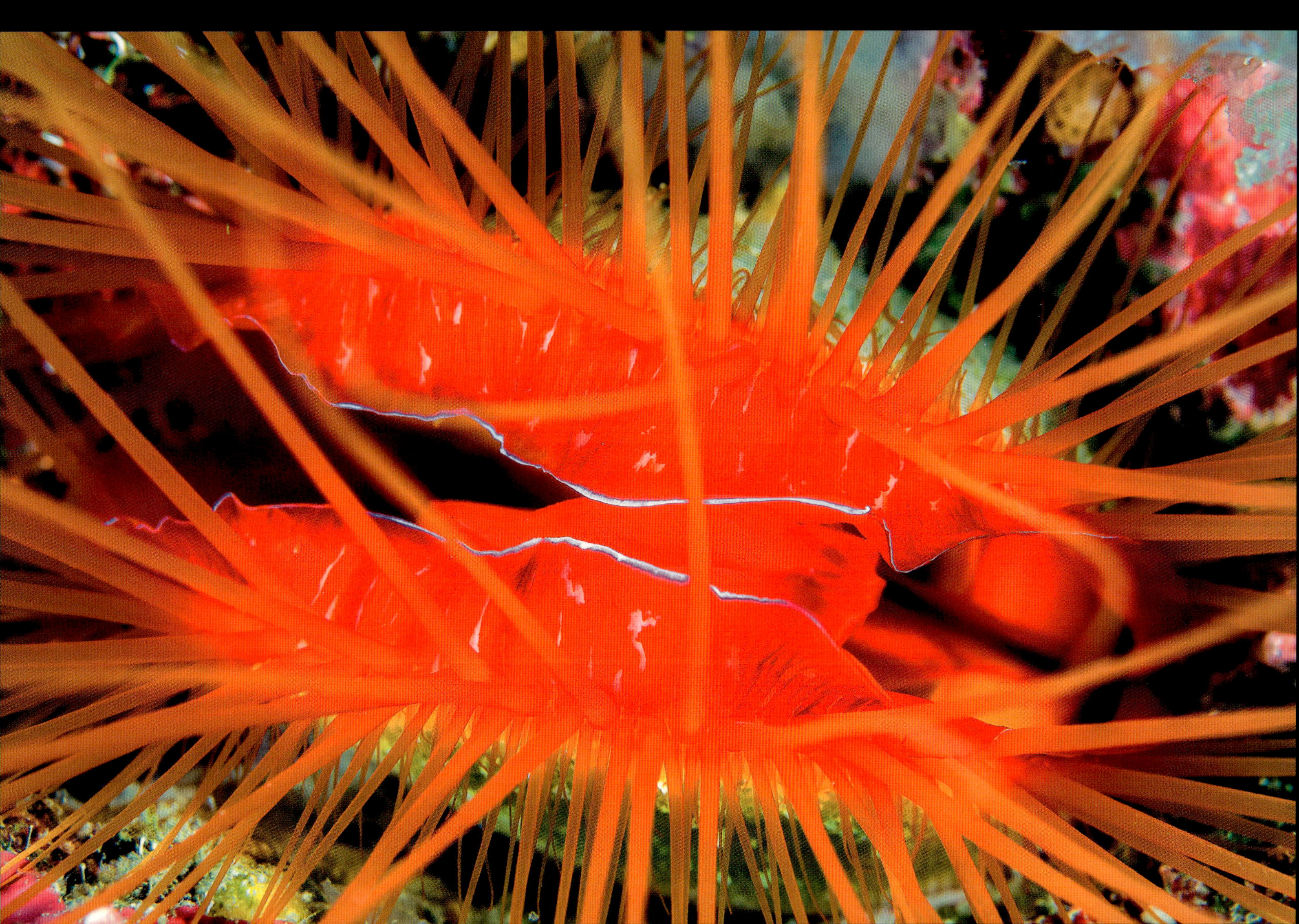

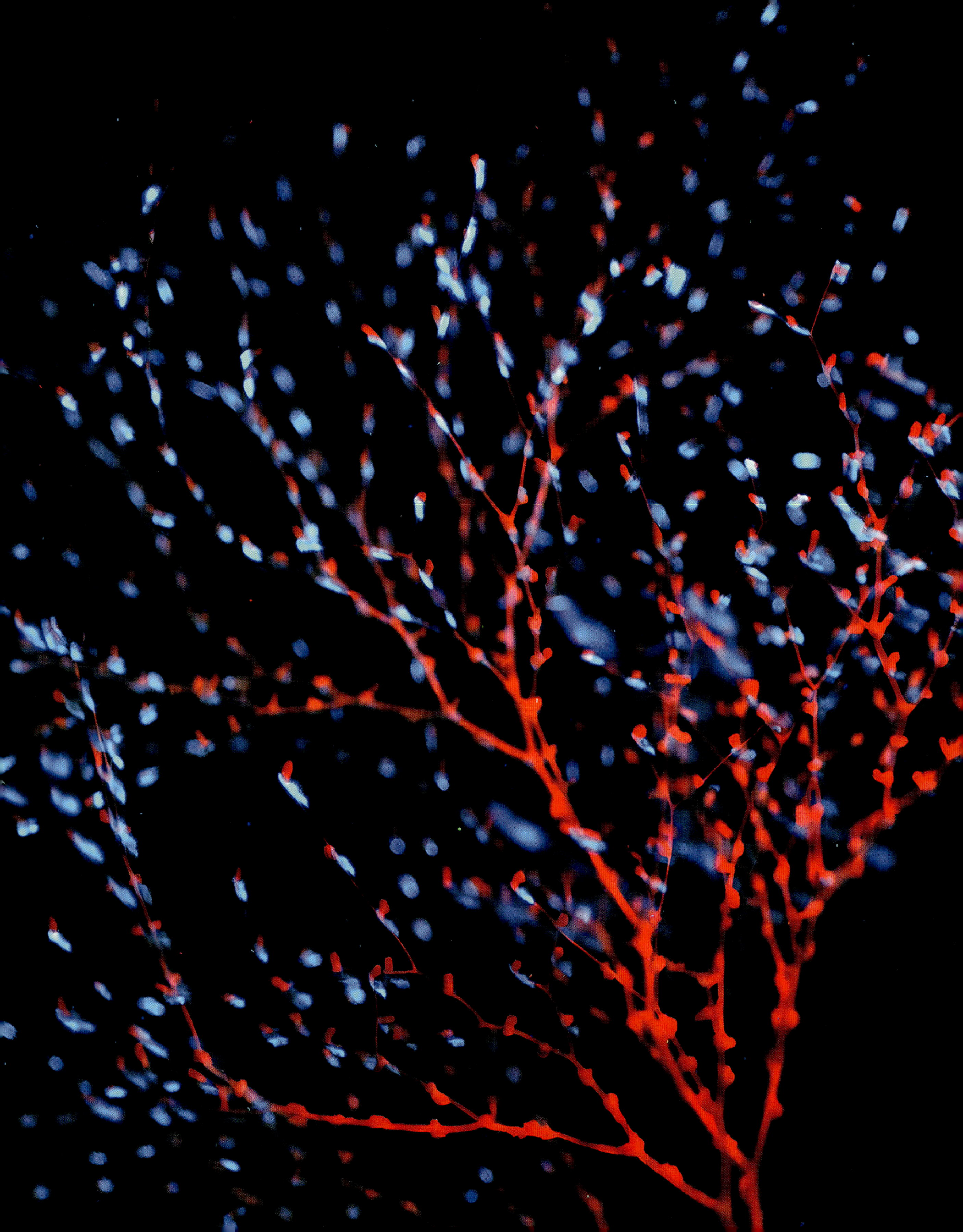

BIOLUMINESCENCE

Making your own light

The processes we've seen so far for generating color worked by selecting or subtracting colors. In the next two sections on bioluminescence and fluorescence, we will see the seemingly magical processes where colored photons are generated by chemical or physical energy originating within living tissues.

Bioluminescence occurs when a chemical reaction takes place within a living organism and releases energy in the form of a photon of light. The light-producing chemical in these reactions is called the *luciferin*, but this general term applies to a number of unrelated light-emitting chemicals. It's like calling something a soup: it doesn't tell you what the ingredients are, but you know the general form. A different chemical is used as the luciferin by bacteria, jellyfish, and fireflies. Having said that, there is something fishy going on in the ocean, because one luciferin named "coelenterazine" is used by at least ten different unrelated kinds of marine animals. In some cases, we've found that the animals are getting their luciferin from their diets, but it has also been shown by us and others that a few animals can synthesize their luciferins by themselves; after all, it does have to originate *somewhere*.

Often we get a chance to see bioluminescence only when organisms bloom in high numbers, and then some of their cohort get swept ashore. That is the case for this wash-up of bioluminescent ostracods (crustaceans) along the coast of Japan.

Organisms (the inclination is to say "animals" but bacteria and other single-celled non-animals can also be bioluminescent) typically control their light emission by using a larger protein to regulate when the luciferin reacts with oxygen to produce its burst of energy. These larger proteins are called luciferases and photoproteins, and they are almost always fabricated by the luminous organism itself, using instructions in its genes. These components are often packed into clusters of specialized cells called photophores. Like tiny flashlights, these can even have lenses, filters, and reflectors to direct the light and optimize its qualities for a particular application.

"Phosphorescence" is an old term that was used interchangeably with bioluminescence in the 1800s, but has its own proper definition today. In a physics sense it is now used to describe things like glow-in-the-dark stickers that you "charge up" with a light and then they glow for a while afterwards. Since this process is a time-release version of fluorescence, we will touch on the mechanism in the next section.

EVOLUTION OF BIOLUMINESCENCE

Although there are a few bioluminescent animals on land and in fresh water, the diversity of luminous creatures is mainly on display in the ocean. In a study analyzing hundreds of thousands of macroscopic animals that live in the water column from the surface to 4 kilometers in depth, we found that 75 percent of them were luminescent. Much of what we say to explain this widespread diversity of bioluminescence is speculative, but there are a few factors that probably contribute. First, most of the ocean is perpetually dim or dark, so any flashes of light send a powerful message, like blowing

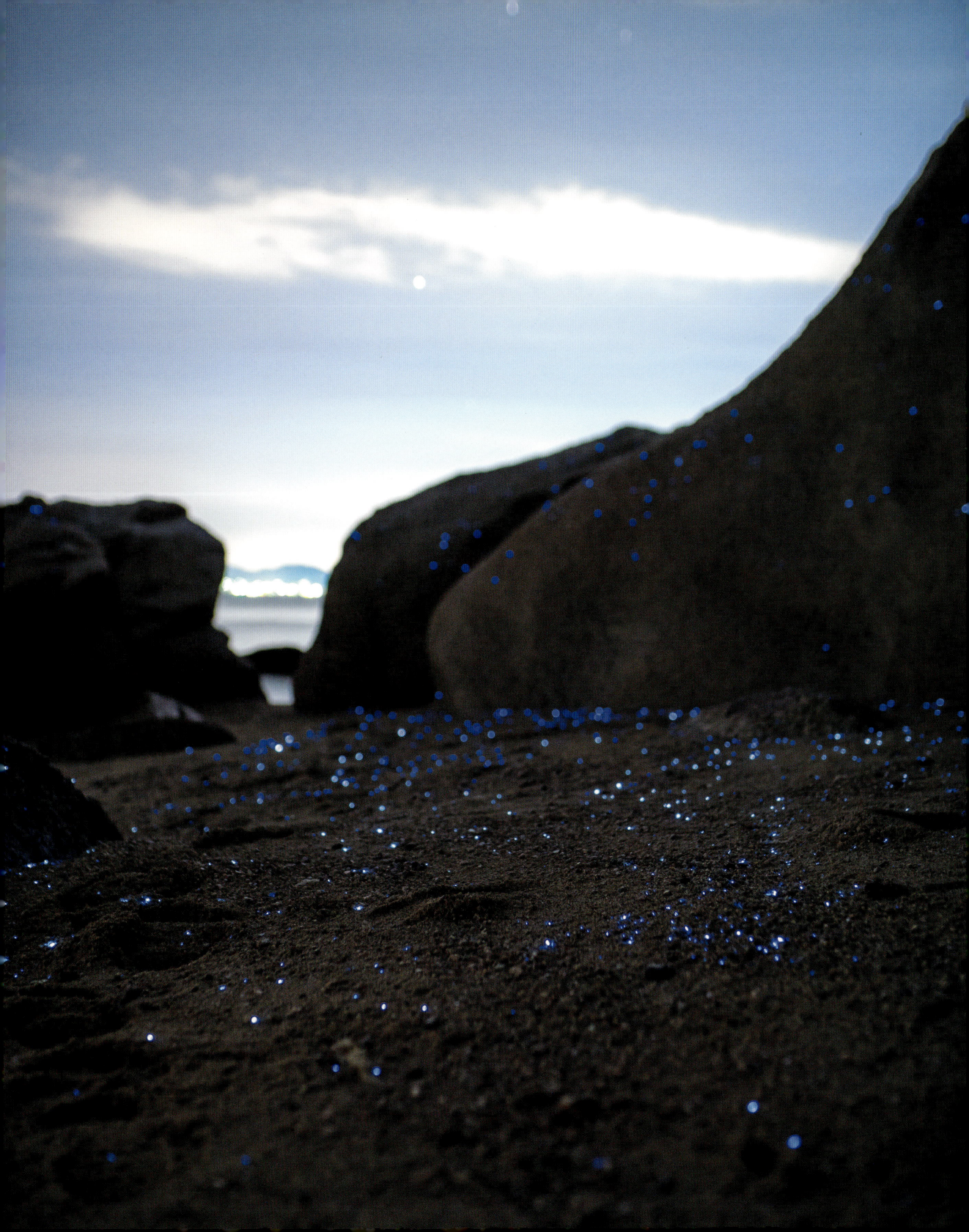

a trumpet in a quiet room. Second, most of the ocean is very transparent to blue light, so a flash can be perceived even 100 meters away, making it extremely efficient. Third, the ocean is the most ancient and stable crucible of evolution, giving creatures many generations to evolve and refine a trait. As for the diversity, the ocean is inherently diverse—not necessarily in the profusion of different species, as with ants or beetles, but at the broad level of categories of animals, like jellyfish, fish, or shrimp. Several major groups of animals *only* exist in the ocean and not even in fresh water: starfish, squid, comb jellies, arrow worms, and sharks (with a few outliers that can venture into fresher areas). Bioluminescence is present in all of these and many, many more. It is quicker to list the few groups of marine animals lacking bioluminescent members (mammals, roundworms, turtles...) than it is to list all those with the ability.

There seems to be a notion that animal bioluminescence is due to symbiotic bacteria. Although there are many kinds of bioluminescent bacteria—and a few of these provide light for fish and squid—most animals do not use bacteria for their glow. They either generate or acquire the luminous chemicals themselves. In fact, the ability to make biological light is thought to have arisen at least one hundred times over the history of life on our planet. This number is hard to nail down, because it depends on accurately identifying luminous species, reconstructing the evolutionary history of all animals, and—importantly—defining what you mean by an independent origin. If something eats a luminous animal, and then reuses the chemicals from its prey to make light, does that count as one origin or two?

We have been studying bioluminescence for decades now, and have been fortunate that some of our discoveries have advanced our understanding of the diversity of bioluminescent animals. In the pages to come, you will see the green bomber worm, the glowing-sucker octopus, bioluminescent sea anemones, the angler siphonophore, and the luminous arrow worm—a few of the glowing wonders that we have documented for the first time.

THE FUNCTIONS OF BIOLUMINESCENCE

One of the most challenging questions in behavioral biology is: What are the functions of bioluminescence? Studying the language of light by poking and prodding organisms is like trying to understand human communication by jabbing someone and recording their yelps and groans. As much as we try, it is not easy to witness natural bioluminescent displays, especially in the ocean: the animals have to be healthy and comfortable, and the process of trying to observe them can disrupt their behavior. "Give us some privacy, please!"

Nonetheless, through the years, scientists have tried to answer the "why" questions about bioluminescence using experiments, observations, and, most commonly, by making up plausible scenarios for others to test. Like Kipling's *Just So Story* of how the leopard got its spots, we tend

From afar, electronic displays, such as on a smartphone, appear white. When you zoom in, it is surprising just how non-white the individual elements are. In newer high-resolution displays, the elements vary in size and number in a way that has been optimized for effective display.

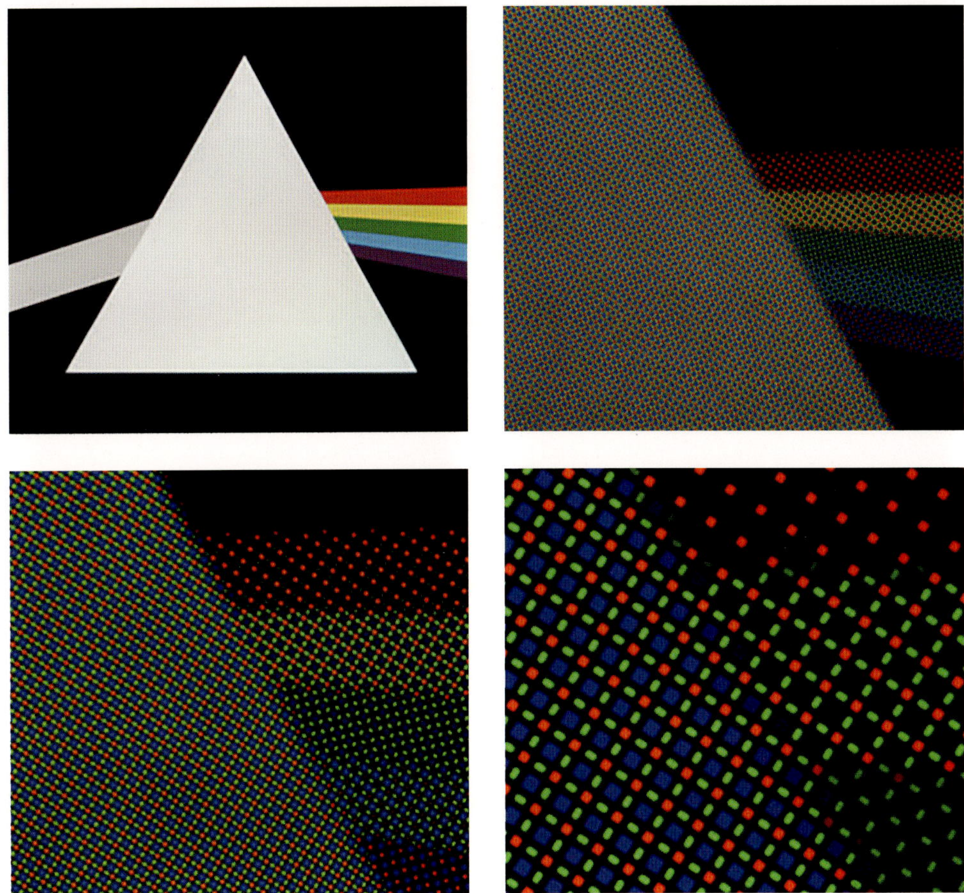

to anthropomorphize and make guesses based on the location and morphology of photophores, even without having seen their light emission outside of the lab.

Putting aside these limitations, there have been many elegant experiments demonstrating the incredible fine-tuning and optimization of bioluminescent displays for particular functions. What is mind-boggling is that some animals—particularly squid but also some fish—may have three different sets of bioluminescent organs that serve at least three different functions in one species.

We generally divide the potential functions of bioluminescence into three categories: defense, offense, and mating. Defensive functions are the most varied and numerous. They include camouflage; blinding or stunning predators; distractive displays such as decoys and smoke screens; attracting your predator's predator; and even what can be thought of as revenge—glowing inside your predator's gut so it gets eaten. Offensive functions are mainly either attracting prey through glowing lures or detecting prey using headlights. Finding mates has been well demonstrated in fireflies on land, but there are only a couple of well-studied examples in the ocean, as well as many cases that seem plausible but haven't yet been tested. Examples of these will be shown in the images that follow.

HOW COLORS ADD

The white light generated from a computer screen is an optical illusion that capitalizes on the fact that humans have three color-sensitive visual pigments. Although these visual pigments, called opsins, have relatively broad ranges of color they can detect, they are different in their sensitivities to blue, green, or red light. When you excite all three color channels equally (physiologically and psychologically equally, and not necessarily equal numbers of photons), we perceive this as white. If you look at your computer or smartphone under a microscope, you'll see that it does not take a lot of magnification to reveal the underlying trickery: what you think is uniform white is actually distinct vividly colored light sources, sufficiently small and closely spaced that they can't be individually resolved.

You'll notice there are no yellow elements in a computer's display. There are at least two ways (actually, an infinitude of ways) that we perceive yellow light: pure yellow photons or a roughly equal mix of red and green photons. Both bioluminescence and fluorescence, which we will see next, typically occur as more or less monochromatic light, with a bell-shaped distribution of wavelengths around the primary color. By far, the most common colors of bioluminescent light in the sea are blues to greens, with a few interesting exceptions. You may be able to guess a couple of reasons why these are the dominant wavelengths: those are the colors that transmit best through the water, so the signal is as strong as it can be. Also, related to the ambient light field, organisms' eyes are most likely to be sensitive to those colors. Because ocean animals may be sensitive to only one or two colors, if you want to communicate, your signal needs to match the receiver's sensitivity. However, just as with birdwatching or catching Pokémon, the most common may not always be the most interesting. Some of the exceptional colors of bioluminescence—the species that emit yellow or red light—are the most intriguing to study.

Variations of purples are some of the hardest colors to define. We can think we are seeing purple by three different mechanisms: you can either have shorter-than-blue photons from the violet end of the rainbow; an additive blend of blue photons with red light from the opposite end of the spectrum (see the image opposite); or traditionally mix subtractive blue and red pigments to make a grape color. This perceptual convergence may not seem surprising at first, but when you add green and red photons you see yellow, and when you mix green and red pigments it yields brown. Between these three scenarios, purple bioluminescence only occurs as short-wavelength violet photons, and even then, only rarely.

The challenge to perceiving violet light is that our shortest-wavelength photoreceptors are maximally sensitive to blue at 445 nanometers, whereas violet light can peak at around 400 nanometers or lower. In humans, the red-sensitive pigment actually has a secondary peak far down the spectrum that gets excited by the violet part of the rainbow. So, we perceive purple from

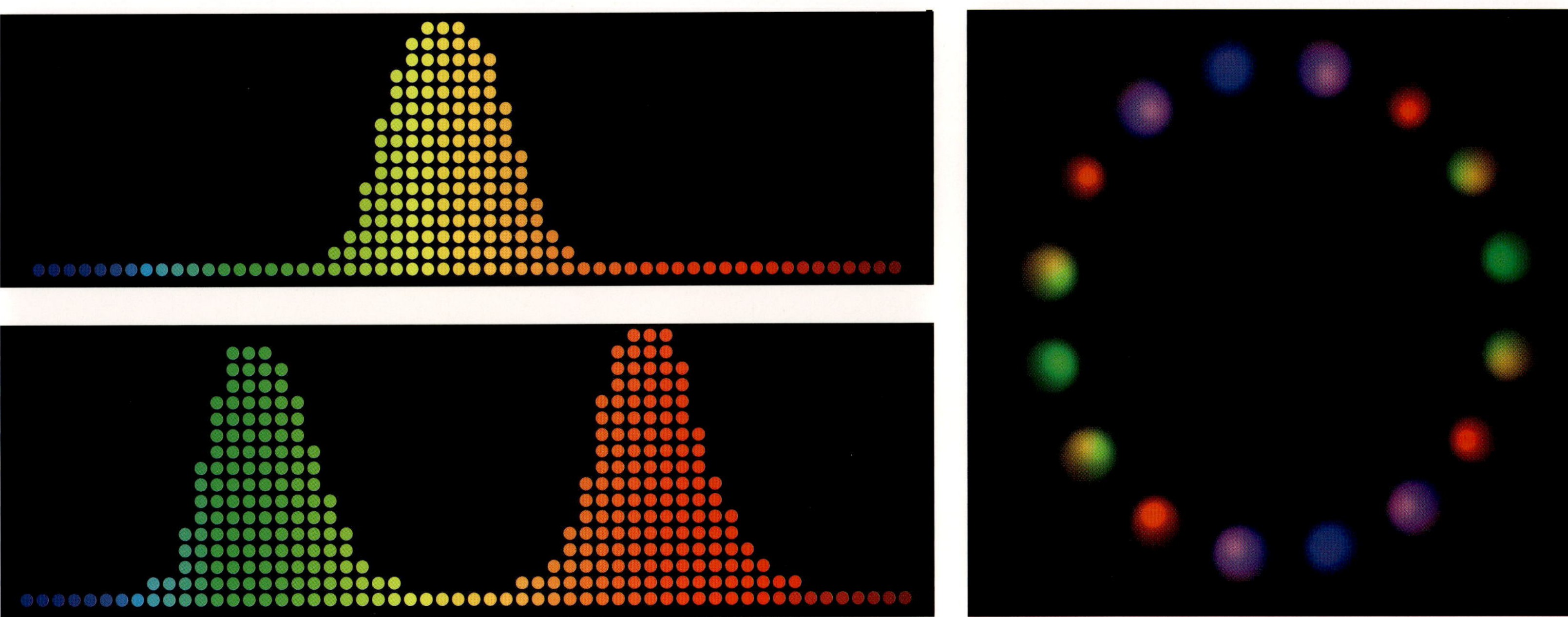

We can perceive yellow light when presented with yellow photons (top), but also when seeing red and green photons without any yellow light present (above). Multicolor LEDs have three closely spaced elements packaged together, and by adjusting the relative brightness you can generate a range of perceptual colors through additive mixing (above right).

two very different sources: super-short-wavelength light or a combination of longer wavelengths. Even cameras are unable to accurately capture this color of light because their sensors are engineered to cover a similar range of wavelengths as our eyes. You will see how this looks for one violet-emitting species later on.

IMAGING BIOLUMINESCENCE

We don't recommend doing it, but if you search the web for images of "bioluminescent jellyfish" you are likely to get nine out of twelve images that are colorized versions of non-luminous jellyfish, like moon jellies, one iridescent ctenophore, and two images of bioluminescent species—none of which are actually showing their luminescence. Photos of animal bioluminescence are highly sought-after but hard to come by.

Even the most basic photographic subjects can pose challenges when you're imaging bioluminescence. We usually try to get as much light onto the camera sensor as possible, which means (if you know photography terms) that we're opening up the aperture as wide as possible. If the lens will support it, we use f/1.4 or f/2, but the trade-off is that the depth of field—the span of distance from the camera that is in focus—is extremely small at these large apertures; often less than a centimeter. We've taken a video of a glowing viperfish where the middle of the body was in focus but the head and tail were fuzzy.

When trying to photograph bioluminescence, either in a shipboard lab or—most challengingly—underwater, we often use a red light to find and frame our quarry. Because most oceanic bioluminescence is blue or green, dim red light doesn't interfere with our ability to record the luminous displays. In fact, having the context of the animal's outline makes the

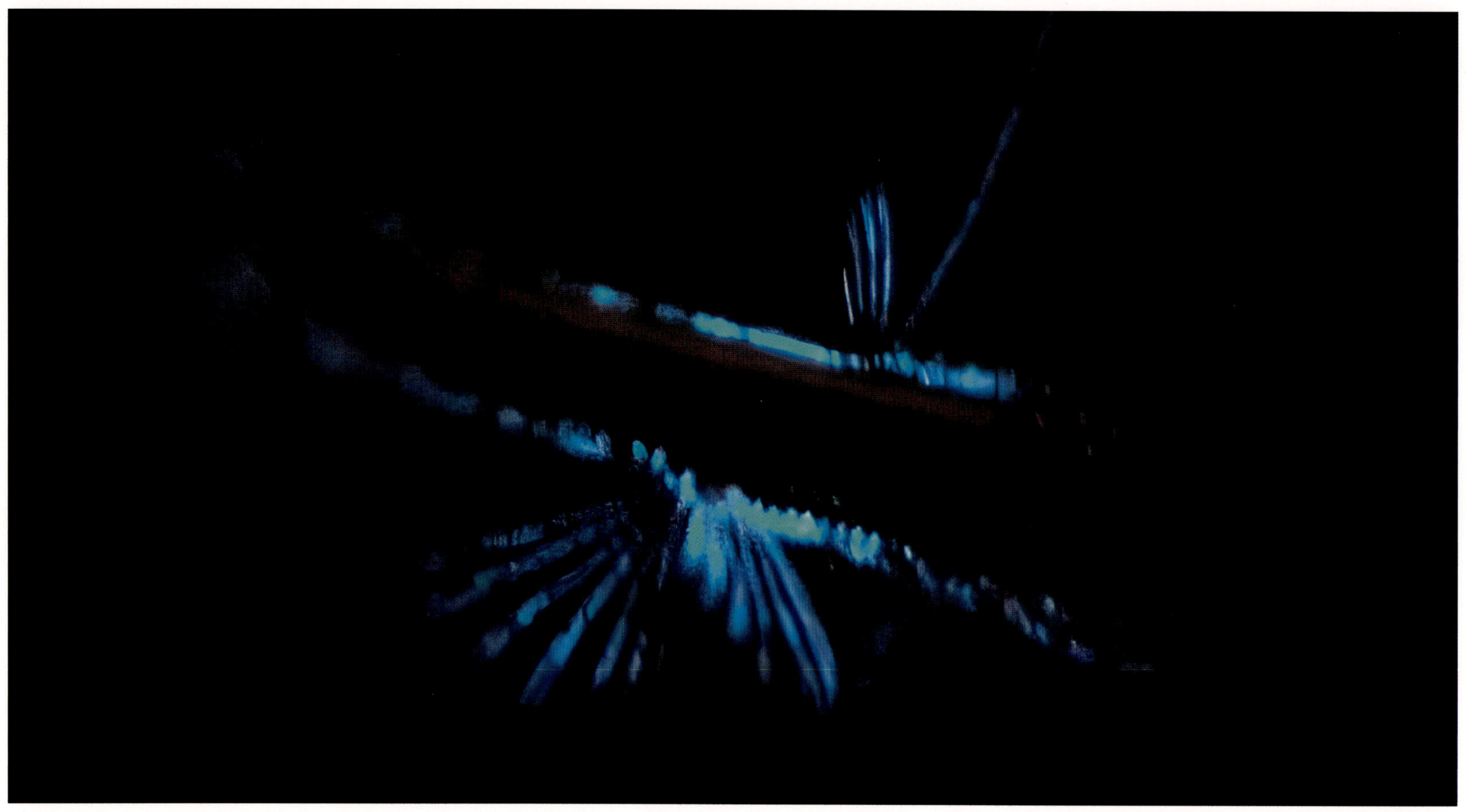

bioluminescence more meaningful than a few specks of blue light would be. Some of the photos in this book are taken with dim red light, providing context to the natural blue or green bioluminescence.

Even the act of focusing is challenging: different wavelengths of light interact with materials differently. When light goes from air to glass or water at an angle, it will bend; the amount of bending depends on the color. This is one way to think about how a prism splits white light into a rainbow. Cameras focus by using glass lenses to bend light. However, red light bends less than violet or blue light, so it is actually impossible to focus both red and blue light precisely—their focal points end up at slightly different spots. Usually this is unnoticeable, but when we work with a shallow depth of field and focus using red light, the blue bioluminescent flashes are a little out of focus!

The fundamental challenge is that bioluminescence is relatively dim in the grand scheme of things. To get enough light, we need to increase the camera's sensitivity (adjust the ISO, in photo speak) or increase the exposure time, or both. When we increase the sensitivity—up to ISO 100,000 for video and 6,400 for still photos—the image can get noisy and grainy. If we increase the exposure time, then any motion turns pinpoints of light into smears and streaks. We don't appreciate this problem when we look at bioluminescence ourselves because our eyes, evolved from the eyes of nocturnal mammals, have excellent vision in low light.

Steve: The viperfish *Chauliodus* is a type of dragonfish that has a bronze (rather than black) body surface. We were chasing this one using the low-light camera on our submersible when it suddenly came to a stop and flipped on all of its bioluminescence. Fortunately, the middle of its body was relatively focused when it froze, allowing us to capture the first imagery of it glowing in its natural habitat. The bioluminescence of this fish is pretty incredible, since it is not just a row of dots along its belly. The light propagates down the fin rays and the long spine on its back.

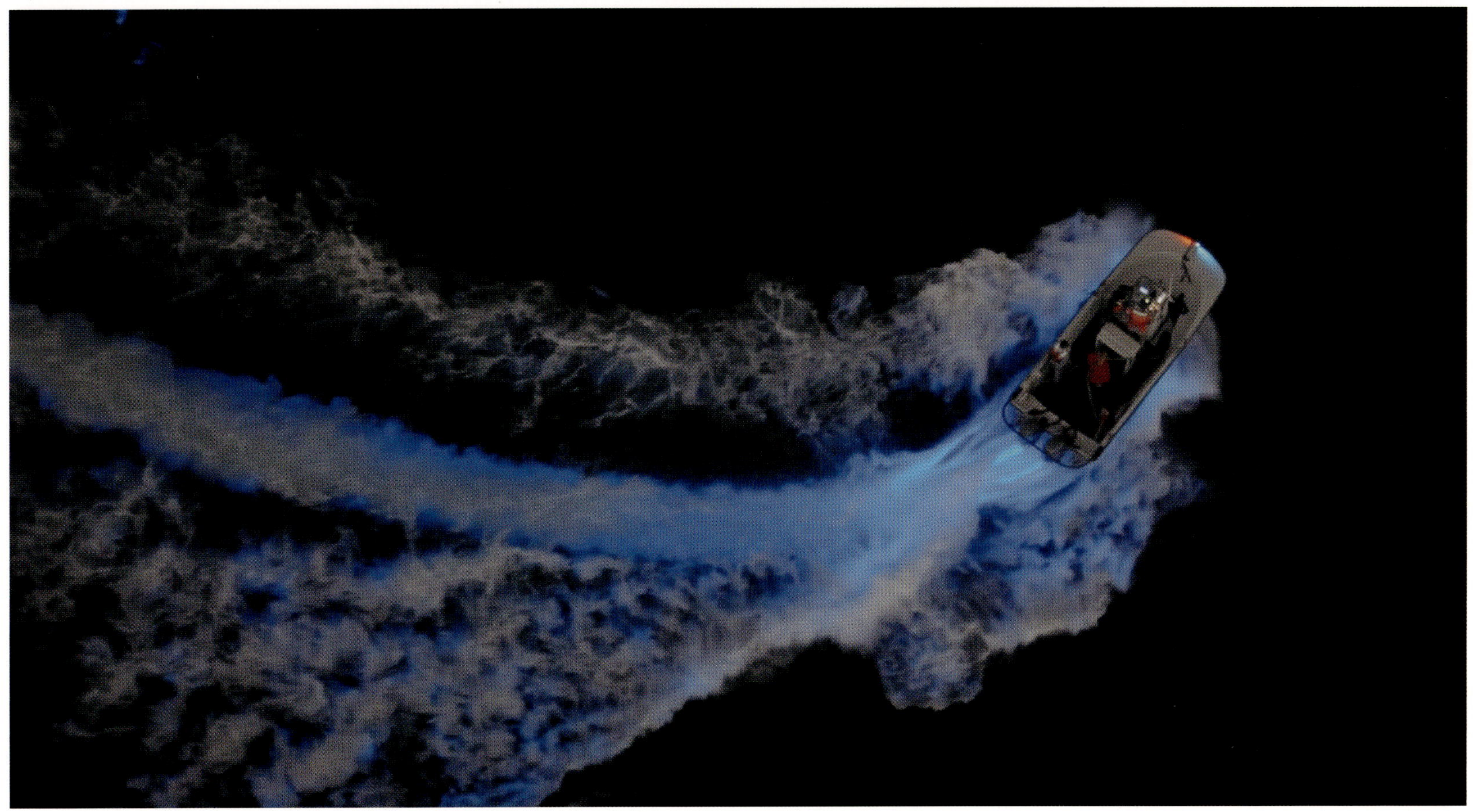

One of the most accessible opportunities to view and photograph bioluminescence is when there is a bloom of dinoflagellates—often called a red tide. These high concentrations of single-celled plankton will produce light when disturbed, as in the wake of a boat, breaking waves, or even footprints in freshly wet sand. In some places, like the bioluminescent bays of Puerto Rico, these events are predictable and become tourist attractions for kayakers to paddle through the glowing waters.

If you are going to try filming bioluminescence or fluorescence, we recommend setting your camera or phone to underexpose relative to what it thinks it should do. Autoexposure is programmed to make an image that has middle brightness—averaging out to gray. Your camera doesn't know that you want a mostly black image with a few spots of light. To overcome this, you need to trick the camera by setting the exposure manually, or adjusting the autoexposure setting way down around -3 on the setting labeled +/-.

Despite these and other challenges, we are immensely grateful for improvements in camera technology that allow us to document and share—in full color 4K video or 24-megapixel still-image clarity— things that we could only see by eye a short time ago. Even a smartphone today packs a powerful camera capable of recording bioluminescence; we've published images in scientific journals that were taken by phone, out of convenience or necessity. So give it a try—on fireflies, fungi, earthworms, or ocean waves—and share the beauty of what you find.

Dinoflagellates

If you have ever seen bioluminescence while walking along the seashore at night (or less romantically when flushing a marine toilet on a sailboat), it is likely produced by thousands of single-celled organisms called dinoflagellates. These "protists" are hard to define, since some are photosynthetic, some are predatory, and others are both. They can also be toxic or not, and bioluminescent or not. A fascinating fact about dinoflagellates is that their light-emitting luciferin has an almost identical chemical structure to their light-absorbing chlorophyll. It seems that they actually interconvert between the two molecules on a day–night cycle. The dinoflagellate luciferin is used only by one other kind of organism, krill, which indicates a dietary origin for the capability.

When dinoflagellates bloom, by dividing and cloning themselves under favorable conditions, they can form dense red tides which turn the water orange or brown by day and sparkling blue by night. These are one of the few bioluminescent organisms you can buy, and many a science teacher has grown these in a small flask and shaken the water in the dark to impress their students. These animals are also responsible for most of the light in bioluminescent tourist destinations, such as the bioluminescent bays in Puerto Rico. We have spent happy nights at sea leaning over the bow of a research ship, watching our glowing bow-wave or marveling at dolphins cavorting with their bodies cloaked in a scintillating aura. One function of bioluminescence in dinoflagellates is thought to be direct startling of their tiny predators, but there is also experimental evidence that at high concentration, their flashes reveal the tracks of grazing predators swimming through them—the burglar-alarm hypothesis. This function requires that the bioluminescence is benefiting not the individual, but all of its numerous relatives living nearby.

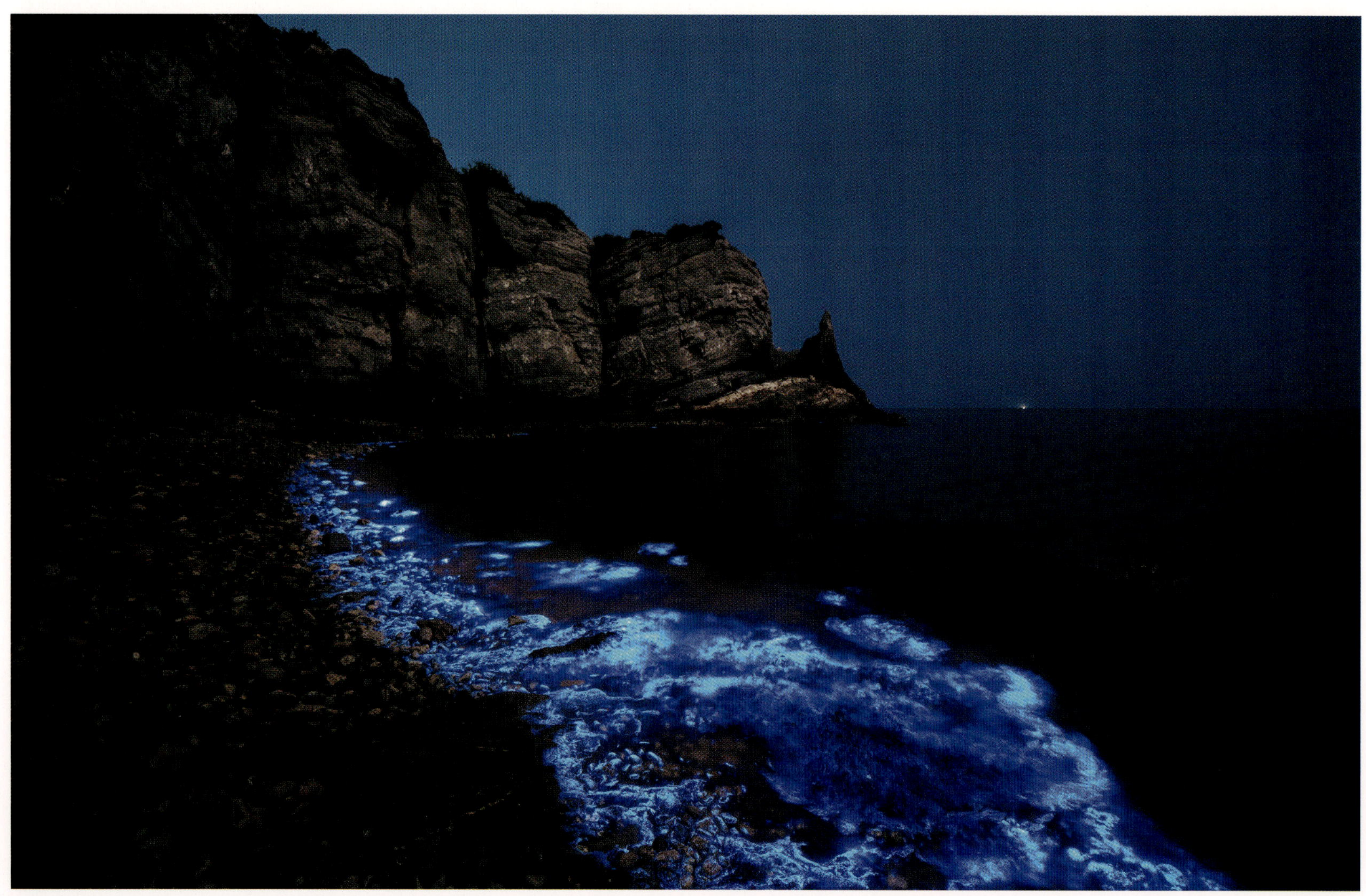

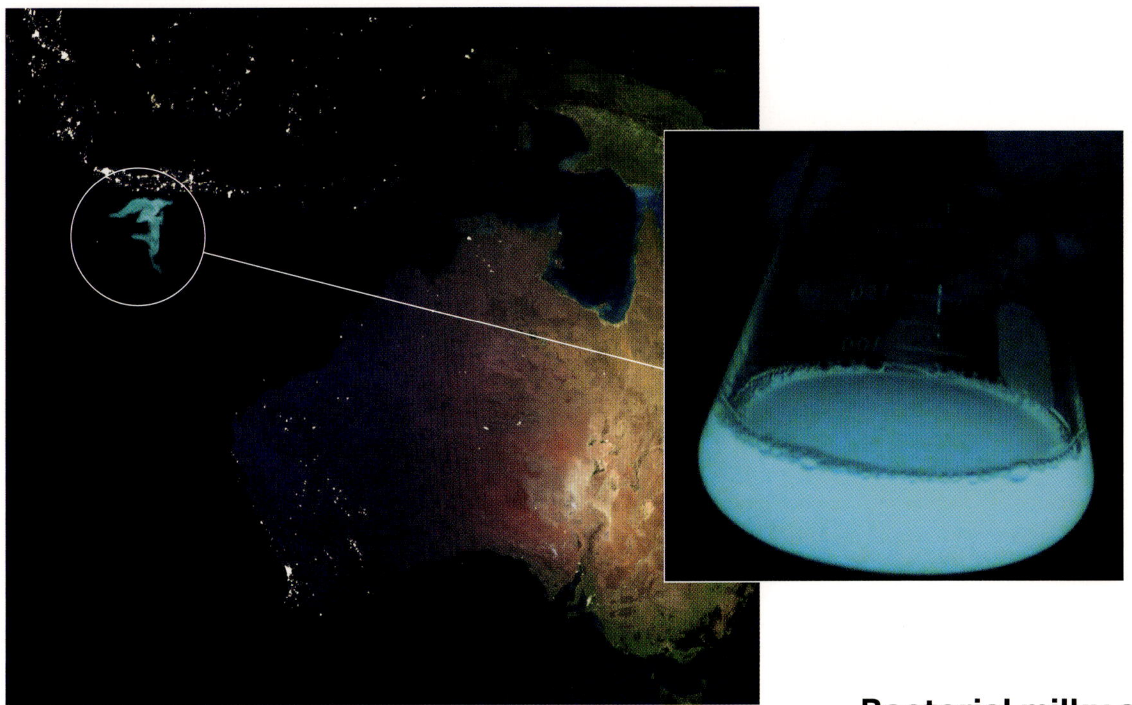

Bacterial milky seas

Although you can make out the pinprick of light made by an individual dinoflagellate, bacteria are the smallest organisms known to make light, and individual cells cannot be seen by eye. When bacteria in seawater grow to sufficient concentrations (about 100 million cells per milliliter, or 5 million cells per drop of seawater), then a transformation happens. Cells signal to each other, acknowledging their good fortune, and they begin to glow. It would be futile for a bacterium in isolation to glow, but together they can transform the water into a diffuse "milky sea."

Incredibly, these tiniest of cells make by far the most extensive bioluminescent displays on earth. Working with atmospheric scientist Dr. Steve Miller and using special sensors on orbiting satellites, we've documented glowing expanses the size of Iceland, which lasted for more than a month. The top-left image is the only composite in the book, constructed to show the true scale of the satellite-detected feature overlaid on images of the globe in day and night.

Bacteria are unique in that they don't need physical stimulation to trigger a flash of light. As long as they and their chemical signaling molecules are sufficiently concentrated, and they have a steady supply of oxygen, bacteria will glow continuously. The result is a surreal surface display that mariners have witnessed for centuries. The true scale of these events—the largest having 100 billion trillion bacteria spanning 100,000 square kilometers—was never appreciated until the appropriate spaceborne technology became available.

Bythotiara

With the lights on, this hydromedusa looks like an old-timey filament bulb, and with the lights off that turns out not to be far from the truth. In hydromedusae and ctenophores, much of the body mass is gelatinous "filler." Most of their cellular activity occurs in the canals that run around, along, and through the body, and these canals are what you can see lighting up in this and many other hydromedusae.

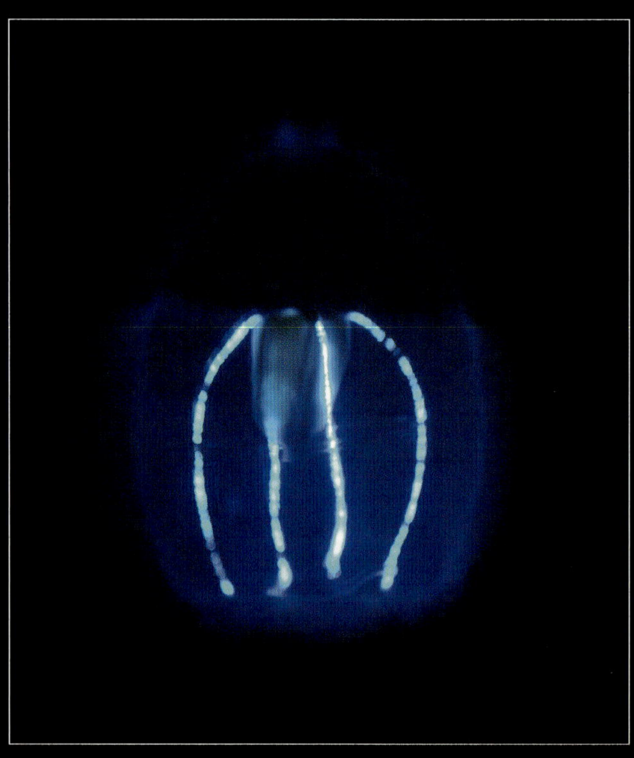

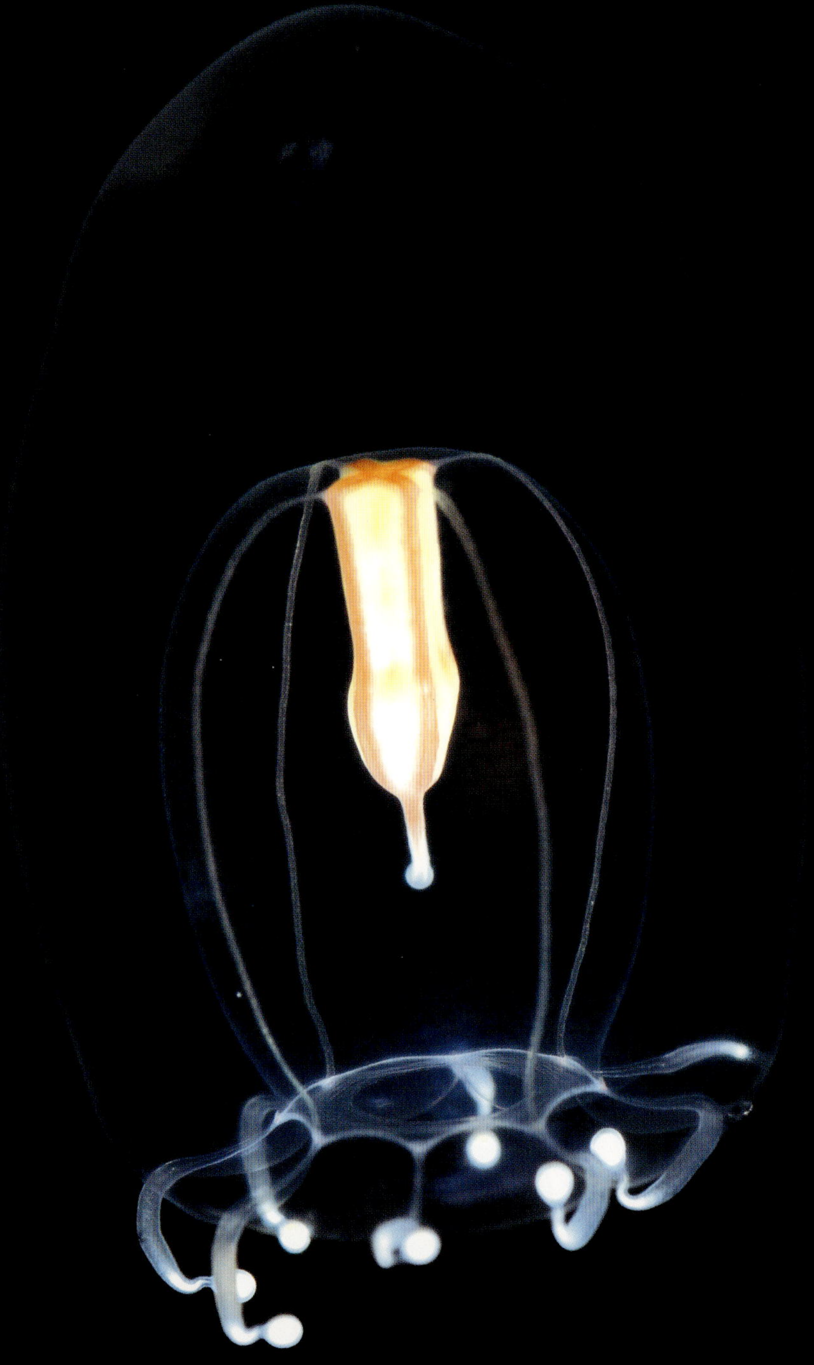

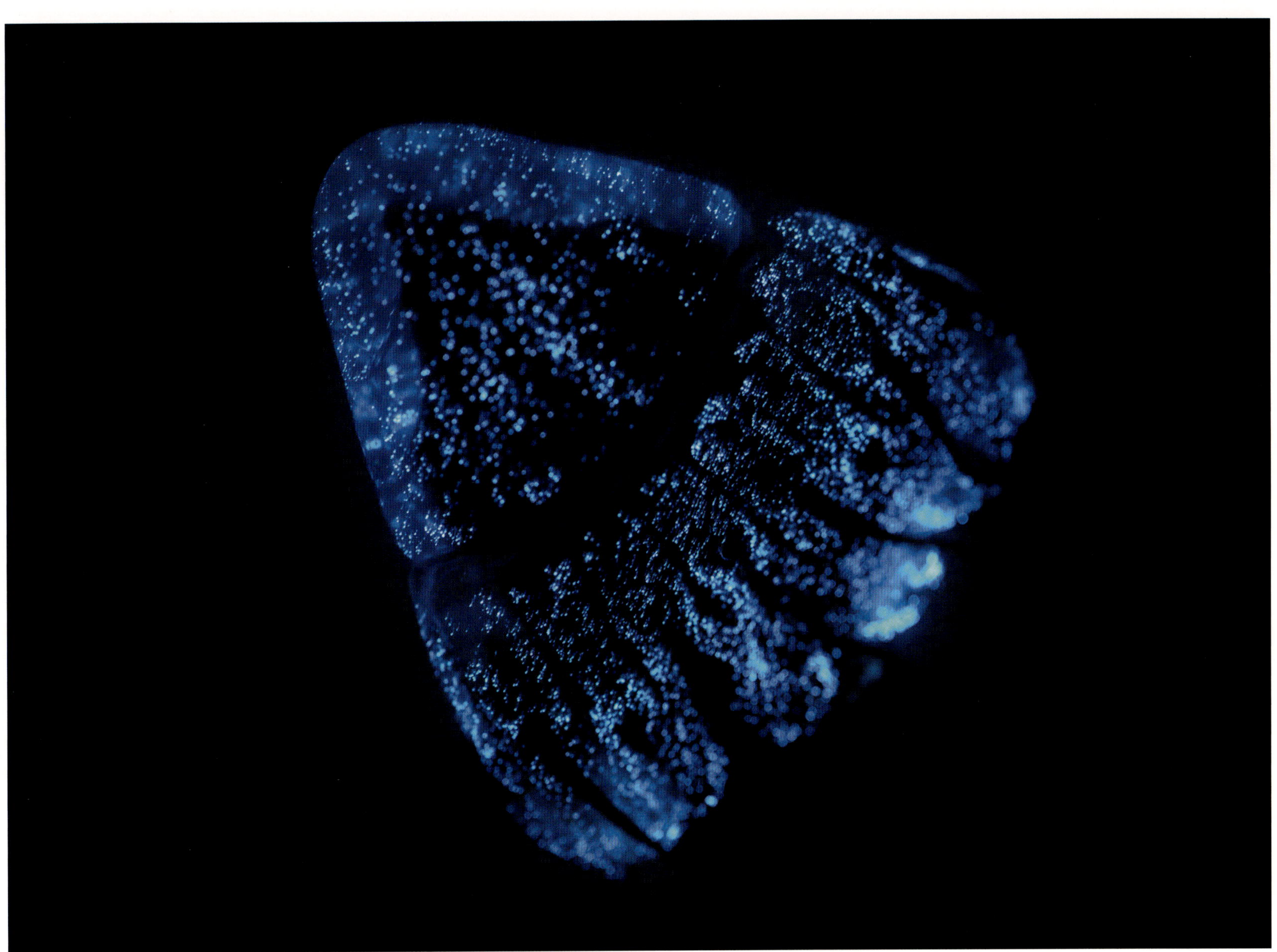

Periphylla periphylla

Periphylla is in the group of Coronate jellyfish ("coronate" means crown, as in coronation), which have most of their abundance and diversity in the deep sea. A few populations inhabit shallow Arctic waters, where the water temperatures are similar to their typical deep-sea conditions. They have a brick-red coloration, as shown in the Pigmentation section (p. 57), but the porphyrin pigments involved may constrain them to life in the twilight zone; the pigment actually becomes harmful when exposed to bright light.

Speaking of bright light, *Periphylla* produces one of the classic bioluminescent displays, with waves of light cascading across the surface of their bell. The luminous granules may also be shed into the water, leaving a sparkling cloud in the medusa's wake.

Atolla

The other very common Coronate medusa in the deep sea is *Atolla*, known as the crown jellyfish. When tapped, these also make bright, energetic displays, which begin as pulses of light and build to a crescendo of pinwheeling spots. We have observed *Atolla* at depth, and in addition to the lights across the surface of their bell, they may also expel a burst of luminous fluid—potentially bioluminescent material left over from their last meal.

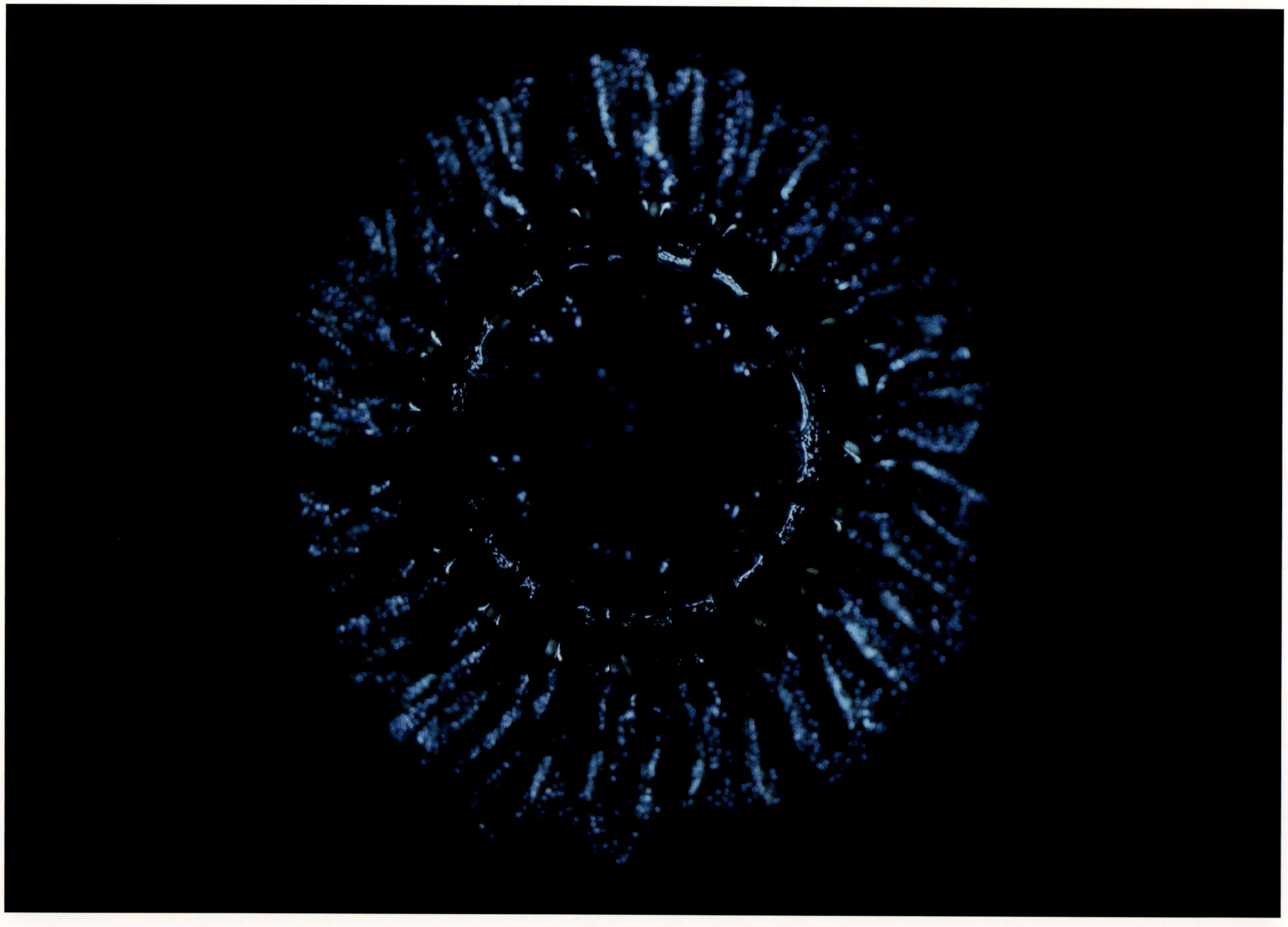

Halitrephes valdiviae

Most jellyfish have bioluminescence that either occurs as lines of light in their canals, like *Bythotiara* (p. 136), or looks like a coating of sparkles on the surface of their bell, like *Periphylla* (p. 137). In *Halitrephes*, the luminescence has a ghostly rippling form, like a two-dimensional lava lamp. If that weren't enough to make this one of our favorite displays, this species emits violet bioluminescence at one of the shortest wavelengths measured. The camera is unable to register this color separately, so in the photos it shows as a deep royal blue.

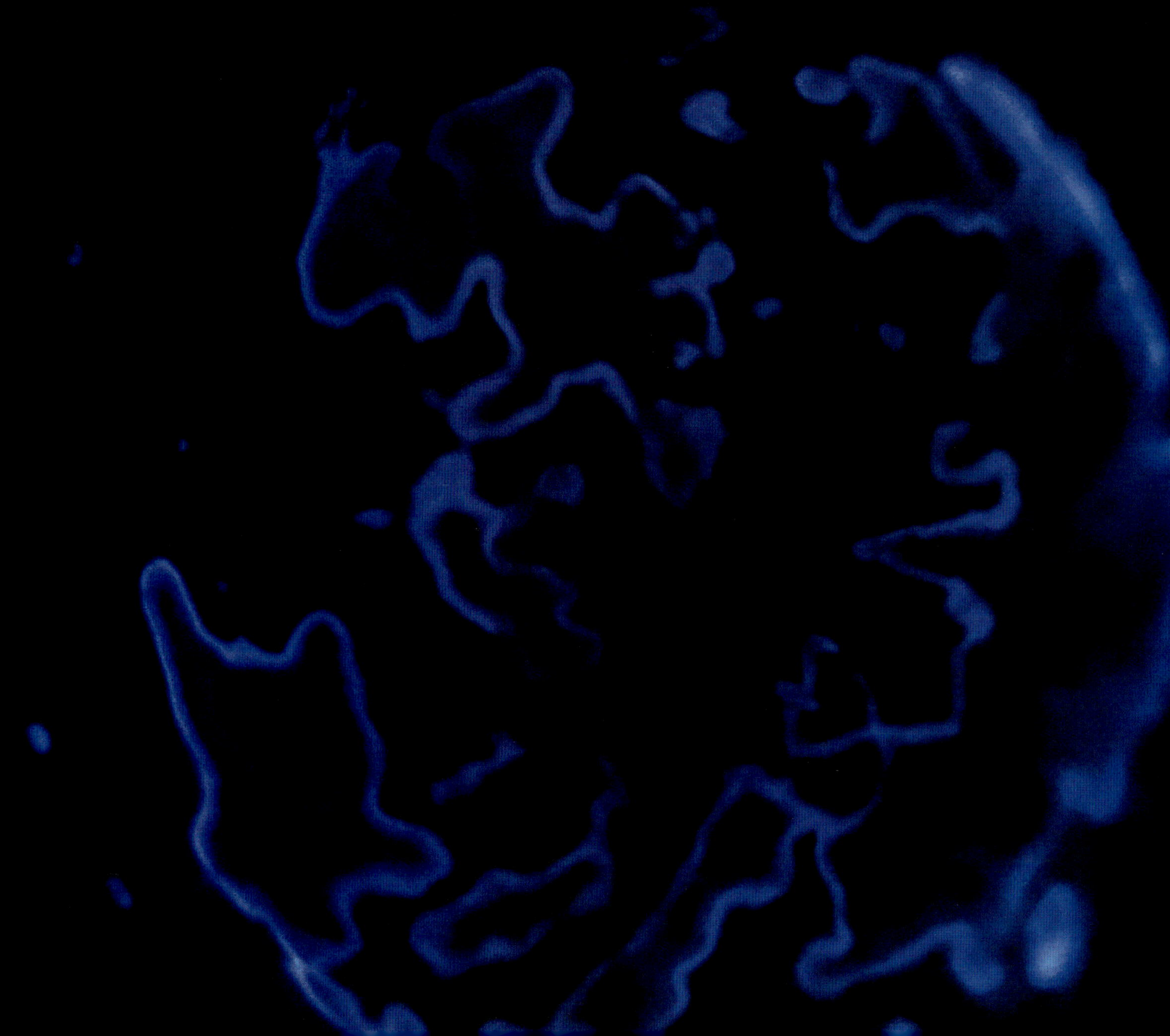

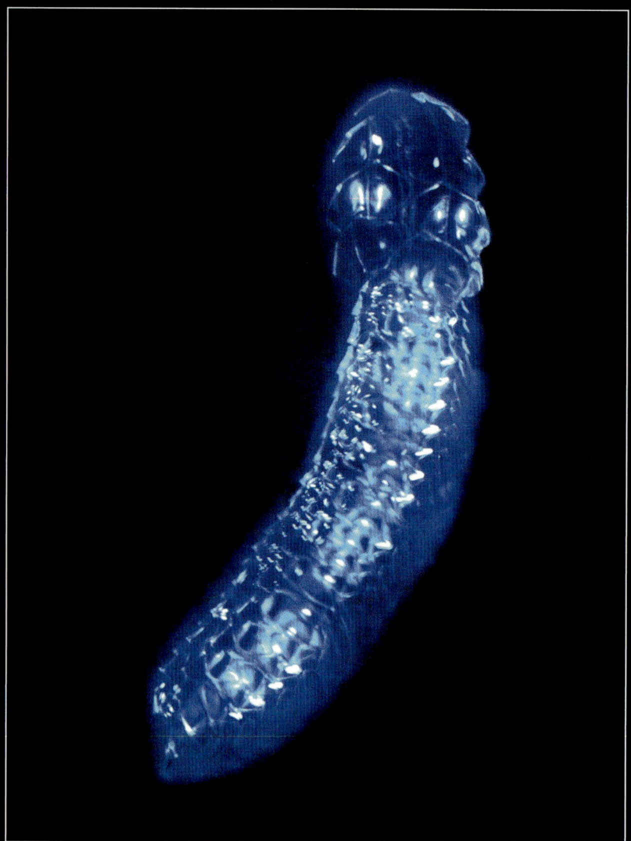

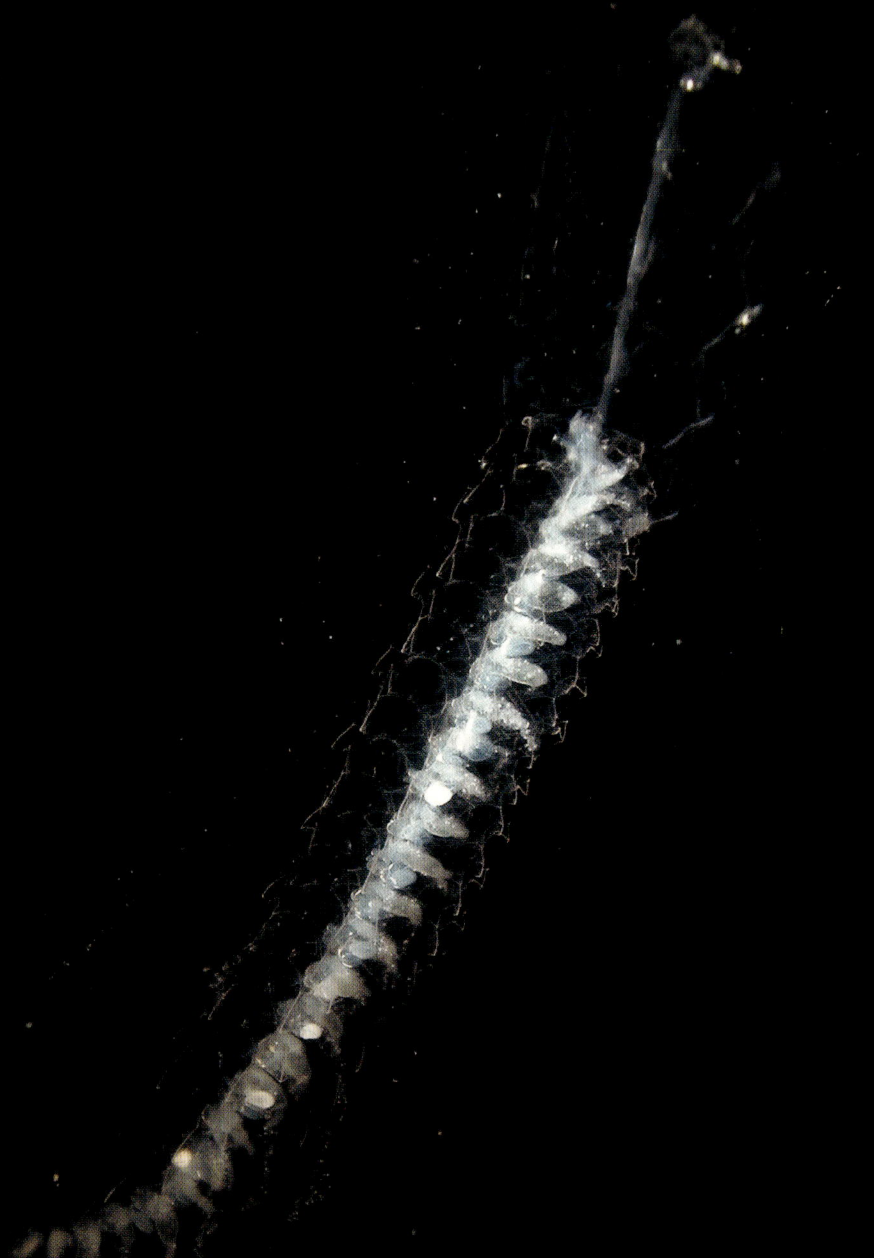

Frillagalma

In addition to their transparent swimming bells, siphonophores are typically sheathed in transparent protective structures called bracts. These overlap down the length of the body like gelatinous shingles. We've found that many transparent siphonophore bits (a very technical term used by siphonophore specialists) have frosty patches on them, which are the locations of light production. Following a disturbance, the bracts may be shed to produce a constellation of distractive lights, although in this particular species the bracts are rather

Rosacea

The siphonophore *Rosacea* has two swimming bells and a long trailing series of gelatinous disks, like the lane lines in a swimming pool. Unlike the point-sources of light in *Frillagalma*, the light-producing cells in *Rosacea* are spread uniformly across the surfaces of the organism. Their glow projects a ghostly form of the whole animal.

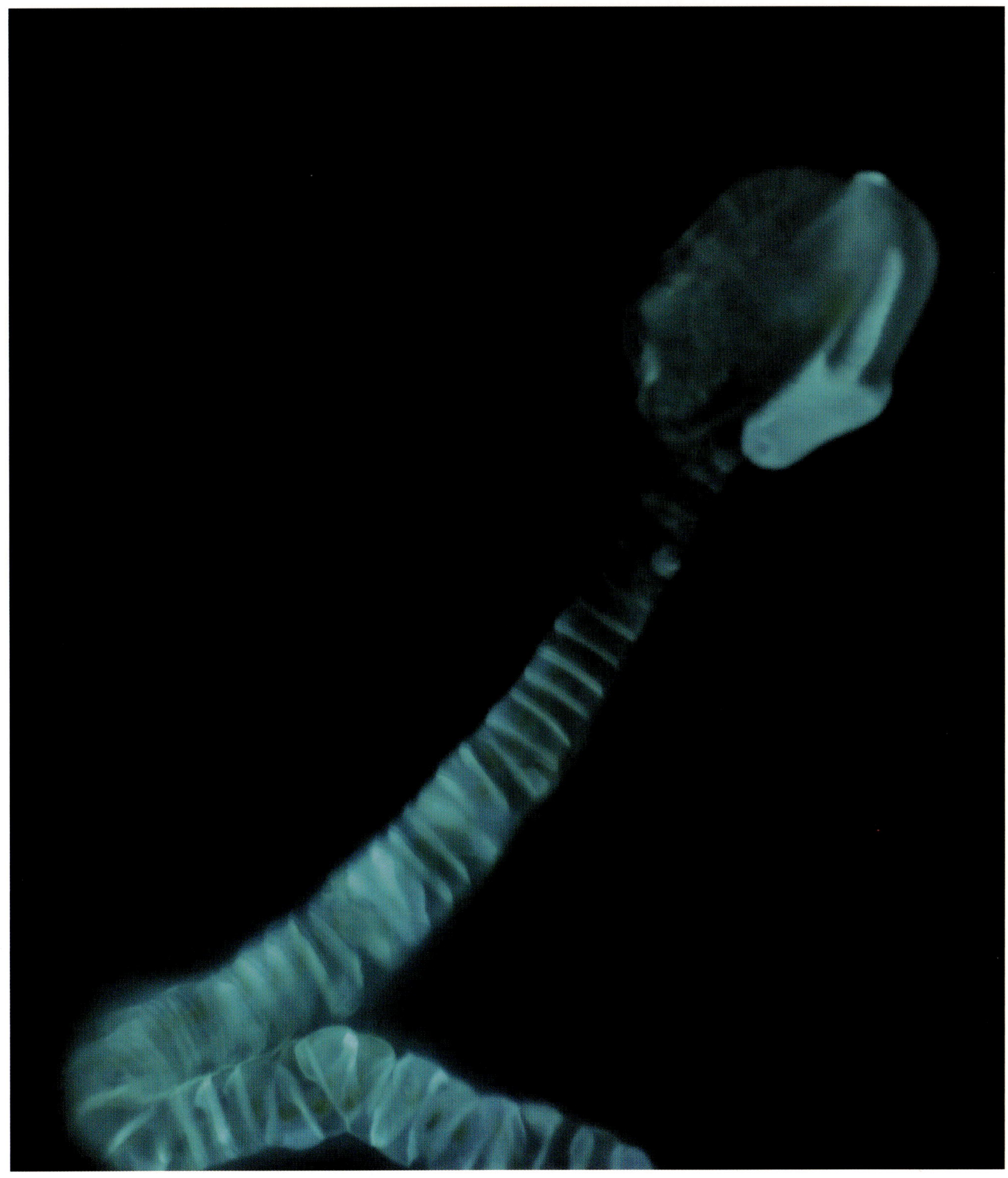

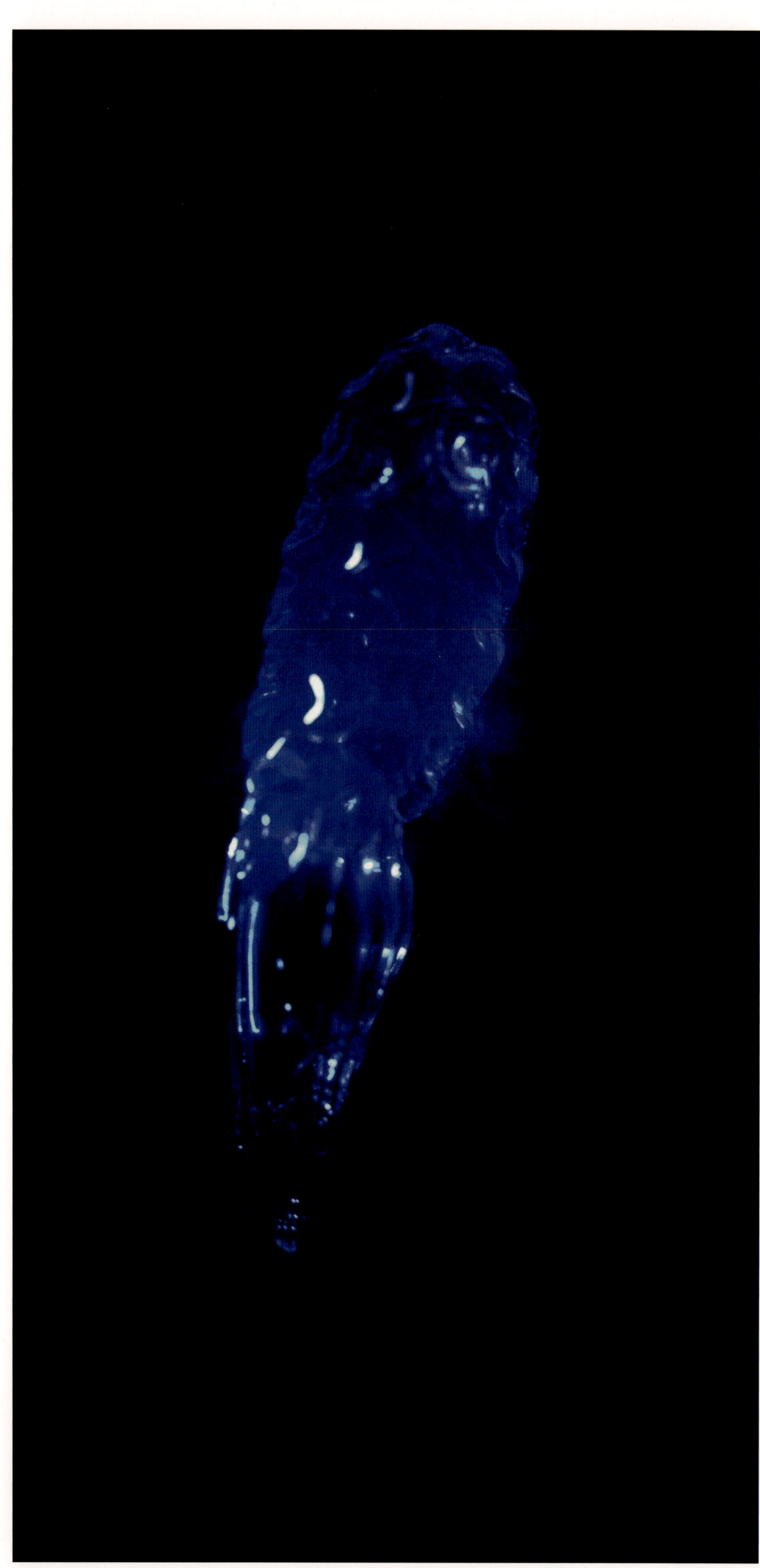

Stephanomia

Siphonophores can be big. They are said to reach lengths of 40 meters—one of the longest animals in the sea—so you can imagine what a display of bioluminescence they can produce. *Stephanomia* has some of the largest swimming bells of any siphonophore—a dozen T-shaped jets interlocked to provide locomotive force. Each is big enough to "swallow" your finger (the actual mouths are in the tail end) and their bioluminescent display is bold and beautiful. Some species have patches of bioluminescent cells on both sides of the opening, so with each pulse, glowing particles come shooting out.

Erenna sirena

Steve: The discovery of bioluminescence in the angler siphonophore *Erenna* rewired how I think about the function and expression of bioluminescence. Species of *Erenna* are deep-living, fish-eating siphonophores that have characteristic black pigmentation in their canals, derived from their black or ultra-black prey. Although their tentacles look like fluffy feathers when extended, they pack a strong sting, and I've had one "welded" to the palm of my hand when its hundreds of stinging cells all fired at once. The easiest way to distinguish *Erenna* species is by looking at side branches of their tentacles, called tentilla, where the bundles of stinging cells are found. In some species these are long and slender; in others they look like beads of a necklace. There were originally three known species of *Erenna,* and we found two more; one which had banana-shaped tentilla with a strange projection. Suspended by a slender filament that was rooted near the base of the banana was a red-coated capsule (shown on the next page), which was filled with bioluminescent chemicals. When we looked back at the other four species, we saw that they all had bioluminescent cells in their tentilla, although only two had traces of red in them. When we observed live siphonophores, they would flick these luminous capsules, jigging their tentacles in the water. It was clear that these siphonophores had lures right next to their stinging cells, to attract fish in the deep sea where meals might be hard to acquire. What remains a mystery is why they coat their bioluminescent lures in a red-fluorescent sheath. When it came time to name this species, we chose *Erenna sirena* because of the alliteration and in reference to the mythical Sirens, who enticed sailors to their demise.

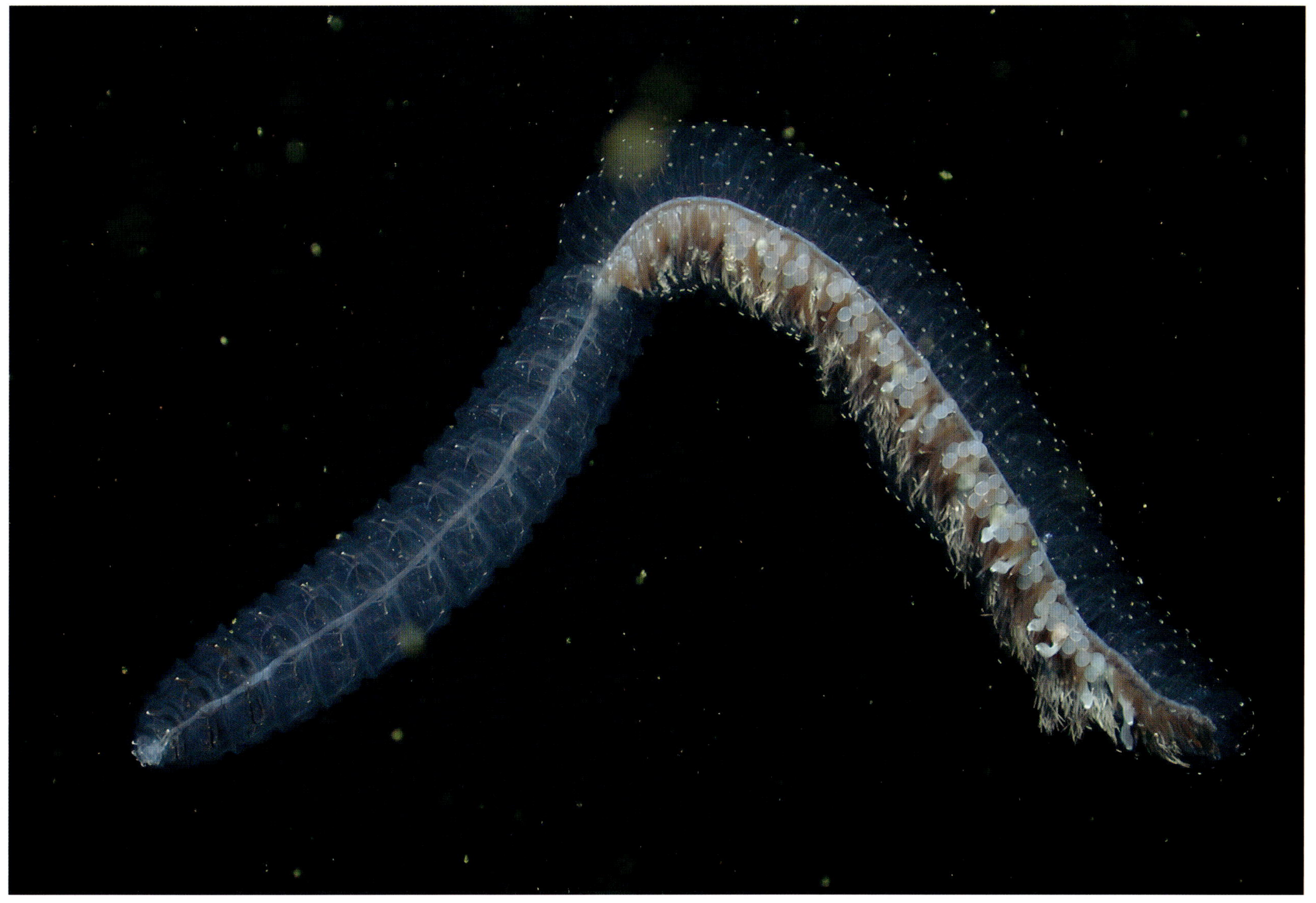

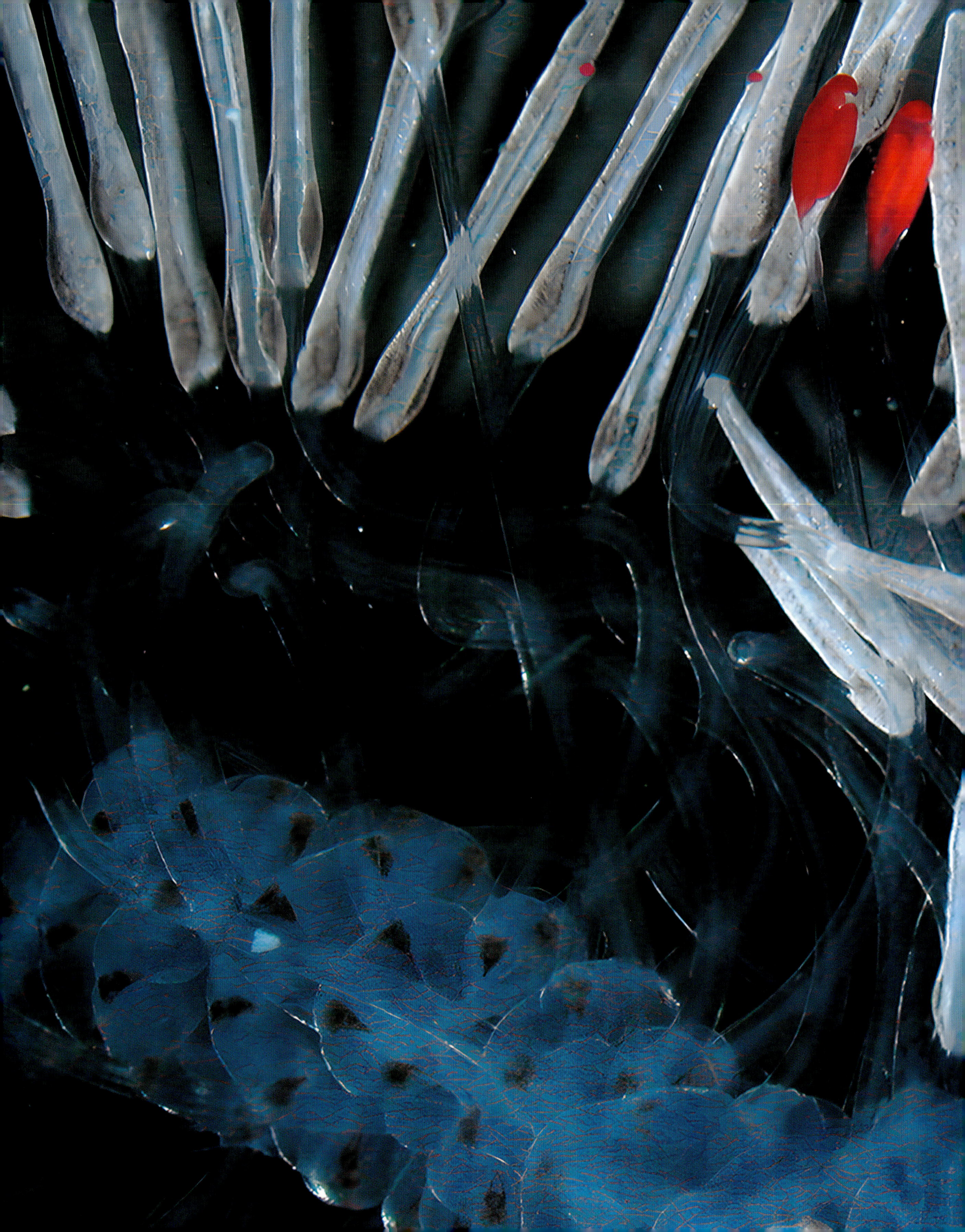

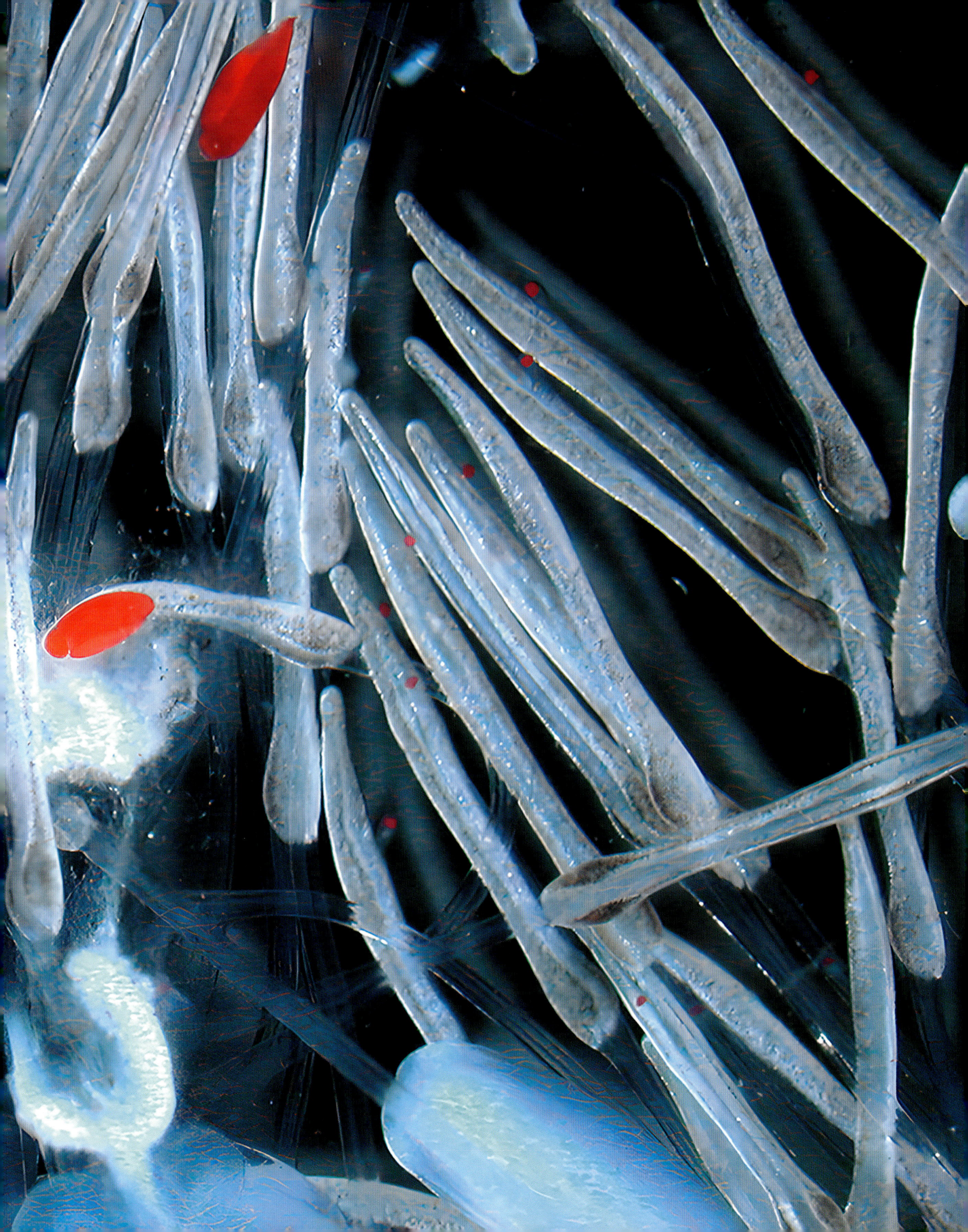

Beroe forskalii

The ciliary combs in comb jellies are arranged into eight rows that run along the length of their body. These comb rows are situated above canals that are extensions of their digestive system and which distribute energy from the stomach out to where it is most needed. The canals are the locations where reproductive cells occur and they also host the bioluminescent cells (you knew we'd get there eventually). *Beroe forskalii*, in particular, does not have simple linear canals. Instead the canals are branched and form a mesh with the canals of adjacent comb rows. Since bioluminescent cells follow the course of the canals, this ctenophore produces probably the number-one-ranked bioluminescent display among marine animals. These photos were exposed for about 7 seconds, so the waves of light are integrated into one complete picture of the mesh, but in actuality the light propagates as a wave across the body from the point of stimulation.

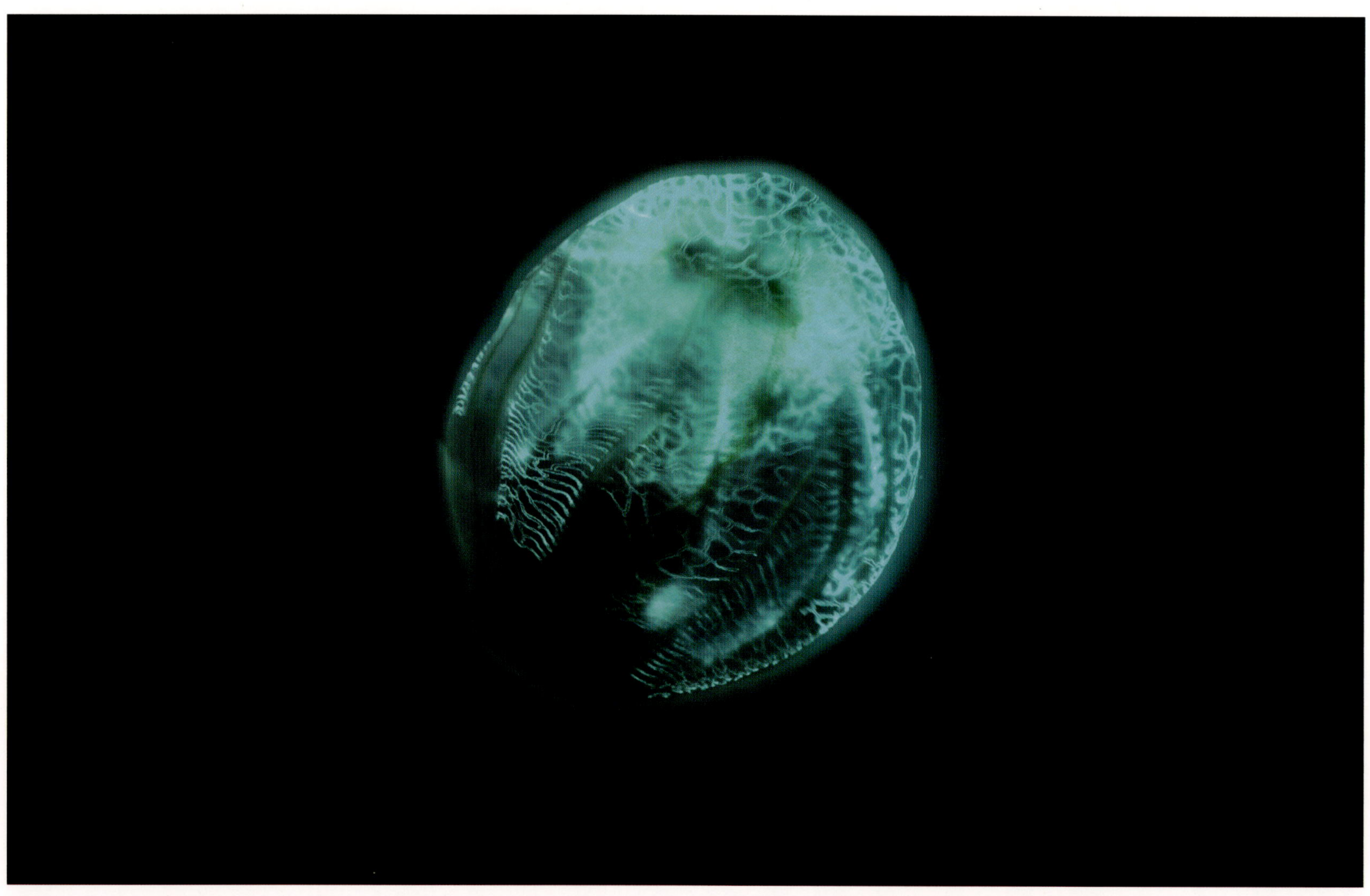

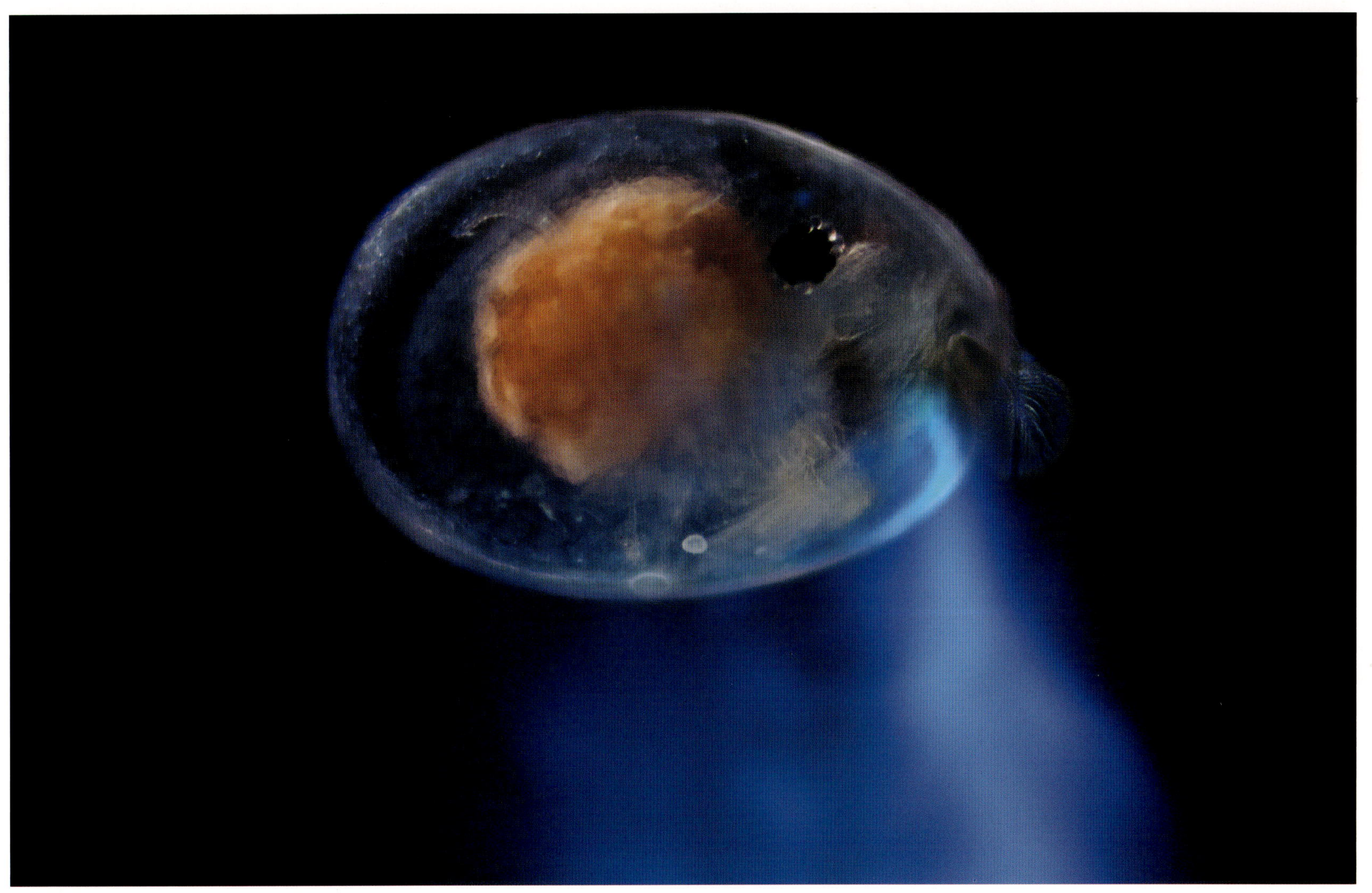

Ostracod bioluminescence and on rocks

Gigantic ostracods were shown in the Iridescence section (p. 118), but most are millimeters, not centimeters long. An amazing fact about ostracods is that they have evolved bioluminescence twice using two different light-emitting luciferins. These were the first bioluminescent organisms studied by bioluminescence chemist Osamu Shimomura, who you will meet again when we talk about green fluorescent protein (GFP).

Although we do not understand the function of bioluminescence in many marine species, we can comprehend it in some ostracods. Under a microscope, these animals look like small shrimp swimming inside a tiny clamshell. They are able to shoot out small puffs of bioluminescent mucus as they swim. One effect is that when a fish eats an ostracod at night, their whole head gets lit up from the inside, making them vulnerable to larger predators. When the ostracods wash ashore, their exuded luminous fluid can set the sand and rocks aglow (shown on the next page).

In species found in Caribbean seas, male ostracods also use light to advertise themselves to potential mates. By controlling the direction they swim and how quickly they eject luminescent puffs, they can encode their identity as a pattern of glowing dots, like Morse code. Closely spaced horizontal orbs may mean one species, an ascending vertical line another, and rapid descending flashes a third. Females, recognizing the appropriate pattern, will emerge from the reef and swim toward the male at the end of the trail of illuminated breadcrumbs. One of the most jaw-dropping displays we've ever seen was a synchronized display in about 3 meters of water: hundreds of males descended in unison, creating a fairyland of glowing spires stretching across the seabed.

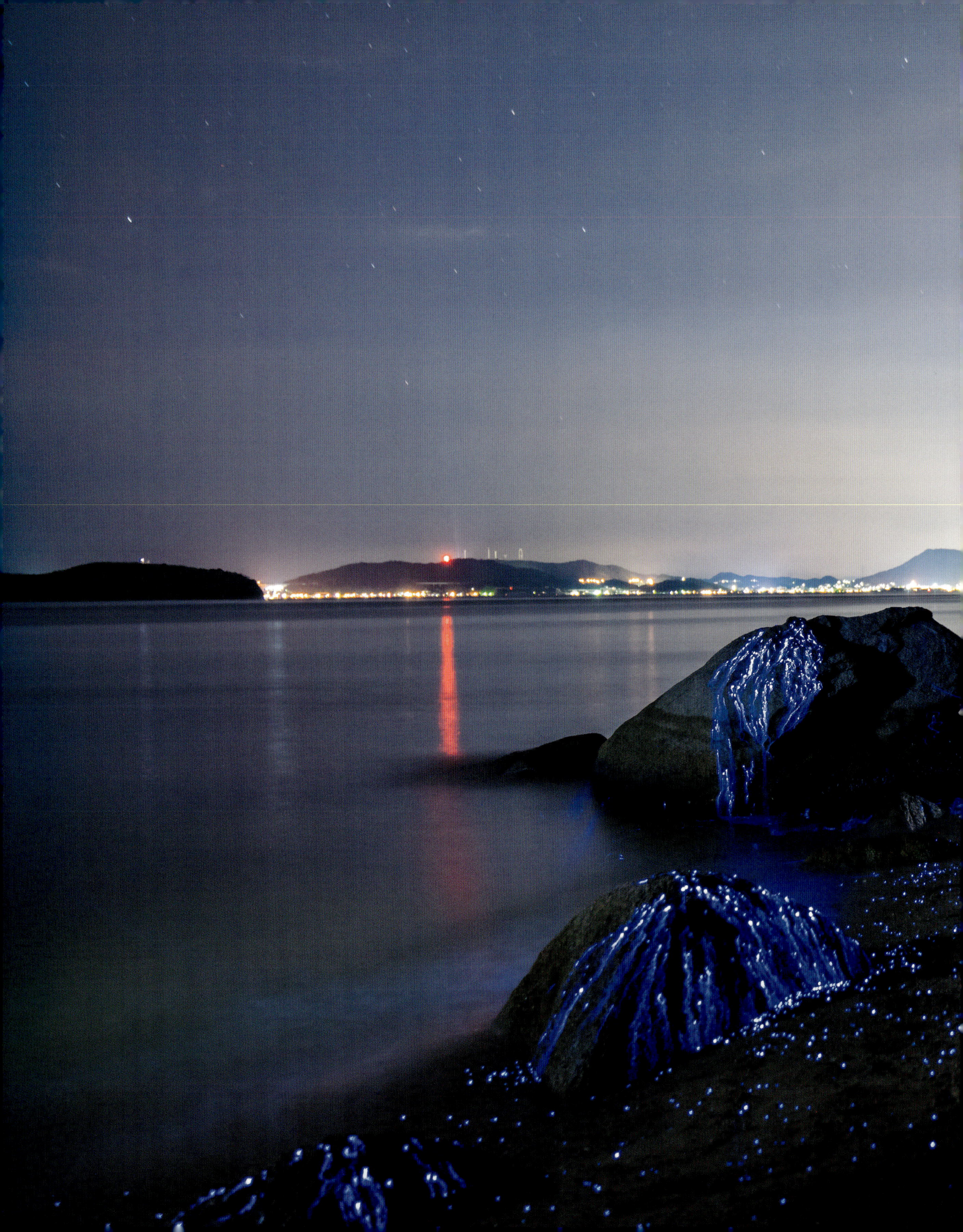

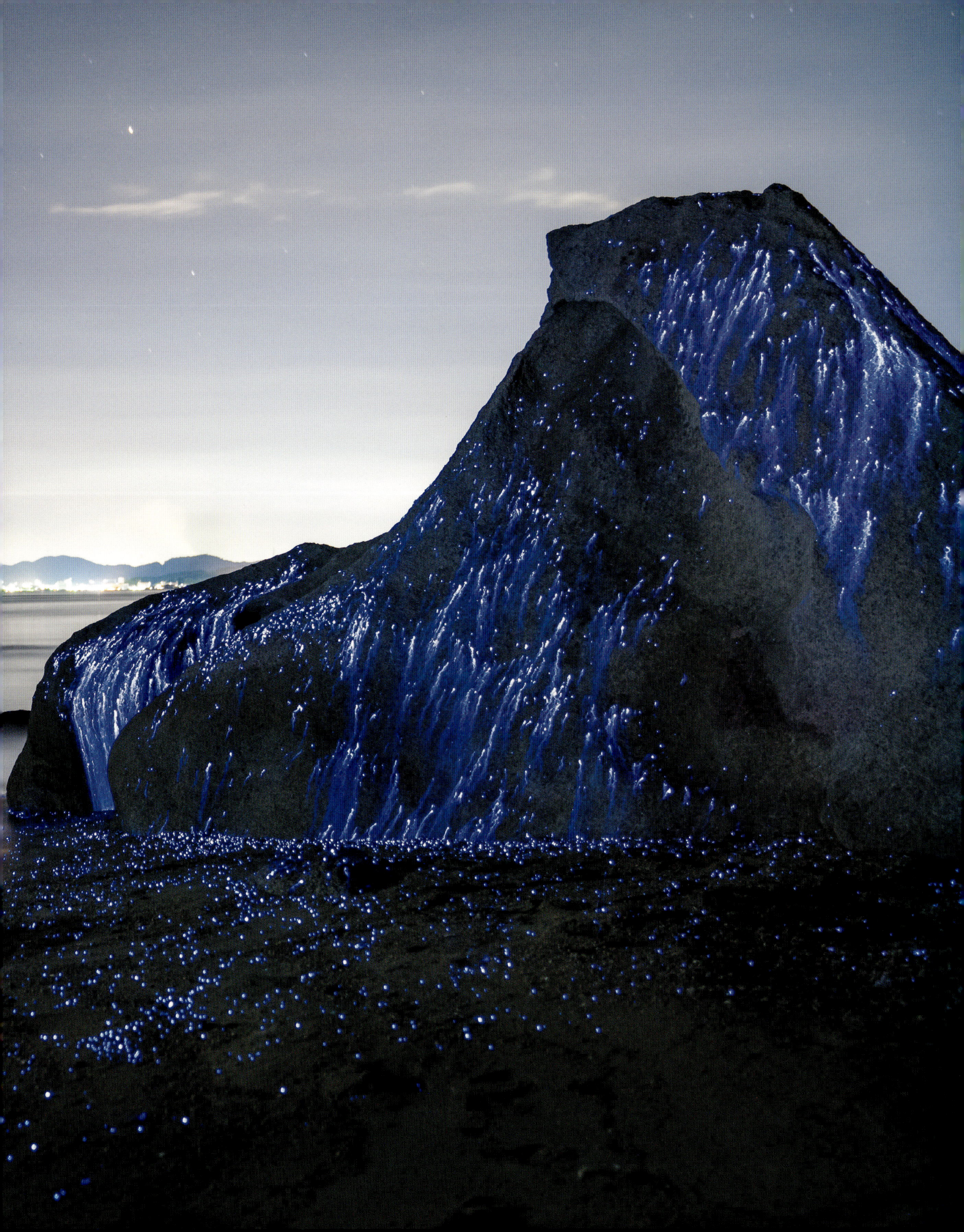

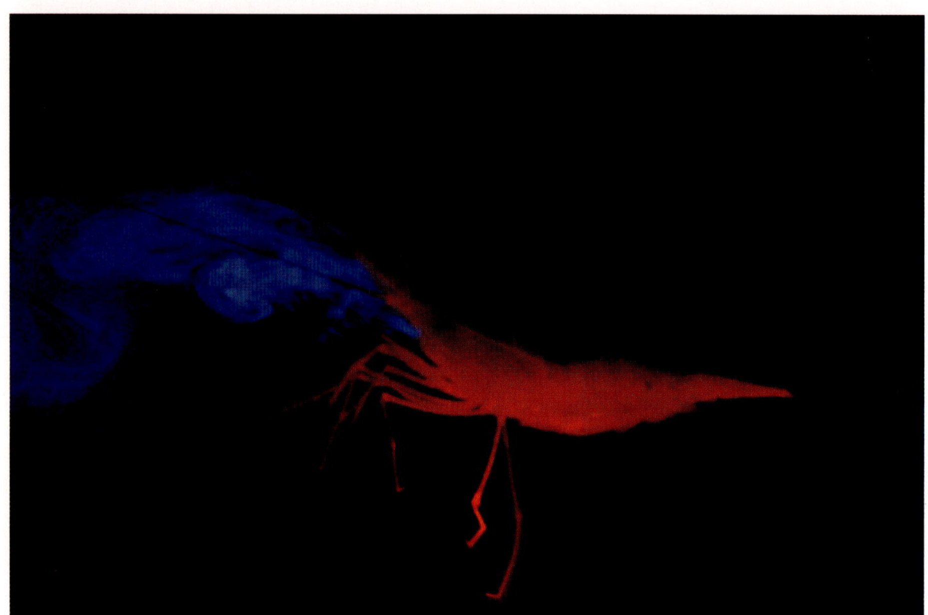

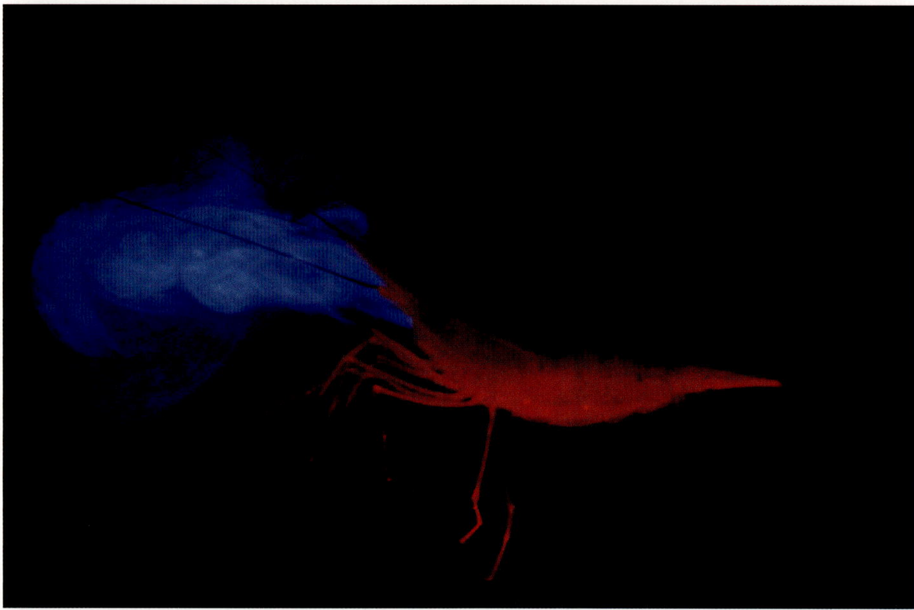

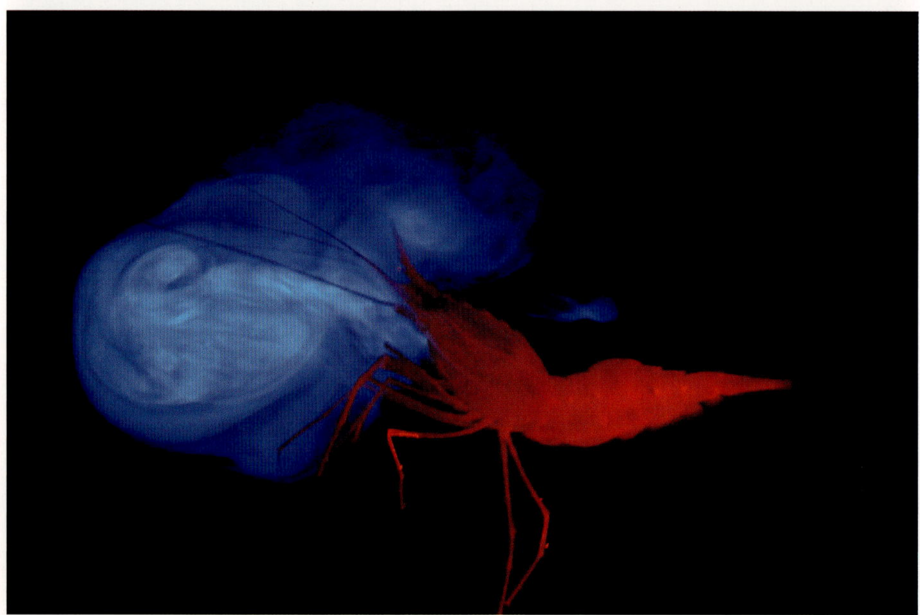

Heterocarpus ensifer

One of the most spectacular forms of bioluminescence is called by the less-than-spectacular name of "spew bioluminescence." In this form of light production, the chemicals to produce light are kept in two separate glands on either side of the mouth. When they are released, it can look like the animal is at the end of a firehose of light. The voluminous and long-lasting display makes for a very effective escape. These three photos show the early stages of a display from this shrimp.

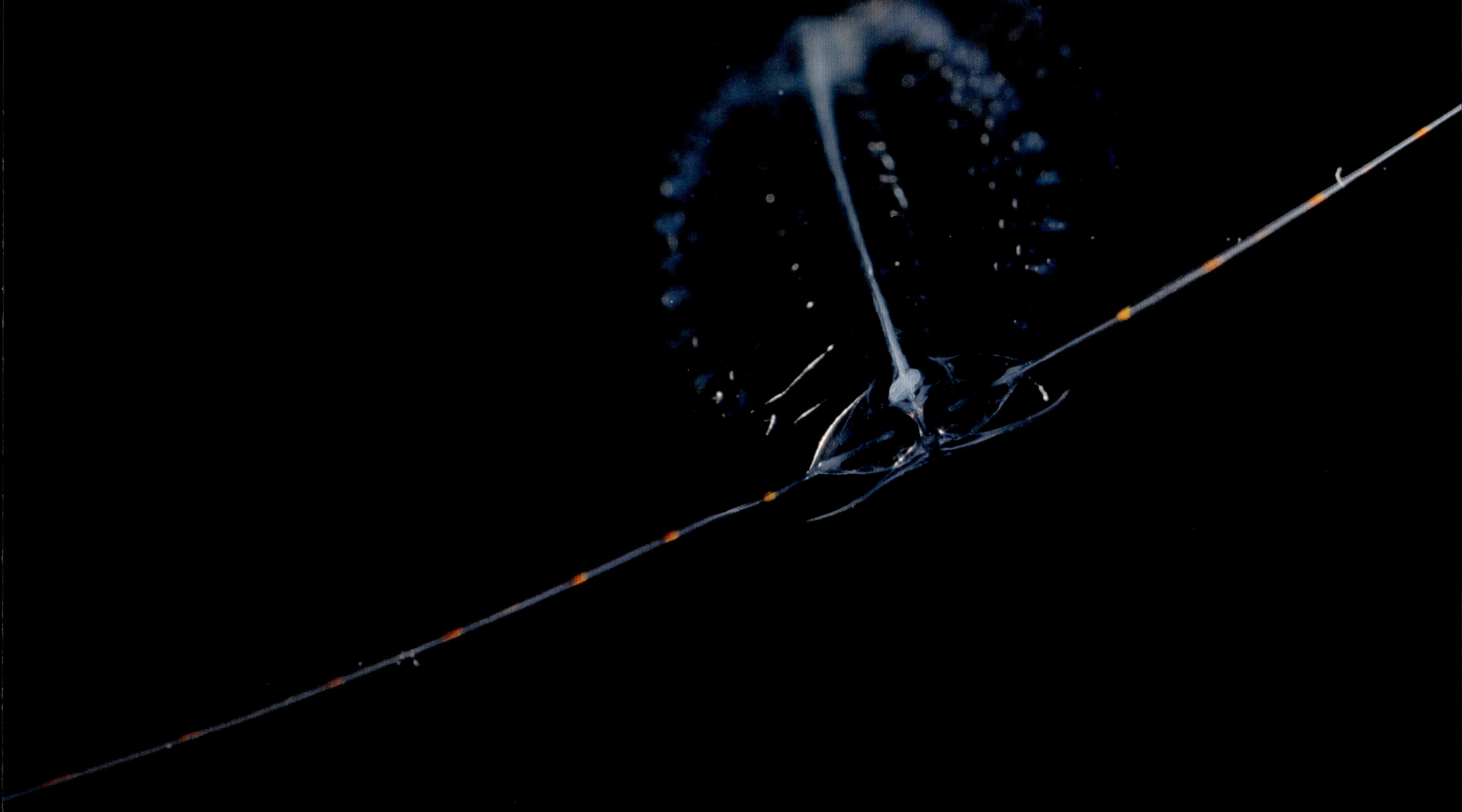

Tomopteris

We have shown *Tomopteris* before in the Transparency section (p. 36), but another special attribute of these swimming worms is that some of them exude mucus that emits yellow light—to the eye it looks like an explosion of golden pixie dust. This color is extremely rare in the ocean, especially in the deep sea where nearly all bioluminescence is blue or blue-green. Curiously, our lab experiments revealed that the yellow coloration comes from a fluorescent chemical that is also found in plants.

Bioluminescent mucus is usually thought to have a defensive purpose, either as a smoke screen or by labeling the predator and thus making it vulnerable to predation itself. If this is true here, the yellow color makes little sense, since light of this color does not travel nearly as far in the ocean as blue light. The functional advantage of this yellow light is a mystery.

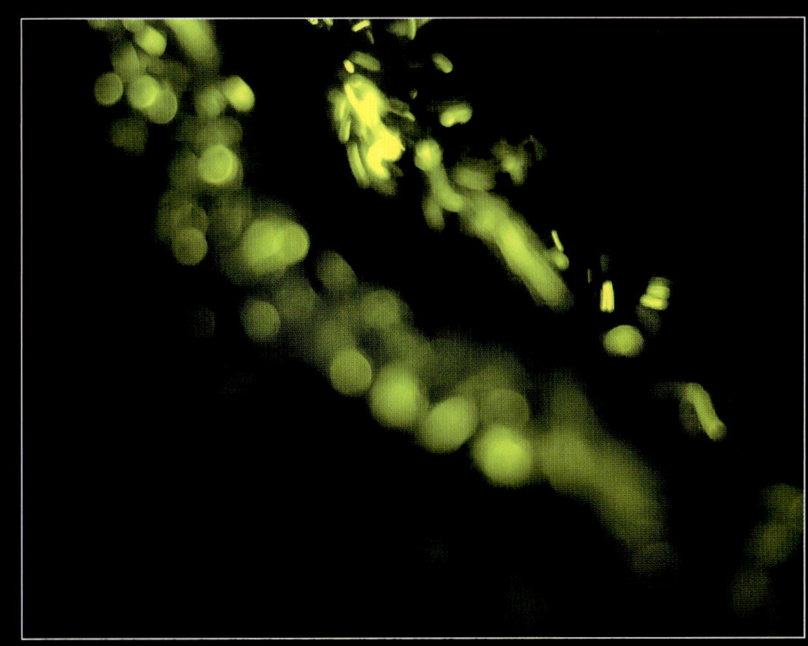

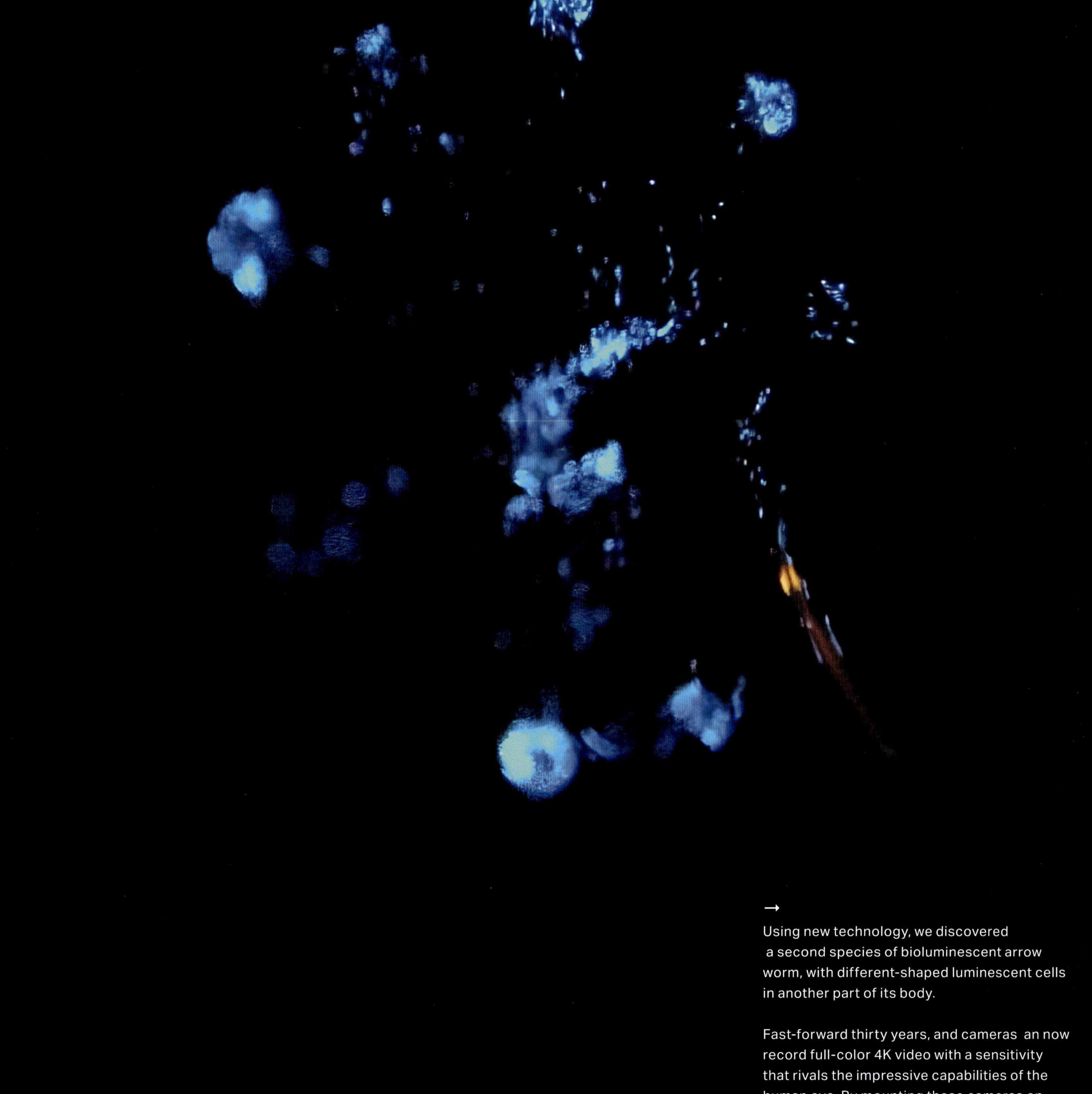

→
Using new technology, we discovered a second species of bioluminescent arrow worm, with different-shaped luminescent cells in another part of its body.

Fast-forward thirty years, and cameras an now record full-color 4K video with a sensitivity that rivals the impressive capabilities of the human eye. By mounting these cameras on remotely controlled submersibles, I could record the same bioluminescent display that I had witnessed by eye decades ago.

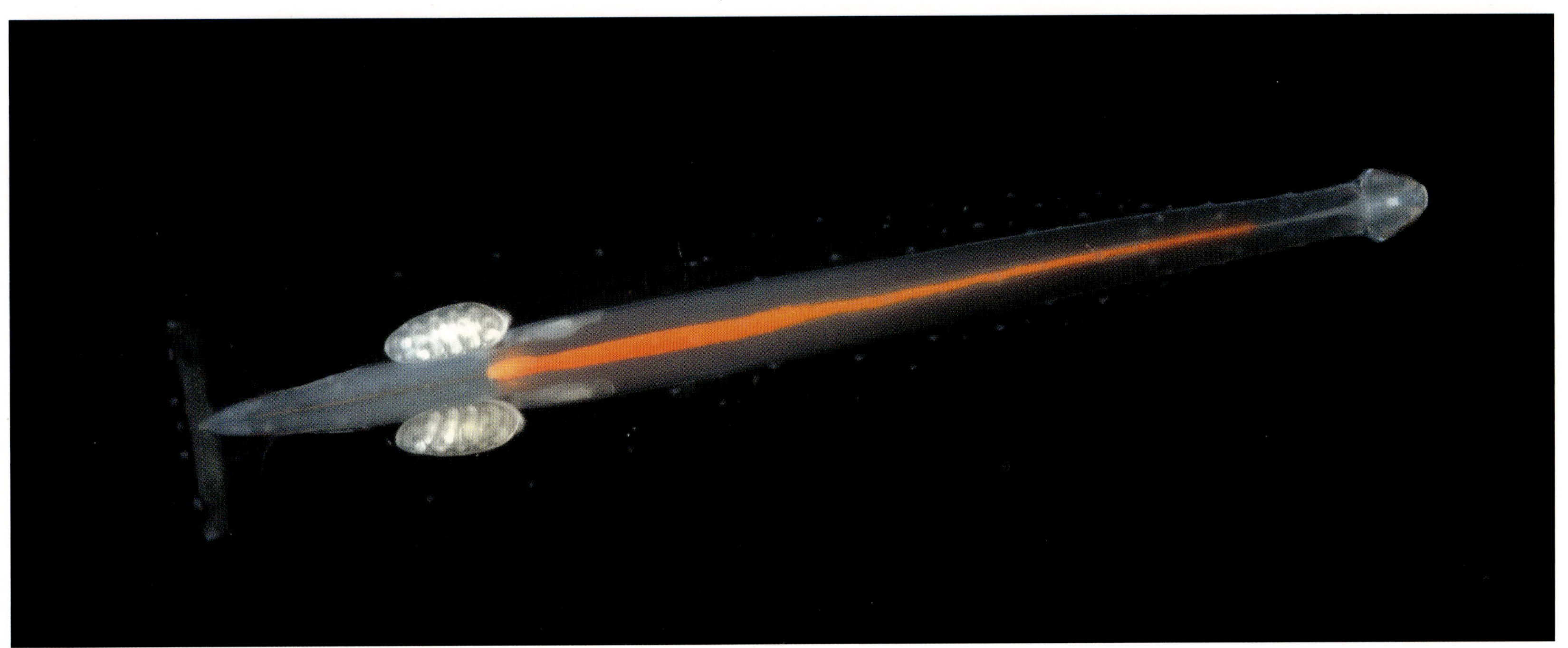

Chaetognath bioluminescence

Steve: On my twenty-fifth birthday, I was crouched in the back of a humid metal chamber, peering through a porthole a "half-mile down" below the sea surface. In the cone of the spotlight outside the pressure-resistant window I watched a menagerie of animals drifting past, but the light was dim enough that I could still see the blue bioluminescent light that they would produce.

A worm the size of a matchstick drifted past as we descended. Sensing us, it darted away, leaving a trail of bright blue pixie dust in its wake. Arrow worms (Chaetognatha) are a group of small predators that grab prey using hooks and inject them with paralyzing poison. They are very sensitive to vibration, and will usually hover motionless in the water until they sense a threat—like our relatively gigantic sub cruising past.

Peter Herring, who is the world's expert on bioluminescence, was in the other chamber of the sub. Peter had written articles cataloging every luminous species that had been discovered, so he was the perfect person to have on call at this moment.

Startled by what I had seen, I called on the headset: "Peter, arrow worms are not by luminescent are they?"

His voice crackled back, "They shouldn't be."

"I didn't think so," I said, shaking my head. I made a quick sketch in my notebook, and the dive continued.

That chance observation led to a three-year quest to figure out just what I had seen out that porthole.

During subsequent cruises with deep-sea physiologist Jim Childress, I used his special net to bring animals to the surface in great condition. People who've studied deep-sea life through the years could typically only look at somewhat mangled specimens—those hardy enough to survive the hours-long trip to the surface in a mesh net. Jim's net sealed the animals in a dark, insulated chamber, so they stayed cold and relatively undamaged on their way to the surface.

I had one clue that would help me in my search for this wormy needle in the oceanic haystack: most arrow worms are completely transparent but the one I saw had a pale orange gut. On Jim's cruises, I would spend hours sorting orange arrow worms into test tubes filled with seawater. Finally, one night as I shook each specimen, a burst of pure blue light filled the tube—one of my first eureka moments studying the deep.

At the time, there were no science-grade digital cameras, so in a darkened room I put my film camera on a tripod and tried to take matching exposures to show both the animal and the glow of its luminescent cells. I had to wait two weeks until we got ashore and the film was developed before I knew whether I had the shot.

This was not a new species—it was described over one hundred years ago and has been seen many times since, but none of the descriptions showed the cells shed into the water during its sparkling escape. These types of smoke-screen displays are fairly widespread among animals, but often the specimens are in no condition to perform when we get them to the surface, so their abilities go unnoticed.

Myctophid

Myctophids, also known as lantern fish, are so common that they're thought to account for over half of deep-sea fish biomass. They comprise about 250 species and have light organs on their sides in patterns that vary by species. In fact, taxonomists use these patterns to determine which species they are looking at. It's been suggested for years that the animals may be doing the same thing—using patterns of lights to decide which animal is in front of them. This is a tempting idea, since it's not clear how else these animals can recognize each other in the dark, but for all the reasons described earlier in this section, we have no idea if this is true.

Like many animals, these well-illuminated fish also have photophores along the bottom of the body for counterillumination, but note the bright white patch just to the right of the tail. This is the sternchaser light organ. We can attest to its effectiveness as an anti-predation device. When sorting through a tub of trawl-collected specimens in the dark, we've been stunned by the extremely bright flash from these photophores, which they coordinate with a burst of escape swimming.

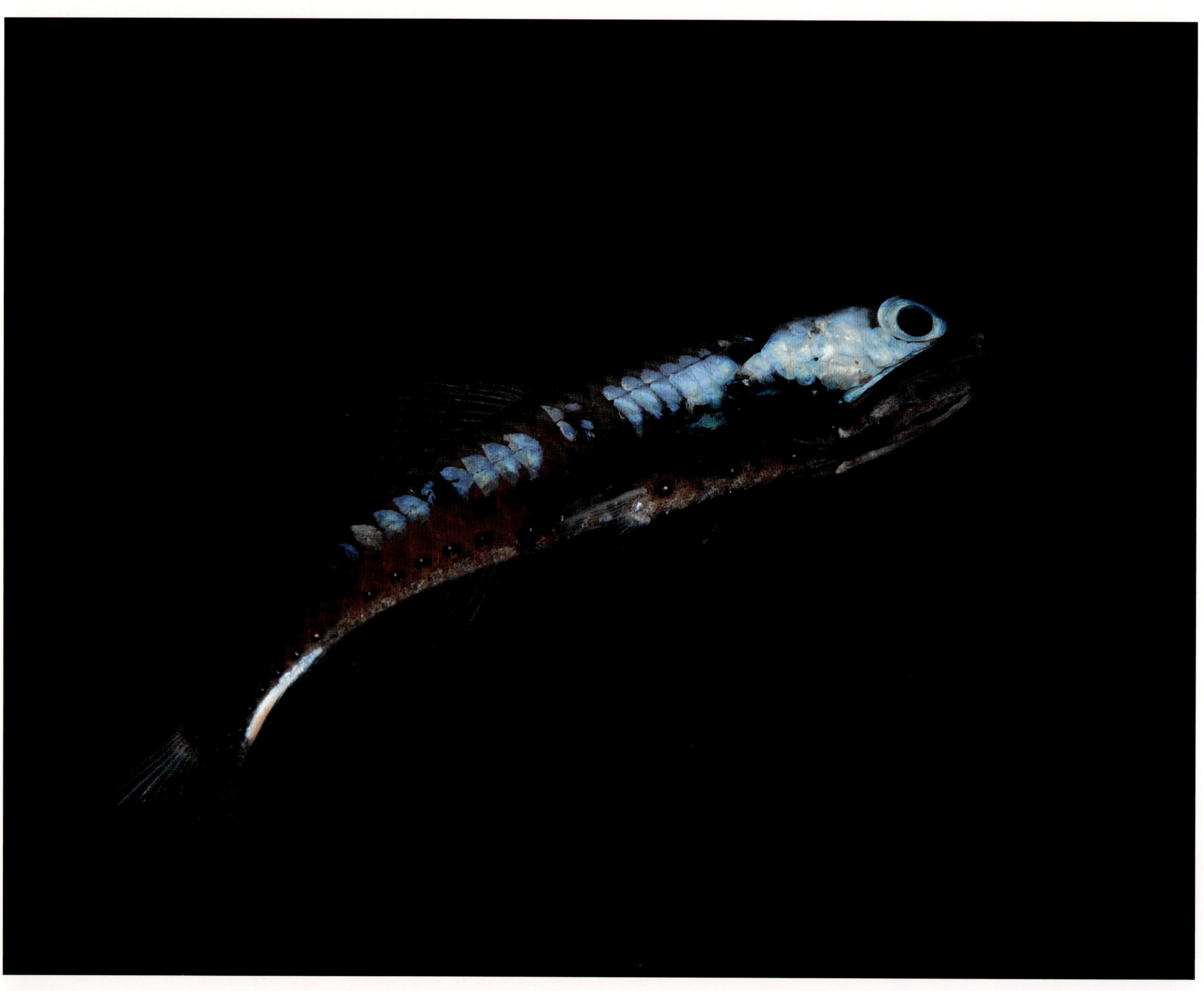

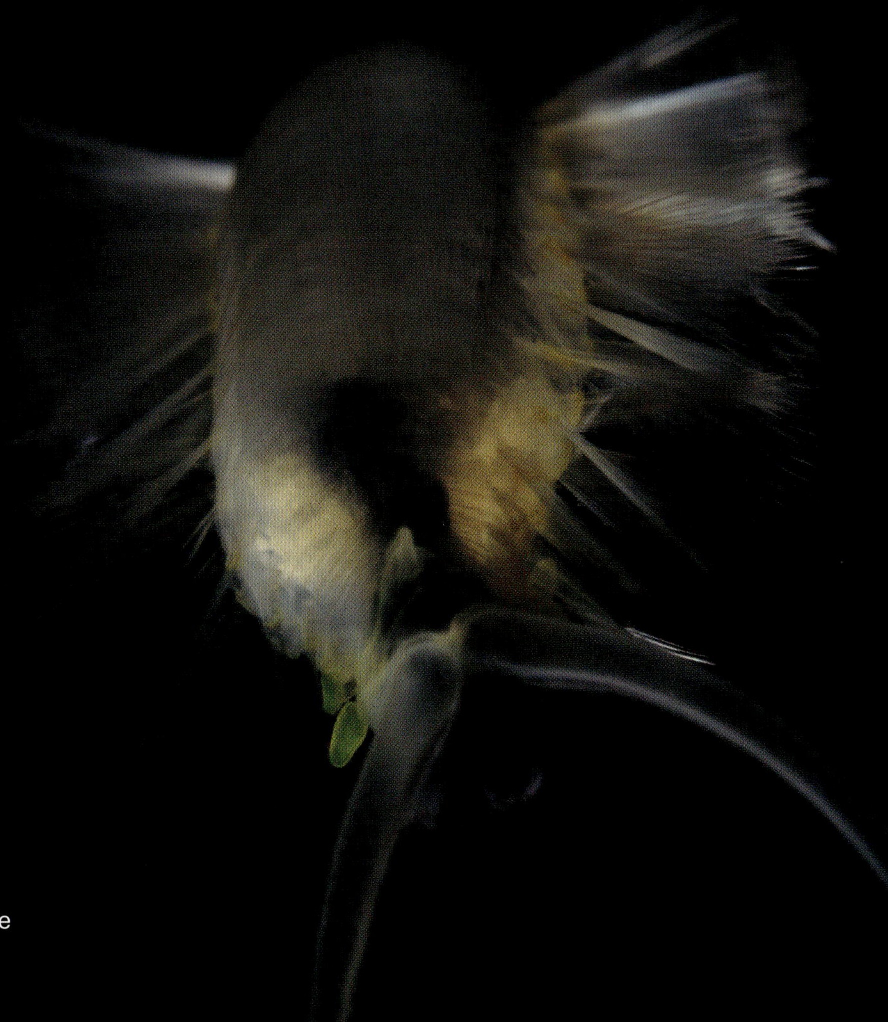

Swima bombiviridis

Steve: When you're fortunate enough to look at deep-sea animals alive over many years, you become attuned to anything suspicious, as you will hear from Sönke's discovery of glowing octopus suckers (p. 167). When I first saw this deep-living worm under a microscope, the suspicious feature was a series of green capsules attached to where its neck might be. I isolated a few and took them into the ship's darkroom. When I pinched them with a pair of fine tweezers, sure enough, a little green flash of light burst forth, earning it the nickname, the "green bomber worm." When it came time to give it a scientific name, we translated its temporary name into Latin to get the species *bombiviridis*.

One gratifying thing about this particular discovery is that, although maybe two people in the world have seen the true bioluminescence of this species, it was the subject of an episode of *Octonauts*, a popular animated show for kids. They did such a good job—you can immediately recognize the true-to-life posture—that now millions of children have learned true facts about this seldom-seen star of the depths.

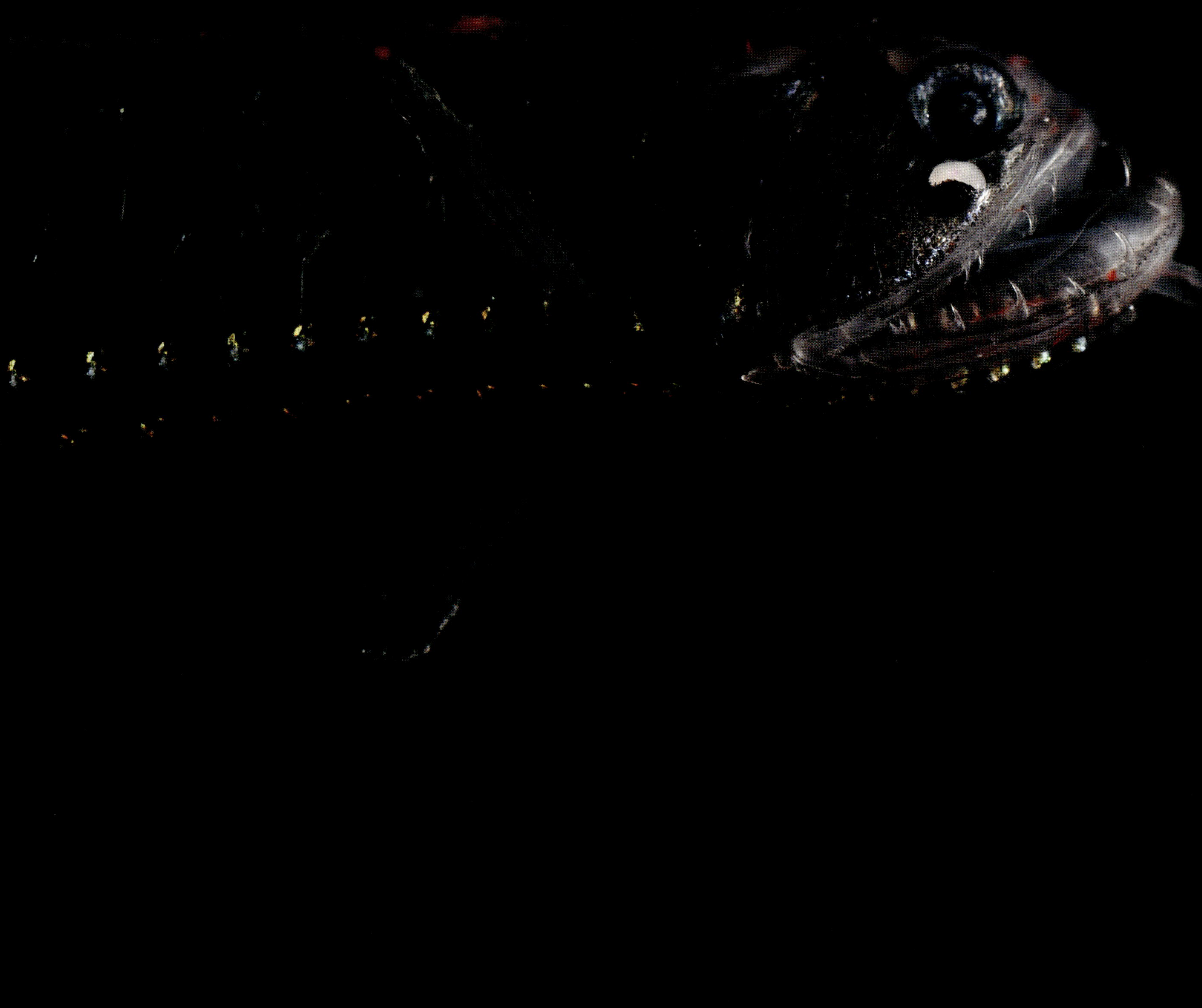

Stomias

This fish is one of many in the family known as the "dragon fishes." They are relatively small, but impressive predators of anything smaller than they are, typically with mouths full of long, sharp teeth. They also appear to use light in multiple ways. This photo shows the bioluminescent organ that hangs from a long stalk below the chin. This organ likely acts not only as a lure, but—because the shapes of the organs vary by species—may also serve as a mating signal for animals that may have no other way of recognizing each other.

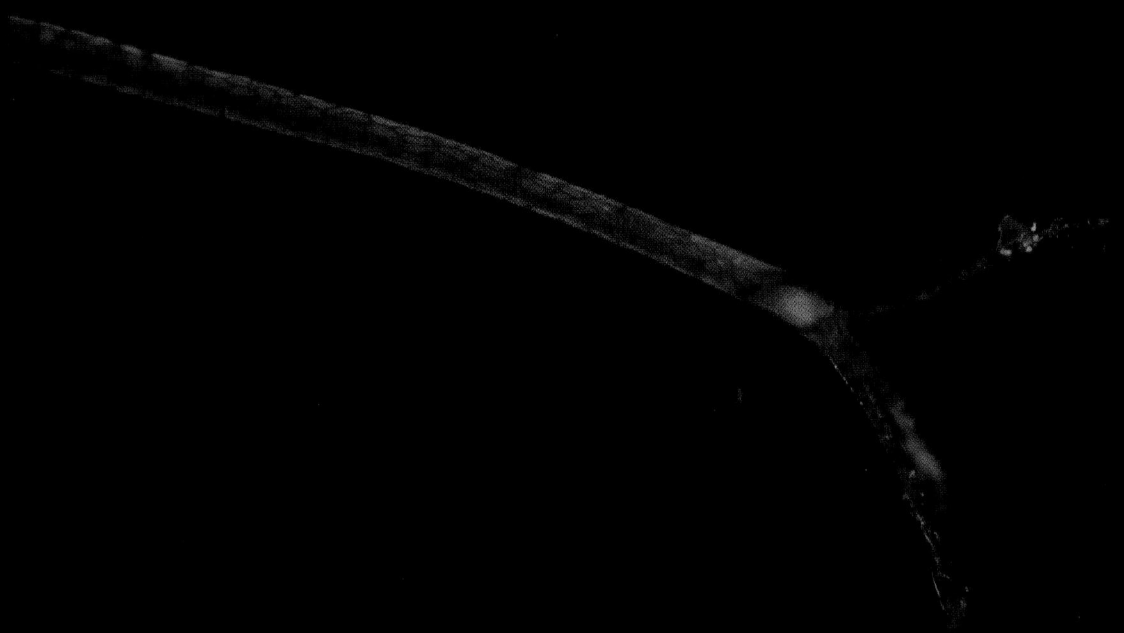

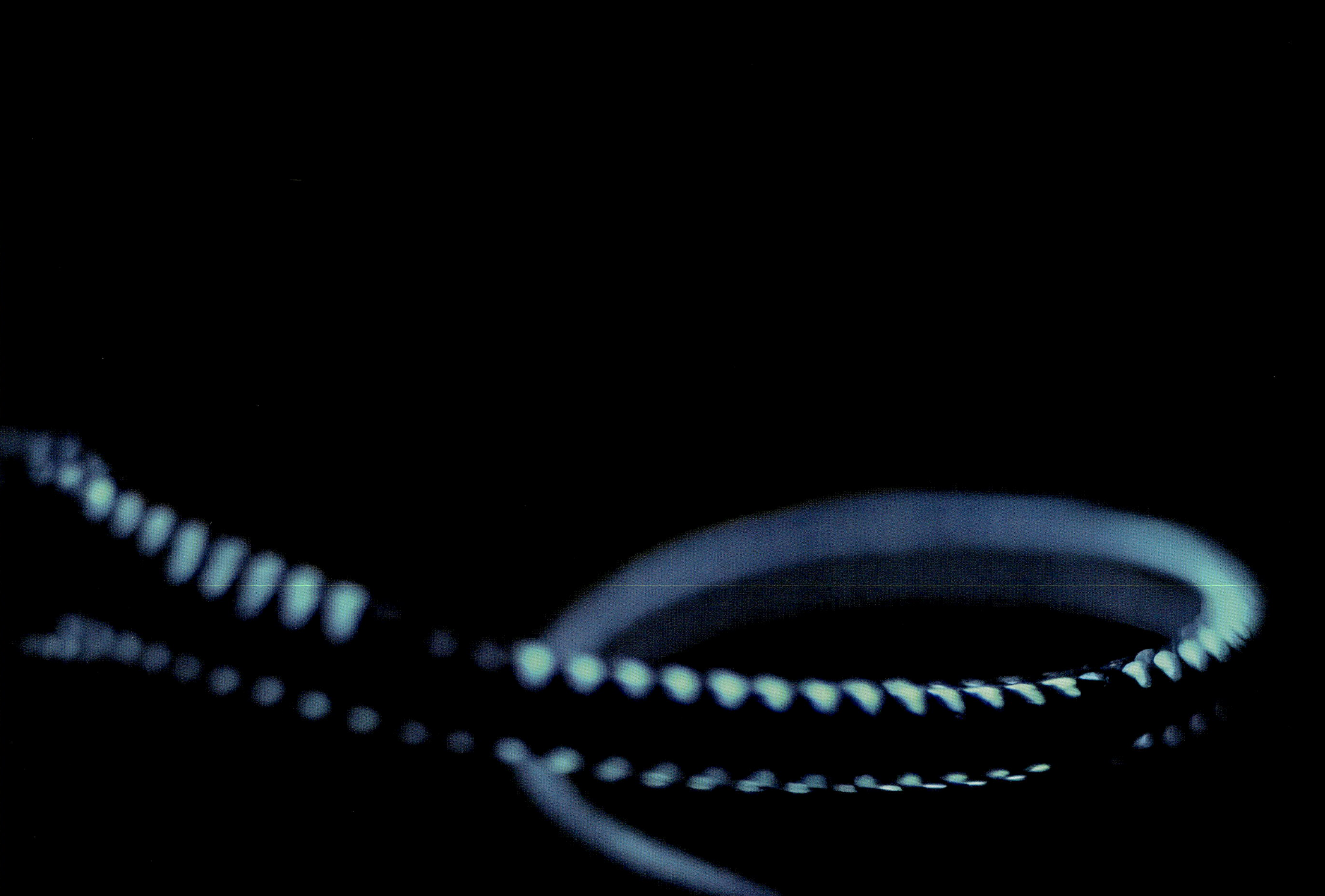

Cyclothone

Bristlemouth fishes like this *Cyclothone* are thought to be the most abundant vertebrates on earth, with about a quadrillion individuals. Indeed, we see these fish in nearly every trawl we pull up, no matter which part of the ocean we're in. The lower side of their bodies (known as their ventral surface) is covered with multiple rows of light organs that are thought to match the sunlight coming down from above, so that the animal can hide its silhouette.

Looking at the serrated jaw, you can see why these fish have the common name of "bristlemouth"—making them fearsome predators if you are only a few millimeters long. What you may not immediately notice is the dark photophore just below the eye. Strangely, it is pointing up in the opposite direction from the rest of its light organs. Several types of deep-sea fish have these isolated lights pointing toward their eyes. The thought is that they use this to monitor how bright their own bioluminescence is. Their goal is to match the light from above with the light coming from their ventral photophores, and not reveal themselves by being too bright or casting a dark shadow.

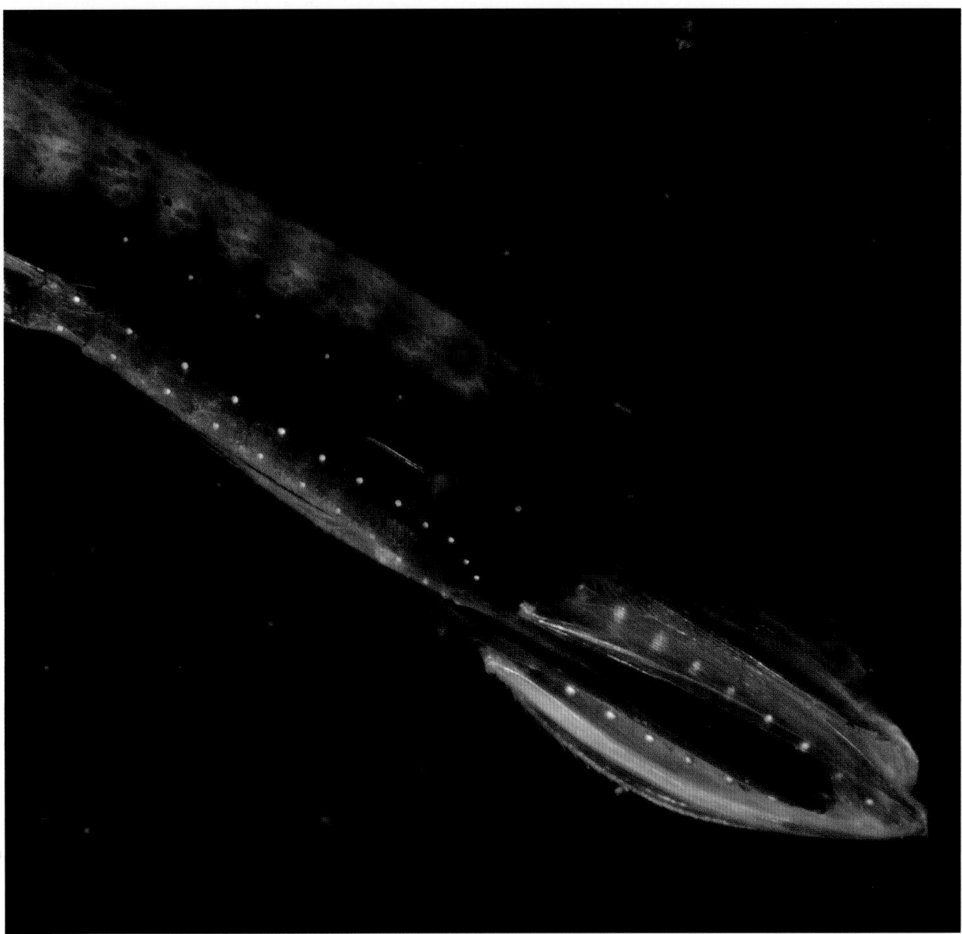

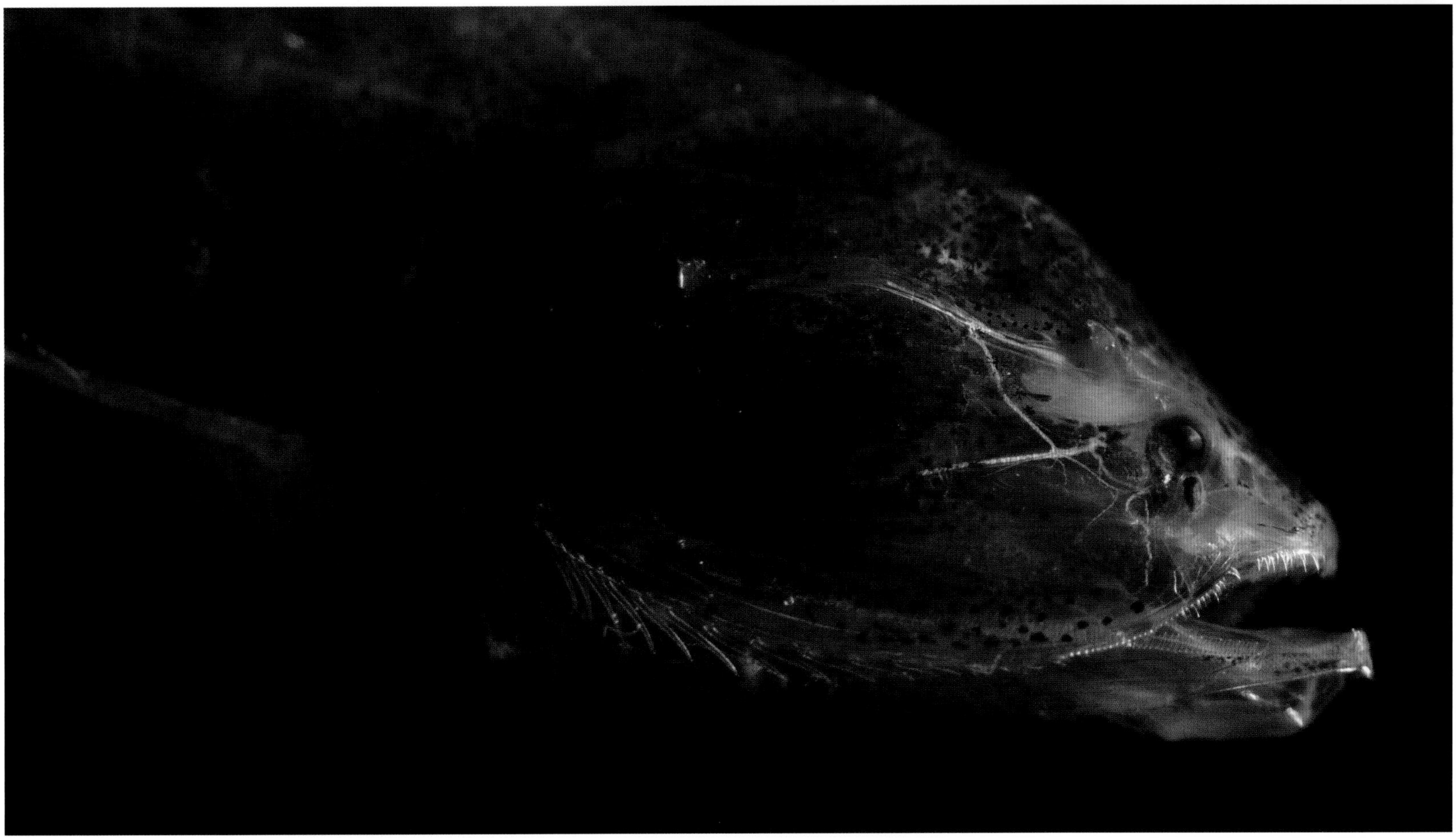

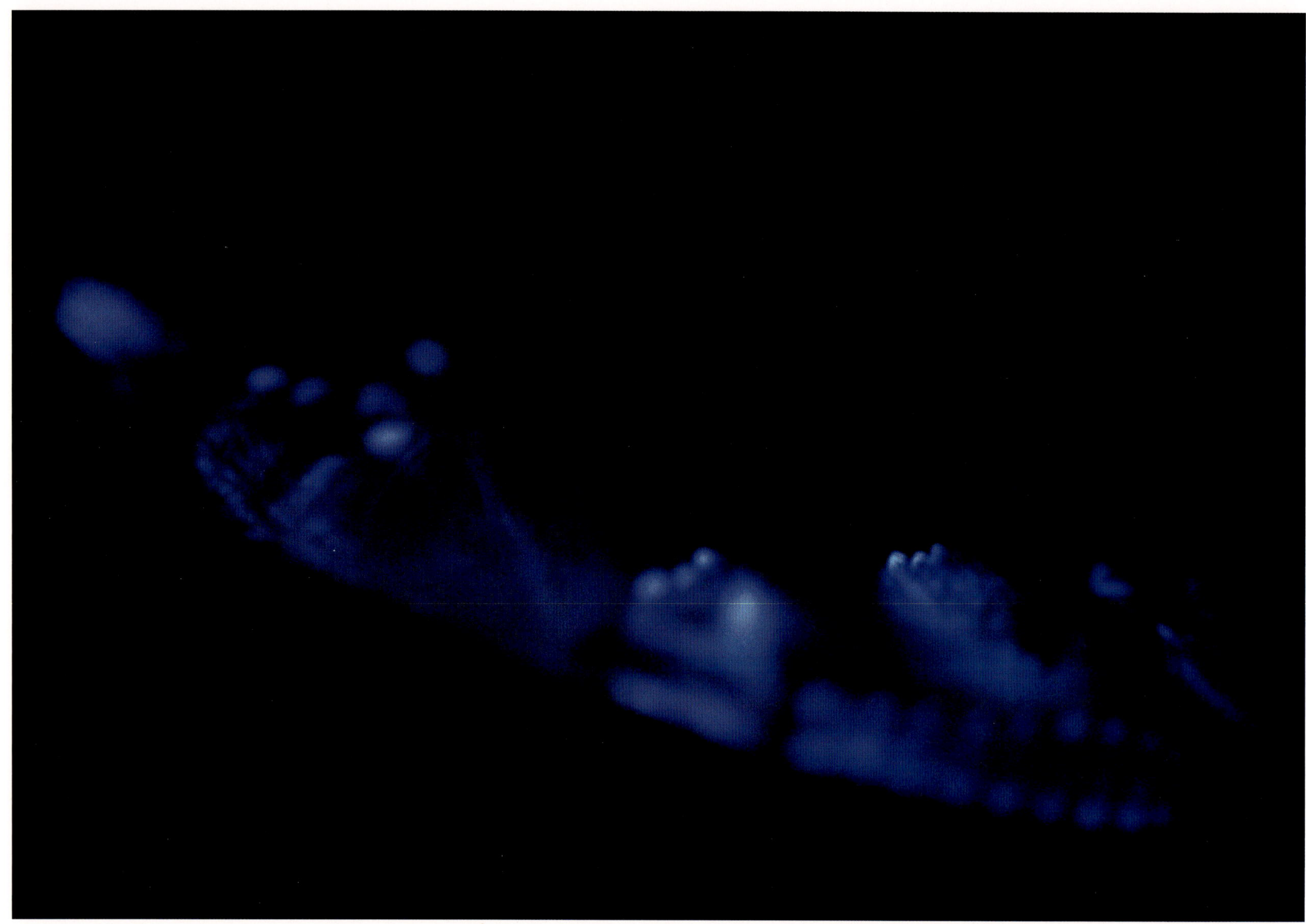

Sternoptyx

Highly flattened and deep-bodied hatchetfish seem to be masters of light. As you saw in the Iridescence section (p. 116), their sides are covered with bright silver plates that help them hide, and they have specialized photophores on their lower surface and strategically placed along the sides, like the running lights of a science fiction spaceship. By adjusting these downward-pointing lights to match the brightness of dim residual sunlight coming from above, they are invisible when seen from nearly all directions.

Anglerfish

These fish, famous from various movies such as *Finding Nemo*, typically have bioluminescent lures on the ends of stalks that hang over their mouths. They will hover in the water column and pounce when prey approach their large mouth. As discussed previously (p. 121), the animal itself has to be quite black so that the lure doesn't illuminate itself and spoil the trick. Although fearsome-looking, these iconic predators are typically the size of a golf ball.

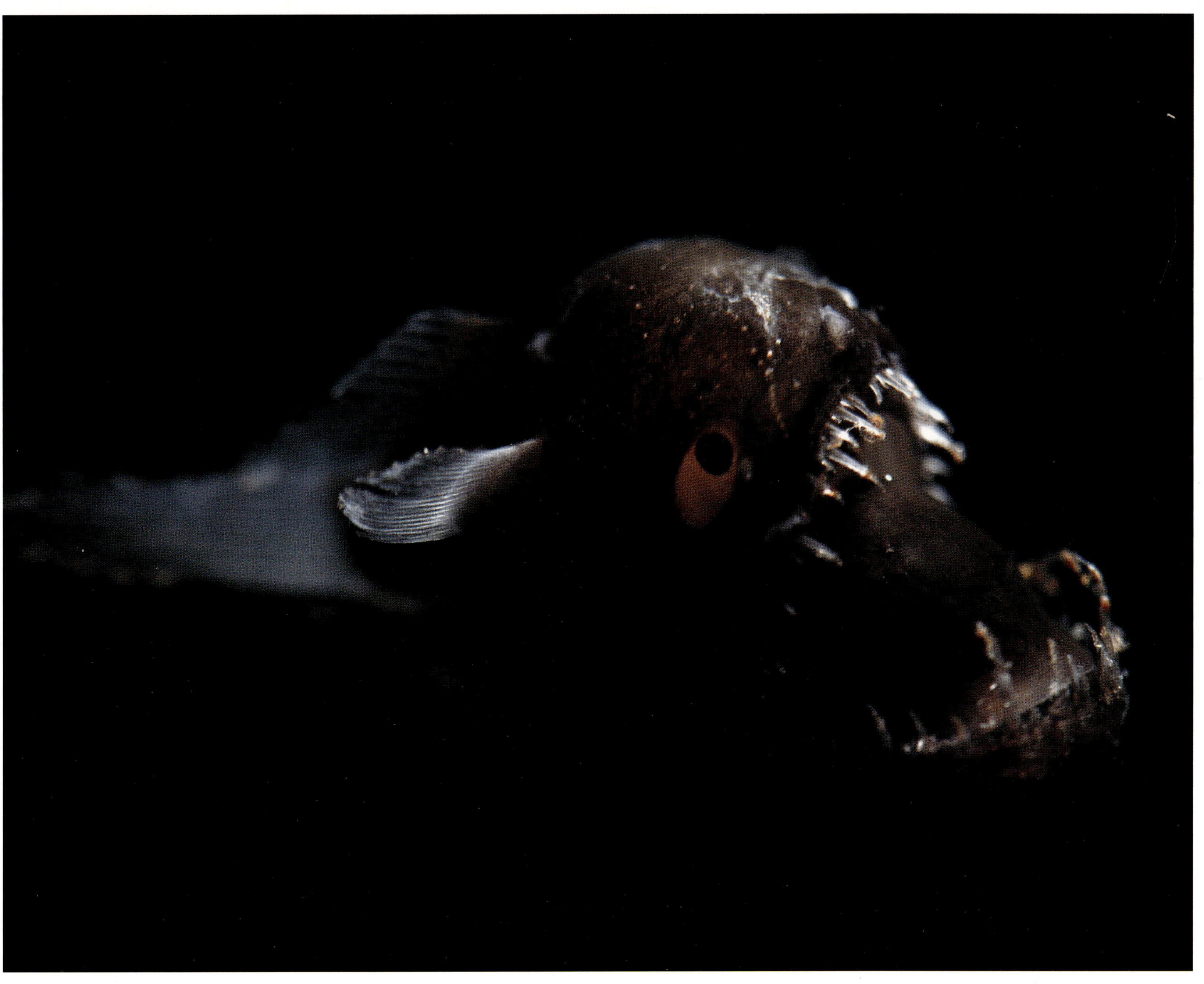

Scyliorhinidae

Like the firefly squid (p. 165), small, shy cat sharks (below) have photophores distributed all over their bodies, which may be used to hide their silhouette. While most animals concentrate their light emission on their bellies, these seem to be able to illuminate in all directions. In the 3D oceanic environment, we sometimes have to let go of our preconceptions about how animals will orient themselves. Different animals hover just as readily head-down or tail-down in the water.

Related to cat sharks are cookie-cutter sharks (right), which have a similarly dense distribution of tiny photophores. These sharks—about the size of a baguette—feed on much larger animals like dolphins and tuna by rapidly striking to take circular bites. The dark unilluminated stripe across their throats is an apparent flaw in their camouflage. It has been suggested that this might actually be a kind of decoy that looks like a small fish and draws larger fish closer to investigate.

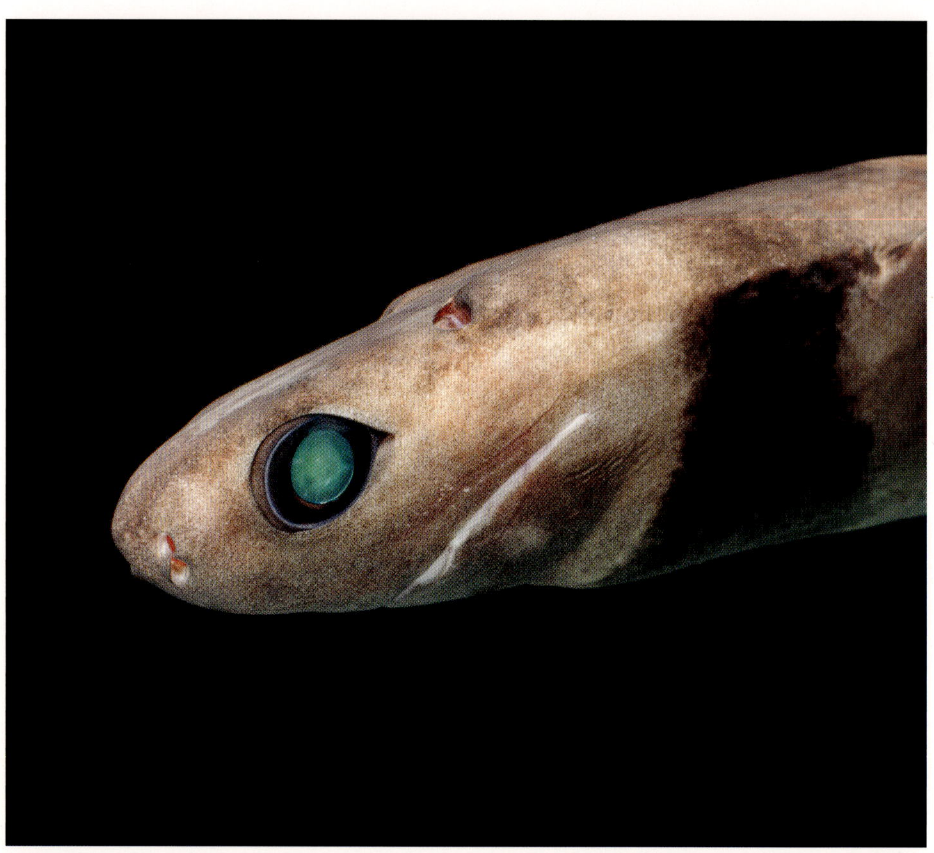

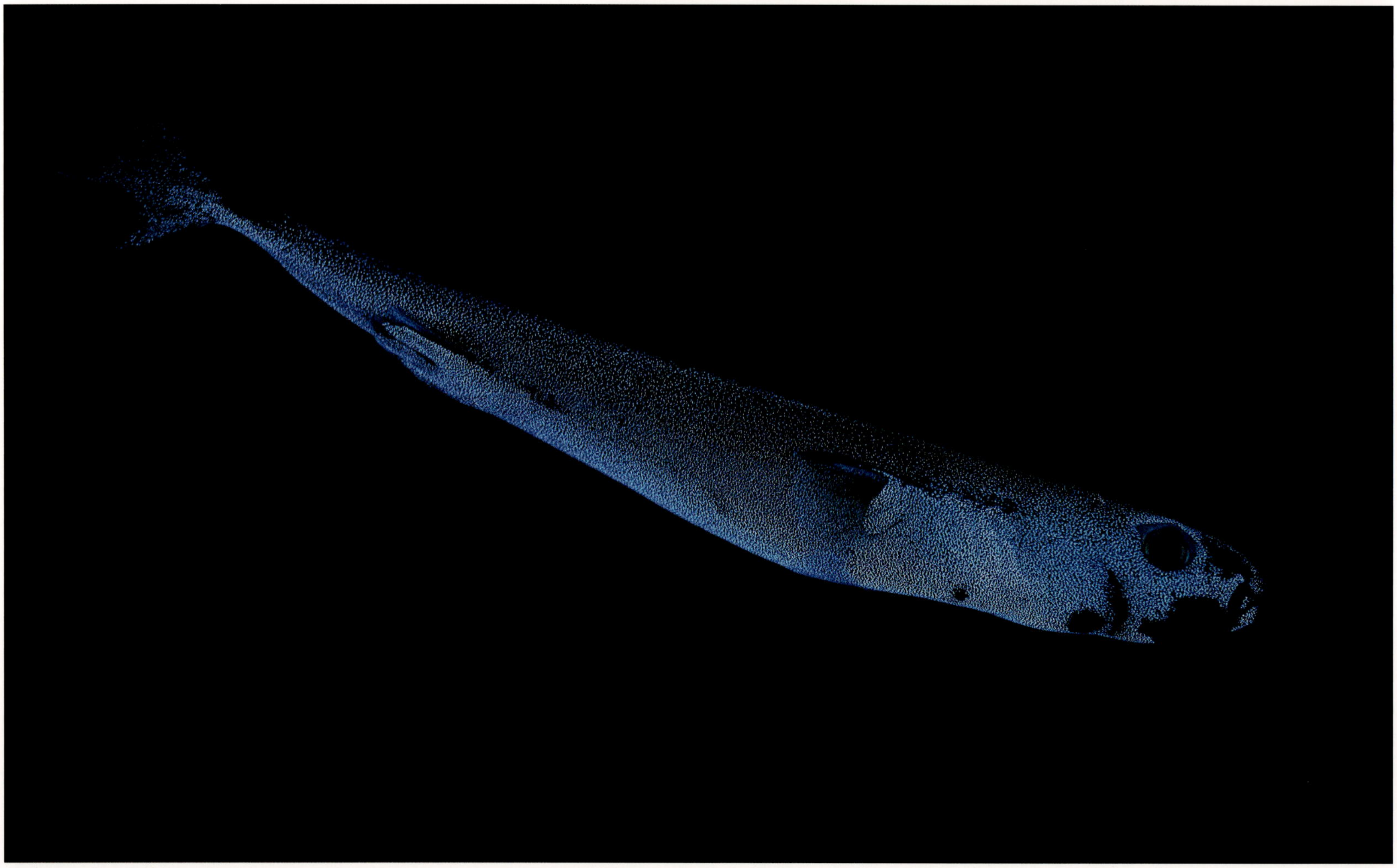

Aristostomias scintillans

Although many fish have light organs under or above their eyes that they use as flashlights, this fish is one of the few known to have a red flashlight. This not only allows them to see the many red items of the deep sea (which are hidden to blue flashlights) but also enables them to do so without being given away by their own light, since so few animals can see red light at depth. Surprisingly, this useful trick seems to have only evolved a few times, but new red-emitting species continue to be discovered. This red light can be such a long wavelength that we've had difficulty seeing it clearly from the live glowing animal. It is not all the way to infrared but is longer than your average red LED.

Note that this photo does not actually show the bioluminescent light. It is one that uses the inherent fluorescence of the bioluminescent light organs to reveal the color and morphology of these photophores.

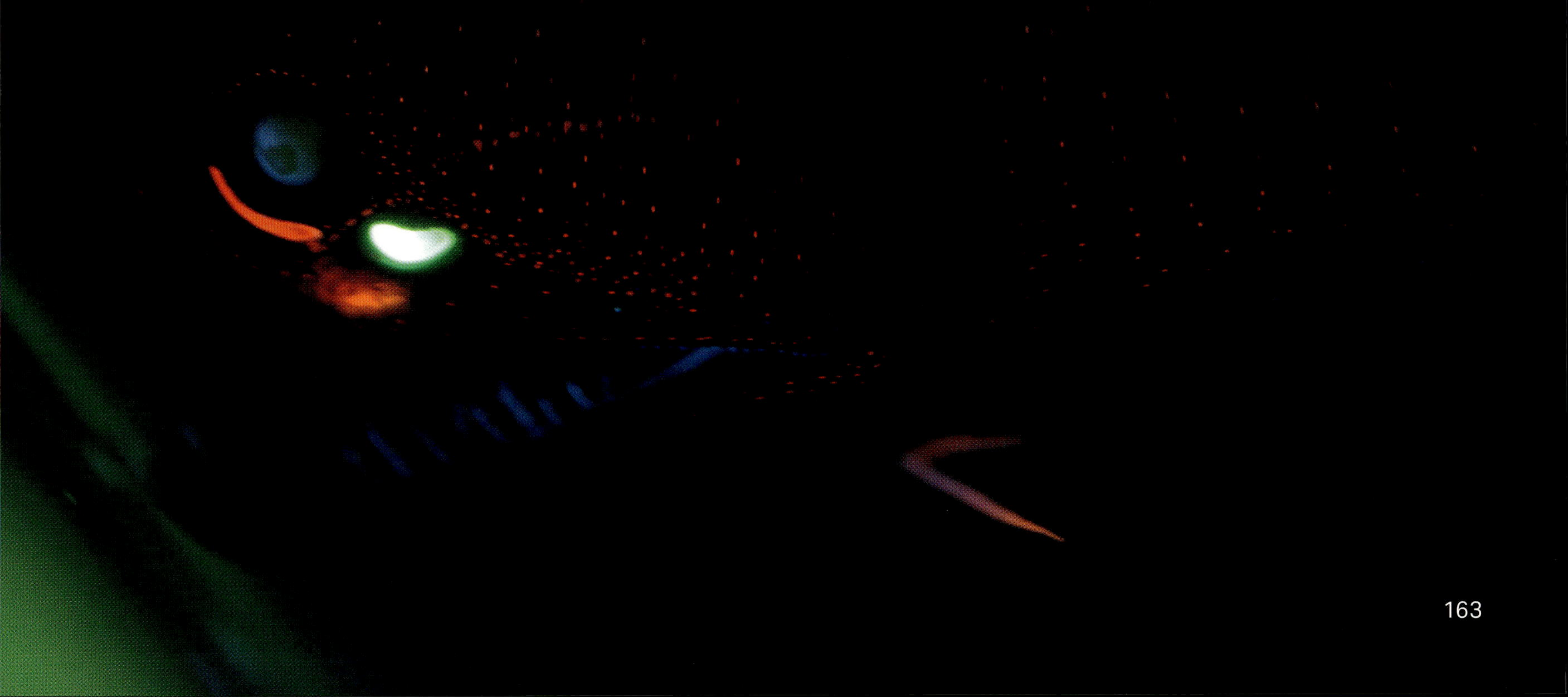

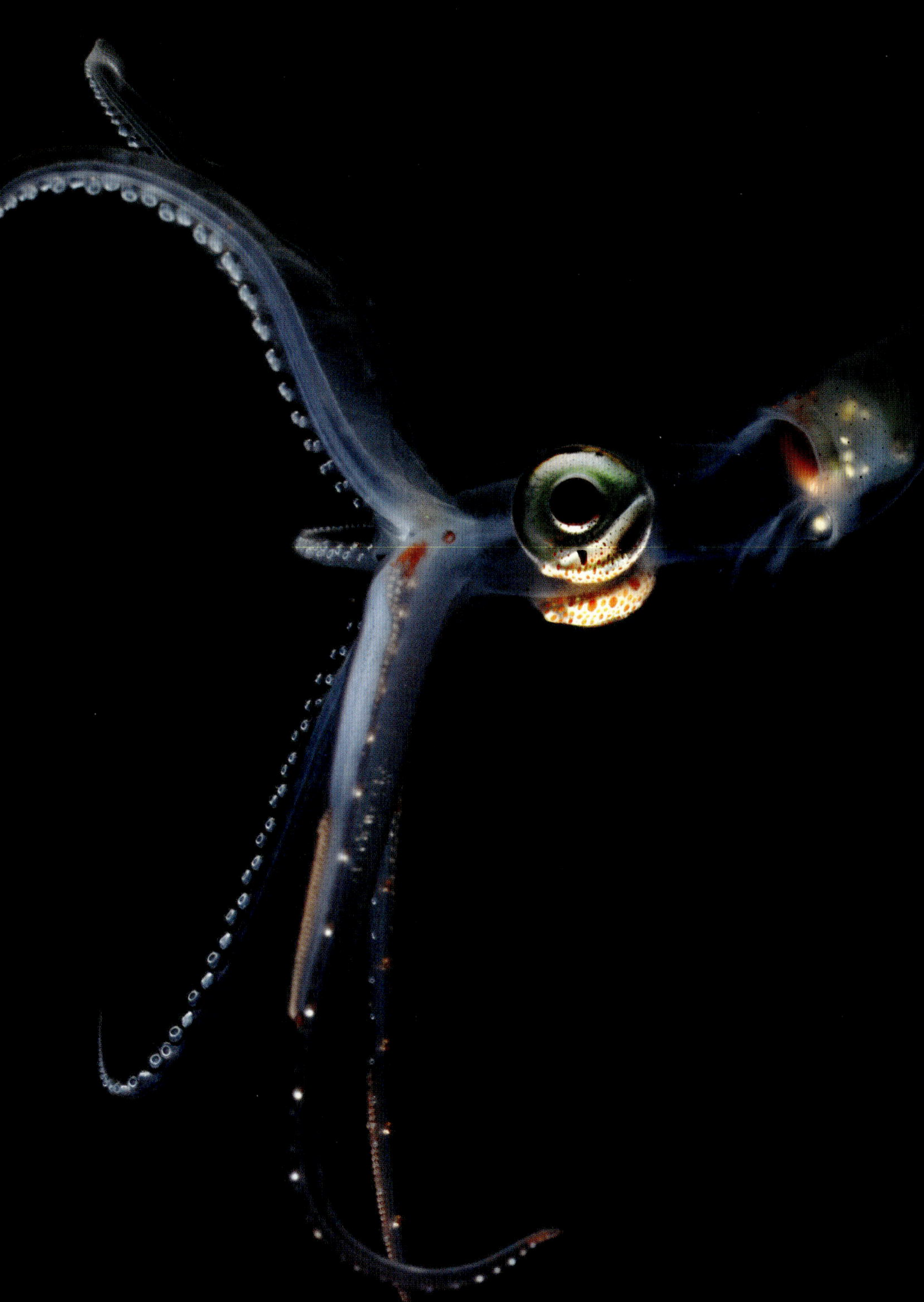

Chiroteuthis

The sword-tail squid has several amazing bioluminescence adaptations. The lower part of its eye is a large light-emitting organ. This serves to counterilluminate the eye, replacing the shadow of its silhouette with bioluminescence, since the eye must absorb light to function. What makes this especially amazing is that no matter how the squid is angled in the water, its eye rotates so that this organ is always pointed downward, not to the side. It is fun to follow these squid with a submersible and see the eye appear to be gyroscopically stabilized as the squid changes its posture.

Chiroteuthis also has a series of light organs along two of its tentacles, which it dangles down to form fishing lures. So far, we have only been able to get a shaky handheld video with a phone camera, showing the flashing tentacles of a specimen that was captured with a trawl net. We are striving to film this behavior in the deep with a special low-light camera mounted to a submersible.

Watasenia

Aptly known as the "firefly squid", this animal has light organs over almost its entire body. Like many animals that are covered with small light organs, it seems to use them to hide its silhouette. It also has bright photophores on its armtips, which can serve another defensive function. While other squids—and indeed all octopuses and cuttlefish—see the world in black and white, *Watasenia* can see color over a small part of its visual field, and we have no idea why.

A vacation suggestion is to travel to Toyama Bay, Japan, for their annual firefly squid festival, where you can see thousands of these flashing cephalopods. The boat departs at three o'clock in the morning, so don't be late!

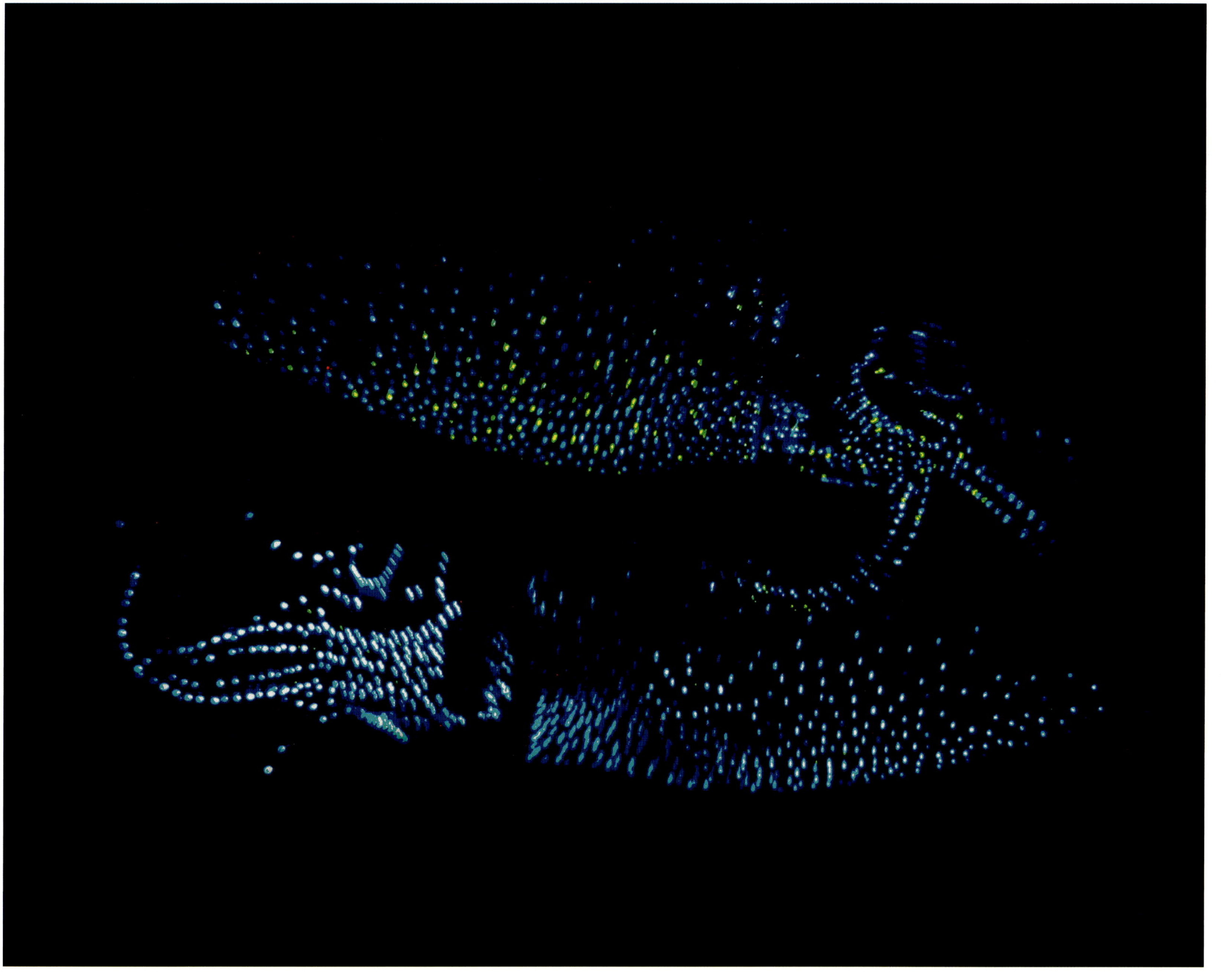

Vampyroteuthis tentacle bioluminescence

The vampire squid is another animal that exemplifies several of the optical adaptations at once. Its chocolate-brown skin provides dark pigments, and two bright photophores near the "tail" may serve as beacons to startle potential predators. When threatened, it will turn its umbrella inside out, adopting the "pineapple posture."

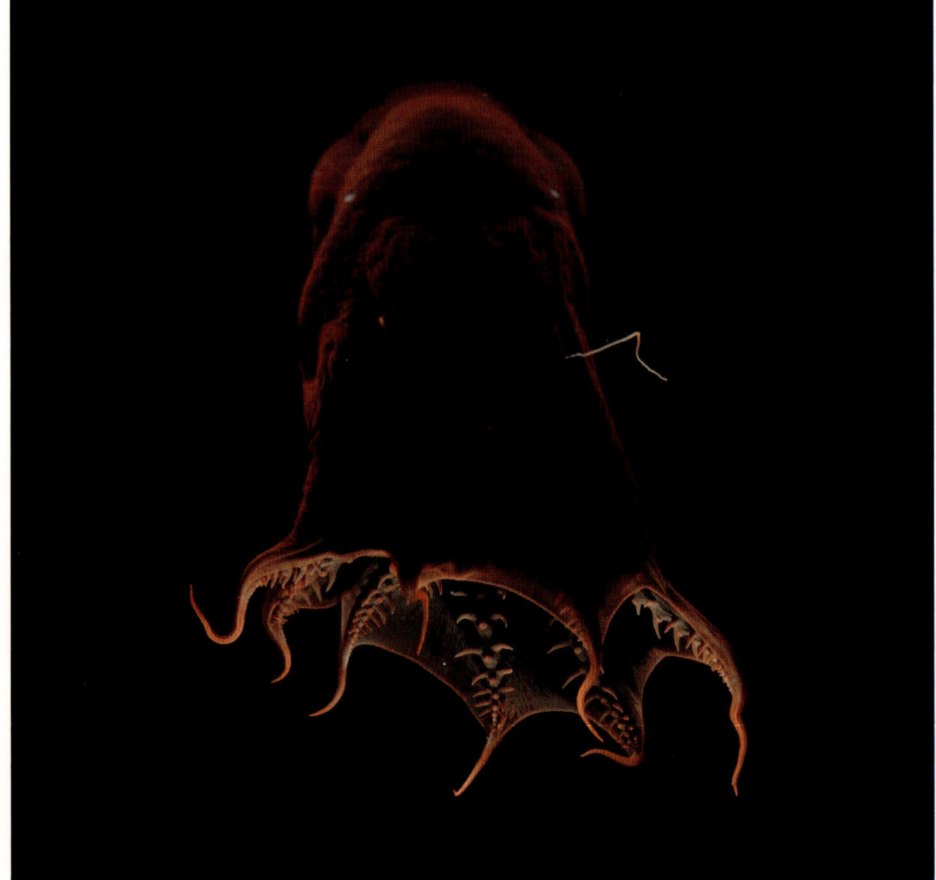

→
Steve: We discovered that there are bioluminescent organs at the tips of each arm. Glowing particles ooze from the armtips, providing a sparkling diversion while the animal makes a stealthy escape.

↓
Remarkably, the light-emitting chemical in the light organs appears to be the same luciferin used by jellyfish. This molecule, as with most sources of bioluminescence, is also fluorescent, allowing us to visualize the luminous tissues under a microscope.

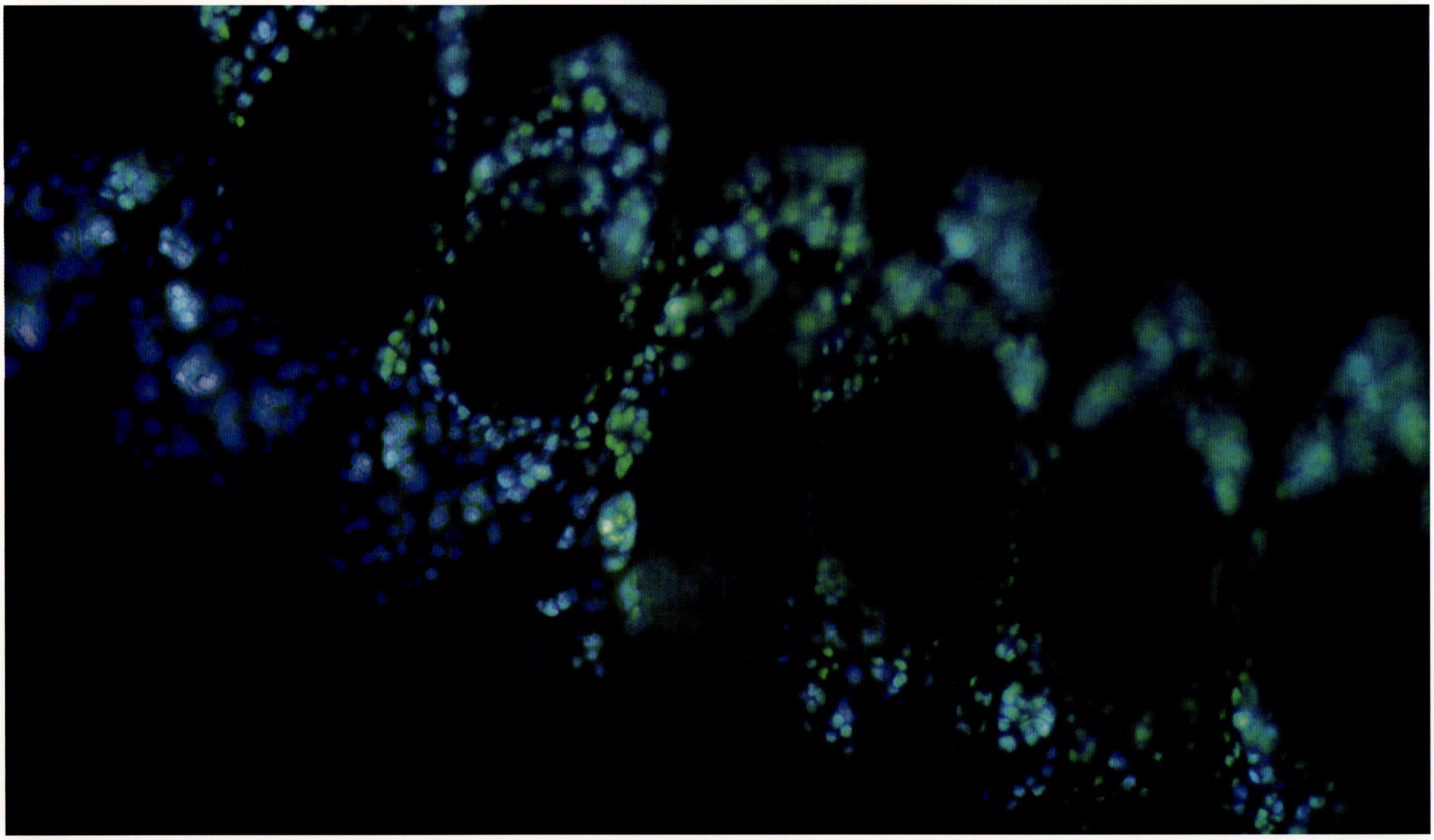

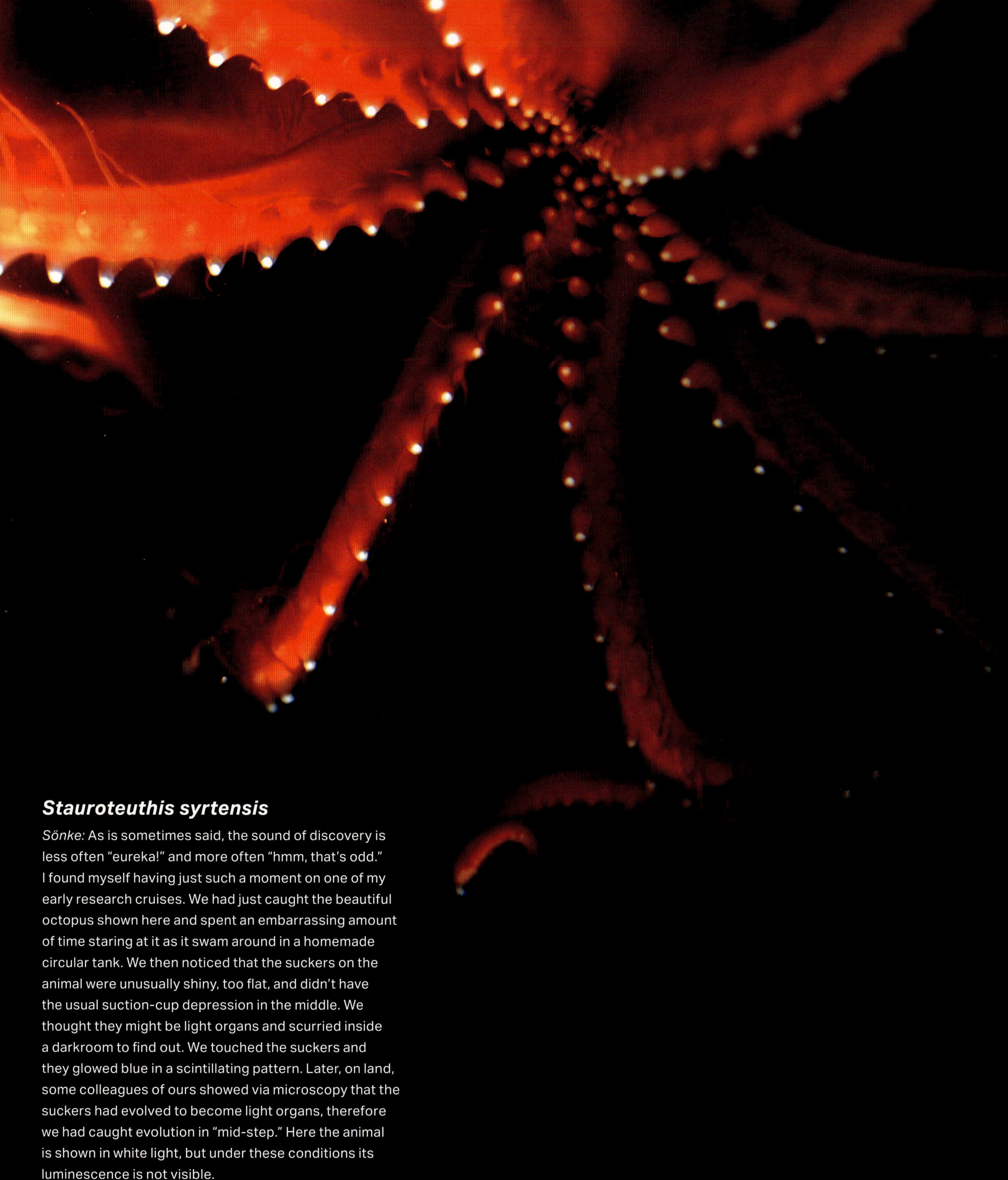

Stauroteuthis syrtensis

Sönke: As is sometimes said, the sound of discovery is less often "eureka!" and more often "hmm, that's odd." I found myself having just such a moment on one of my early research cruises. We had just caught the beautiful octopus shown here and spent an embarrassing amount of time staring at it as it swam around in a homemade circular tank. We then noticed that the suckers on the animal were unusually shiny, too flat, and didn't have the usual suction-cup depression in the middle. We thought they might be light organs and scurried inside a darkroom to find out. We touched the suckers and they glowed blue in a scintillating pattern. Later, on land, some colleagues of ours showed via microscopy that the suckers had evolved to become light organs, therefore we had caught evolution in "mid-step." Here the animal is shown in white light, but under these conditions its luminescence is not visible.

Pyrosoma

In general, anyone wishing to study bioluminescence needs to pay close attention, because most of the emitted light comes in quick flashes. This colonial tunicate, however, can glow for many minutes. In addition to their long-lasting light emission, pyrosomes have the interesting trait that they will start to glow if you shine a light on them. In fact, we've set up a row of eight pyrosomes next to each other, carefully shined a light on one end, and watched the glow be passed along until the whole line was lit up. Although their luminescent ability is well known—they were given their name meaning "fire body" more than 200 years ago—there is still a debate in the scientific literature about whether bacteria are involved in light production.

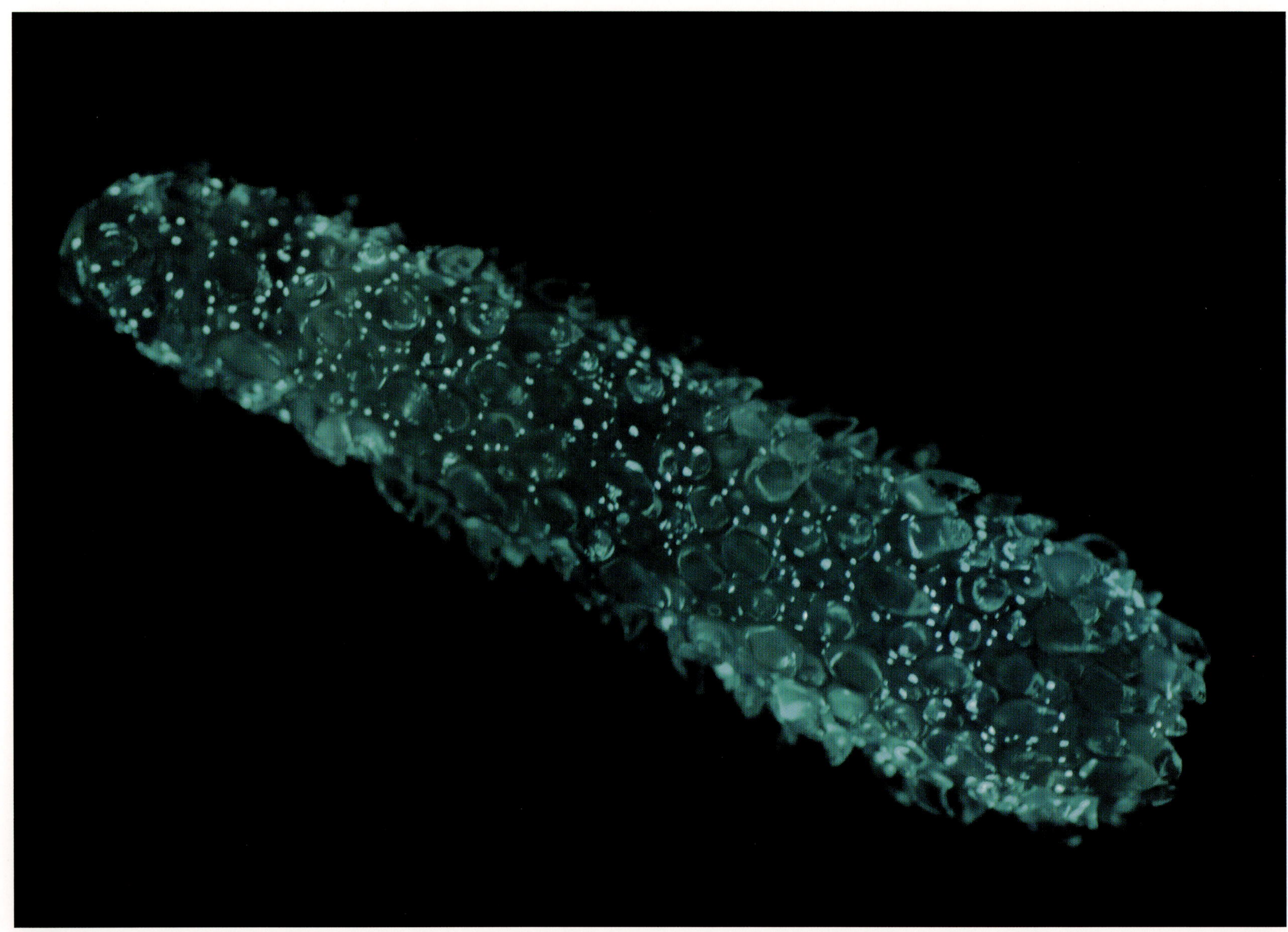

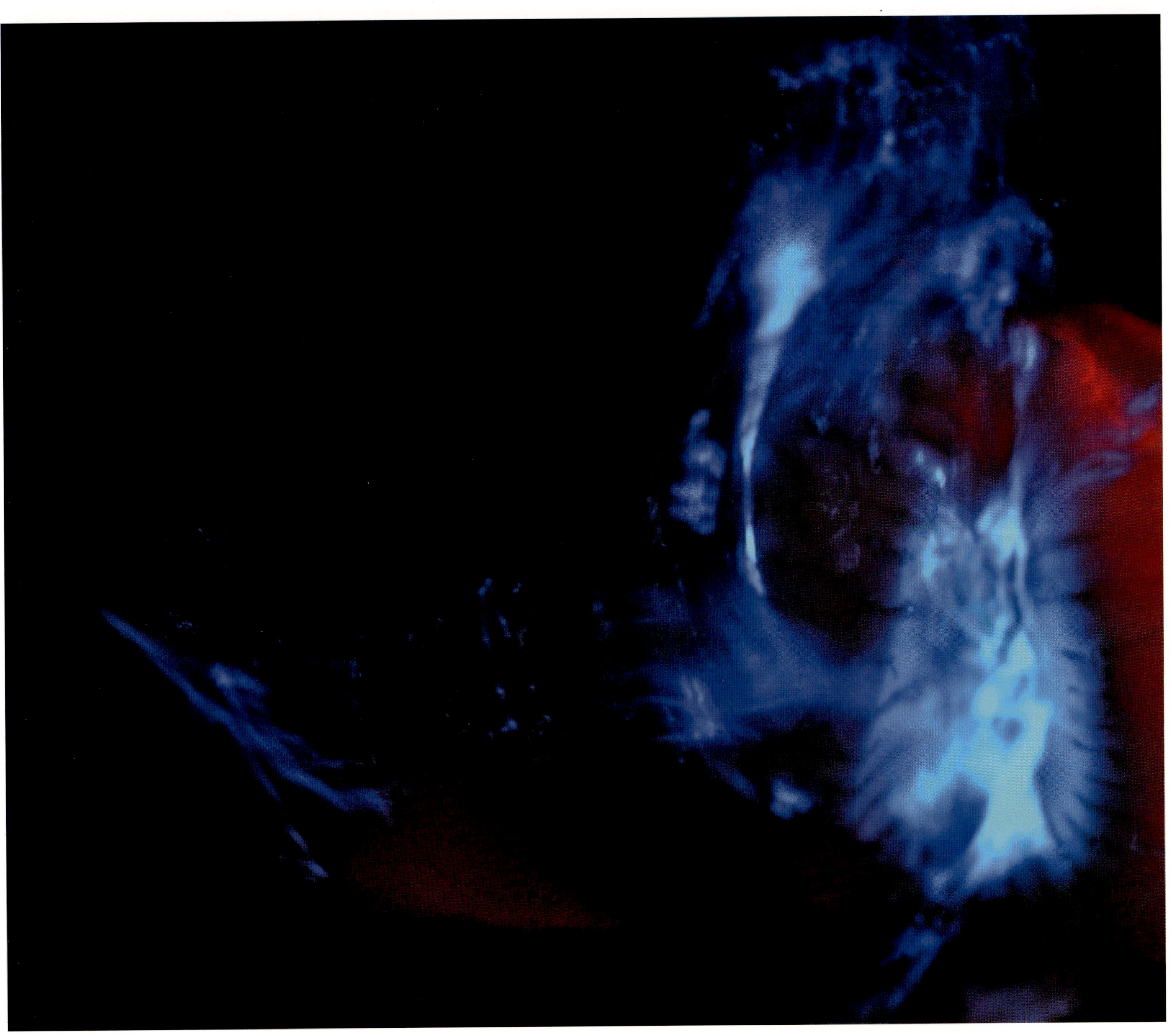

Actinoscyphia

These are also known as "Venus flytrap" anemones, and any photo of an undisturbed animal makes it abundantly clear why. They do look like huge deep-sea Venus flytraps. They're fairly common on the deep seafloor and can be found wherever there's a hard surface (a rock, dead coral, or even a shipwreck) that they can attach to. When disturbed, they emit copious amounts of blue bioluminescent mucus, which is likely intended to label the disturber, and thus serves a defensive function.

Bioluminescent anemone on hermit crab shell

Not all sea anemones settle on geological features. In some ways, the story of the deep benthic life is a story of finding suitable surfaces. Since much of the deep seafloor is a soft ooze of material that has rained down from above, durable substrates may be hard to locate. In this photographed specimen, illuminated by our red light, two anemones have settled on the shell of a hermit crab. Since the crab is also using the shell for its house, this dead snail is quite useful! This arrangement is mutually beneficial for both the crabs and anemones, as they can provide each other with different forms of protection. The anemones are small and their tentacles are retracted so they are hard to see, but they are given away by the bioluminescent splotches they produce.

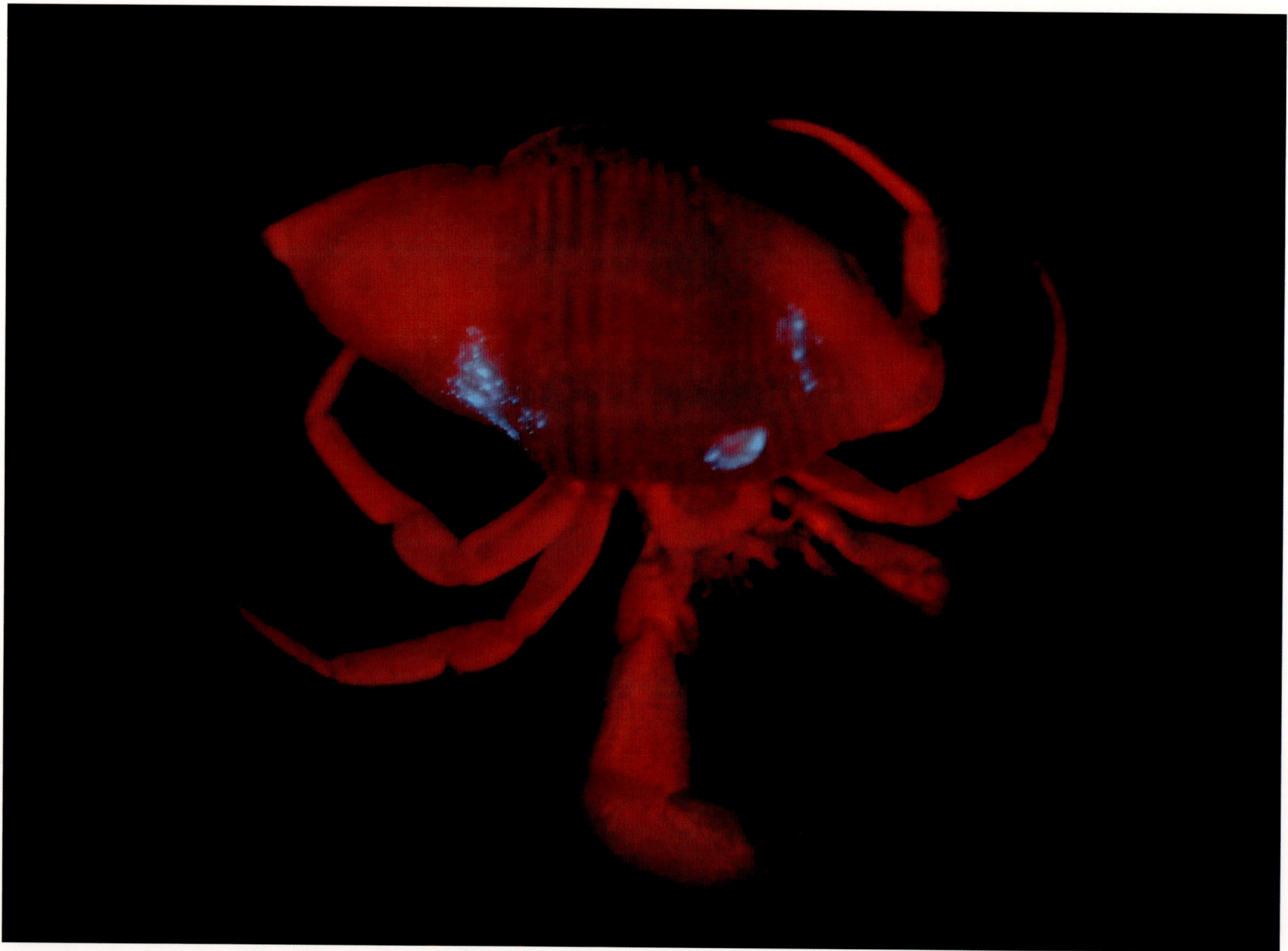

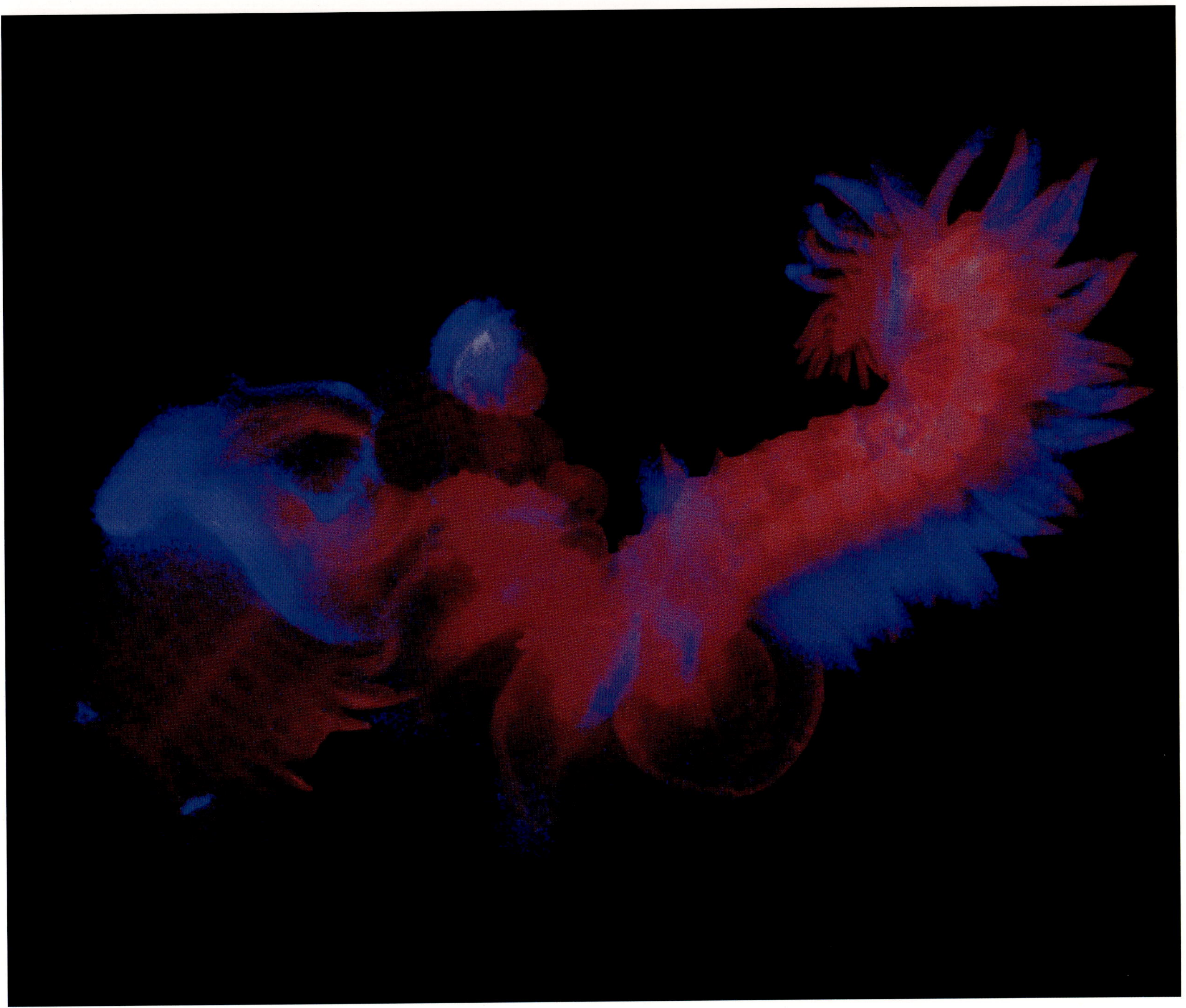

Chaetopterus variopedatus

Parchment-tube worms are found most commonly in their cluster of tubes in the mud of shallow sand flats. Each end of the tube sticks above the mud, but the rest of the animal is completely buried. The worm usually shares this dark and likely dreary burrow with a few small crabs. Unusually, for an animal that has almost no one to signal to, this worm has a truly impressive bioluminescent display that consists of both light organs and luminous mucus. Depending on the species, the light can be violet, blue, green, or even yellow, and despite several attempts, no one is sure why the light is emitted at all. The glowing mucus is ejected from the tube like a volcano of light, and may deter settlers from taking up residence in their tube.

Umbellula

Although we have mostly focused on bioluminescent animals that swim or drift in the water column, there are a number that live on the deep seafloor. One of these is the soft octocoral *Umbellula*. This odd-looking animal-on-a-stick is a type of sea pen, related to corals. It has a set of feeding polyps on a stiff support that keeps the living part of the animal above ground and up in the current. It's possible that the blue bioluminescence found in patches among the polyps may lure plankton, but it may also serve a defensive purpose—or both. Sea pens are often bioluminescent and frequently use GFP (p. 180) to change the emitted light from blue to green.

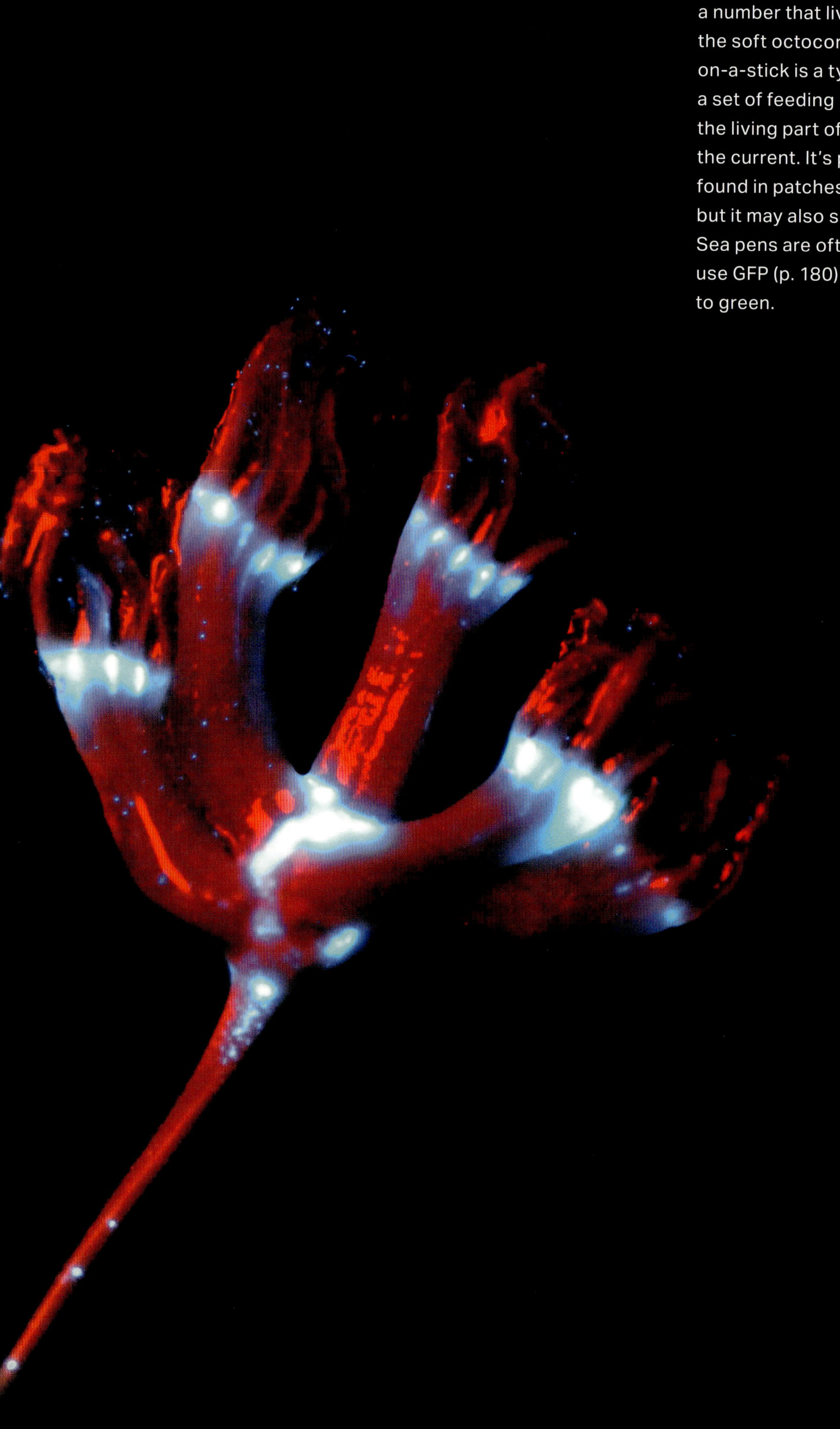

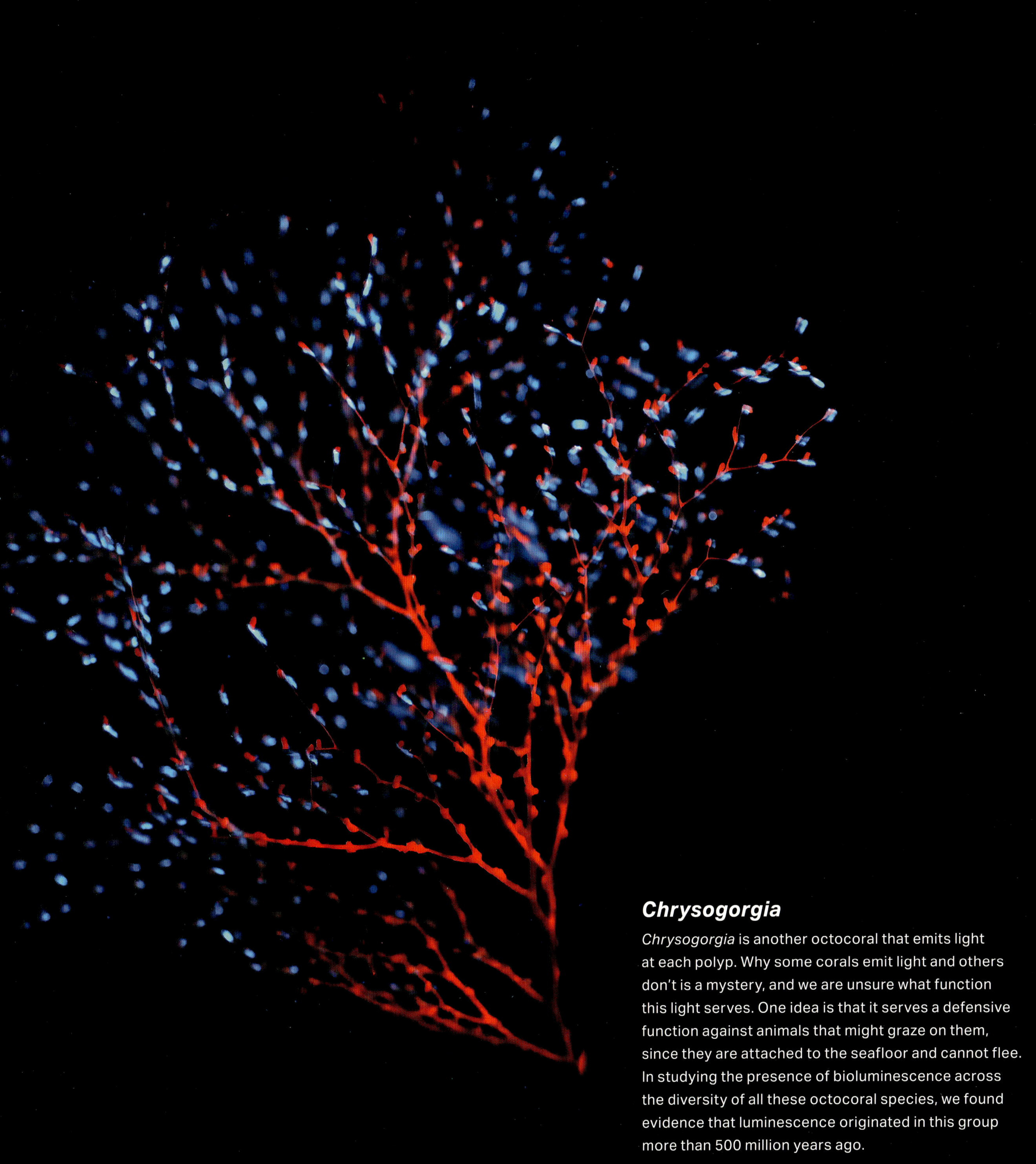

Chrysogorgia

Chrysogorgia is another octocoral that emits light at each polyp. Why some corals emit light and others don't is a mystery, and we are unsure what function this light serves. One idea is that it serves a defensive function against animals that might graze on them, since they are attached to the seafloor and cannot flee. In studying the presence of bioluminescence across the diversity of all these octocoral species, we found evidence that luminescence originated in this group more than 500 million years ago.

Ophiocanthus ternispinus

Most of the surface of the earth is underwater, and most of this seafloor is about 4 kilometers deep. In one of life's puzzles, many of the animals living on deep ocean sediments seem to be either brittle stars, sea stars, or sea cucumbers—all of which belong to the exclusively marine group known as echinoderms. Both deep and shallow brittle stars can be bioluminescent, and because their slender arms can quickly regenerate, it is thought that sacrificing a glowing part of their armtip while retreating to their burrow is a reasonable way to survive a predation event.

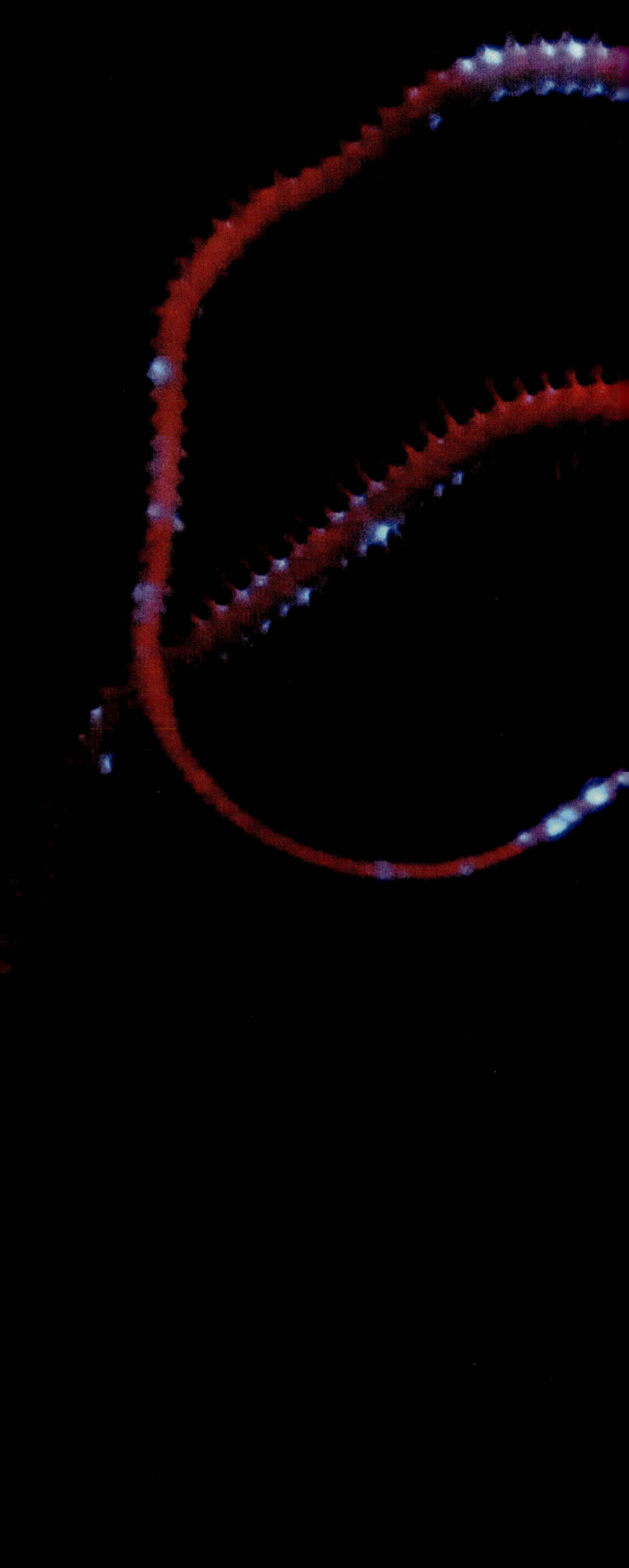

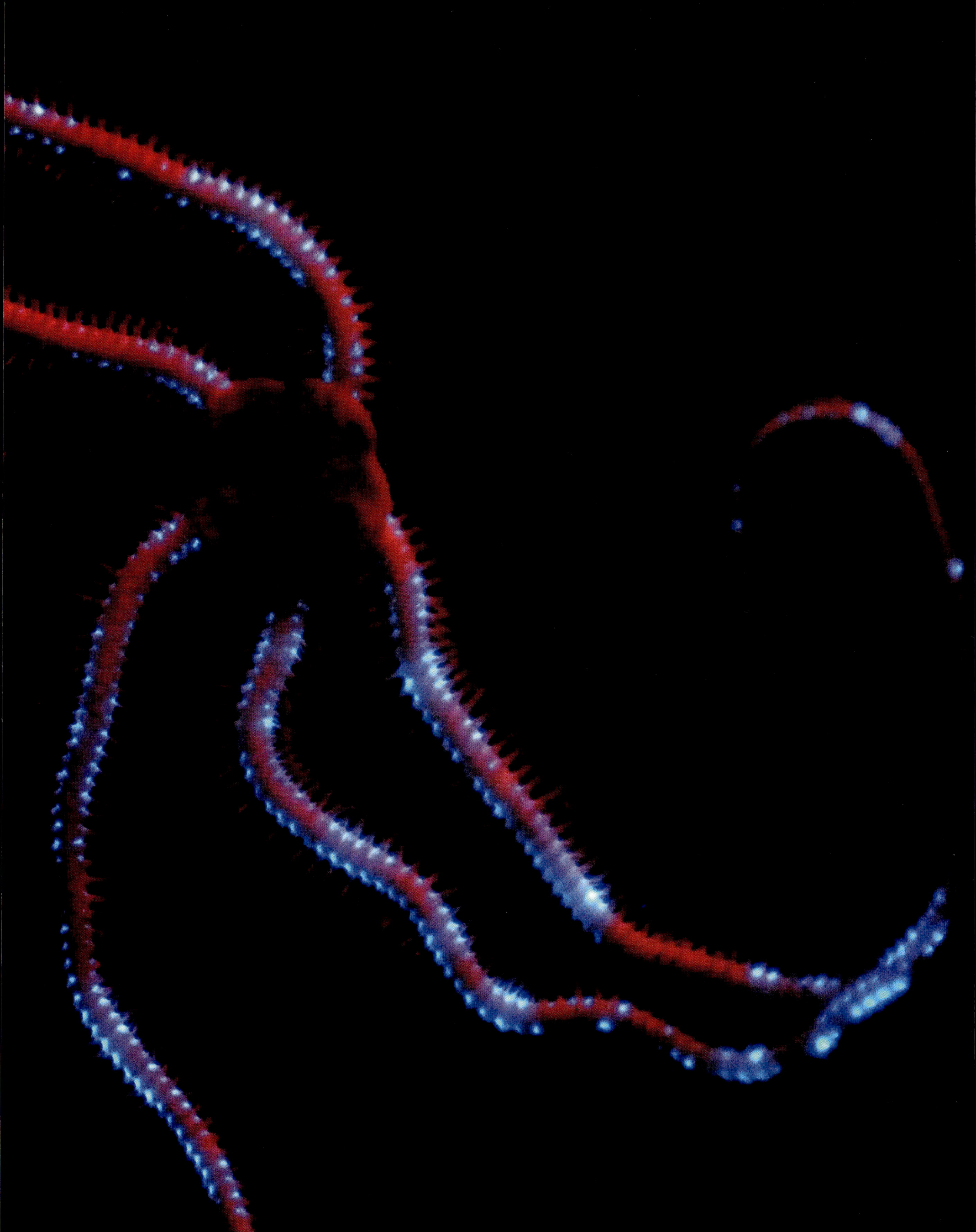

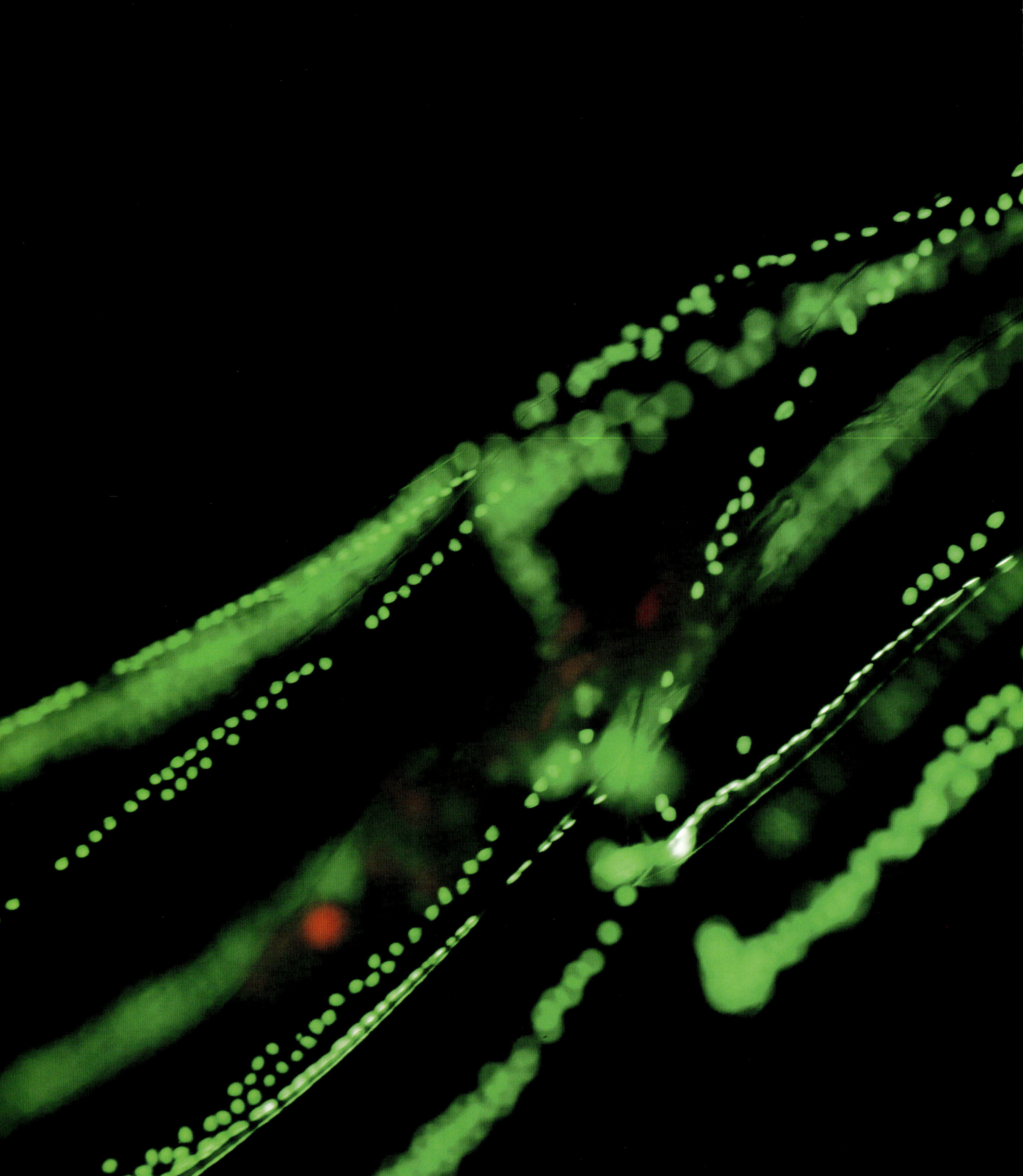

FLUORESCENCE

Making light from light

We end our book with fluorescence. In an ocean full of optical delights such as iridescence and bioluminescence, fluorescence is both delightful in the beautiful colors it can produce and mysterious in that we don't know what—if anything—most of these colors are for. It's therefore fitting to finish with a subject that has so much potential for discovery.

Fluorescence is a process in which some of the light that is absorbed is re-emitted, but at a different wavelength. ("Biofluorescence" is just a misleading word for fluorescence and shouldn't be used.) Light is energy, and a rule of science is that energy can never be destroyed; it can only be turned into another form of energy. In some cases, such as the roof of a black car, the energy of the absorbed light turns into heat by making the molecules of the black paint move and vibrate more quickly. In other cases, found in specialized microscopy techniques, the energy of the absorbed light creates sound, actually ringing the material like a bell. In fluorescence, however, some of the absorbed energy is emitted again. Because it is only part of the energy, the emitted light has to have less energy than the light that initially struck it, and therefore it has a different color. Photons of different colors have different energies that go down in the direction of the rainbow (ultraviolet > violet > blue > green > yellow > orange > red). The fluorescence of a given material also only tends to happen when the light that strikes it is of a particular color. Finally, the emitted color of the light is typically one to two colors down in energy on the rainbow. Therefore, some materials absorb blue light and then emit green light. Others absorb green light and emit red light. There are even materials (e.g. "blacklight" posters) that absorb the ultraviolet light that we can't see and emit multicolored light. Some laundry detergents contain fluorescent chemicals that convert ultraviolet light to visible light, making your socks truly "whiter than white."

It's important to realize that fluorescence—unlike bioluminescence—cannot make light. In fact, because only some of the energy of the original light is emitted, the entire process always takes light away. Typical photographs of organismal fluorescence, including most of those we show in this book, maximize the appearance of fluorescence by illuminating the organism with the light that creates fluorescence (known as the excitation source) and then viewing the object through a filter that only lets the emitted colors (known as the emission spectrum) through. In our photos, a yellow filter blocks blue excitation light, but lets green, yellow, or red emitted light pass.

These imaging conditions can make it appear that fluorescent objects or organisms are producing their own light, which they never are. The instant we turn off the excitation source, the fluorescence disappears. The blue-blocking filter also gives the impression that fluorescence is always obvious

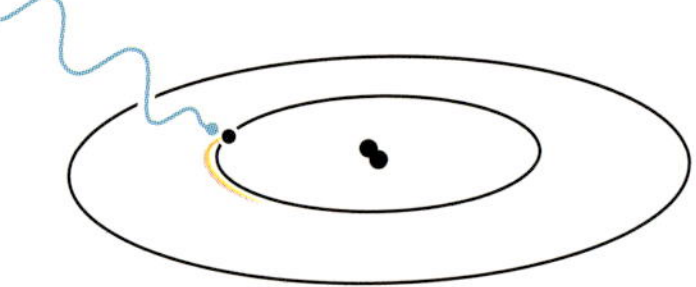

Photons are a manifestation of energy, and electrons orbiting around a nucleus represent different levels of energy themselves. Fluorescence happens when light strikes a molecule, as shown.

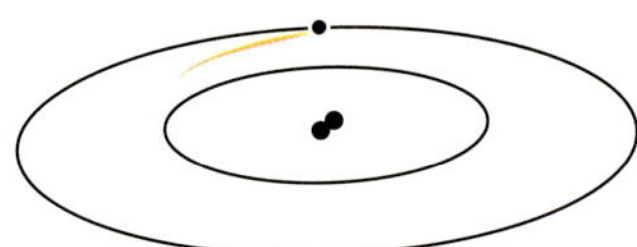

Then the energy transferred bumps an electron in that molecule up to a higher orbital.

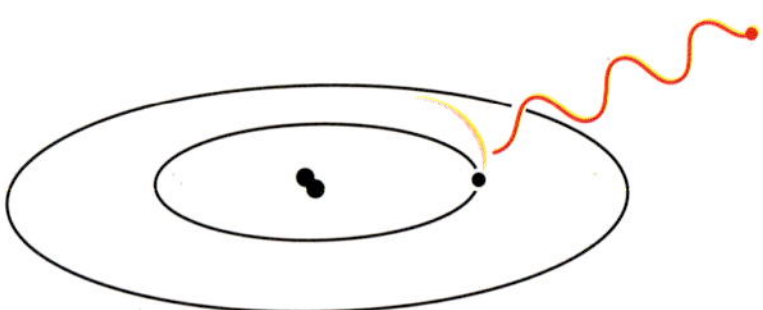

The electron is not stable in this new state, so it almost immediately falls back down to the previous orbital, re-releasing the energy as a new photon.

Because the energy coming out is a bit less than what went in, the new photon appears as a longer wavelength of light, shifted toward the red end of the spectrum. Note that high-energy ultraviolet light can excite fluorescence, but excitation can also be blue, green, or even longer wavelengths. When the electron is excited by a chemical reaction, the rest of the process is essentially the same, but the light is called chemiluminescence, or bioluminescence if it occurs in an organism.

Excitation leading to light can even come from sound (sonoluminescence), heat (incandescence), or breaking of chemical bonds (triboluminescence). To see triboluminescence at home, crunch a Wint O Green Life Saver in front of a mirror, or peel off a layer of duct tape in the dark, and you will see a dim glow along the seam where the tape is separating.

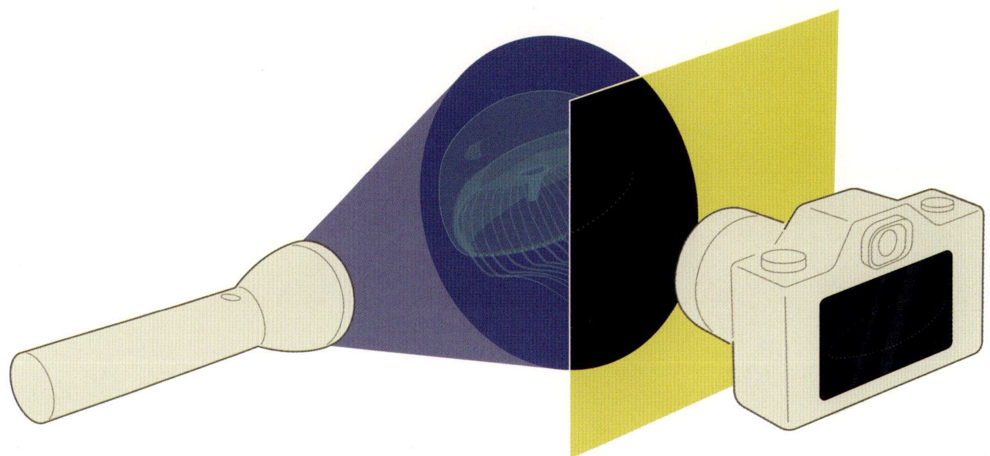

To photograph fluorescence, it is helpful or even necessary to block out the bright excitation source, revealing just the fluorescent emission by itself. To do this we illuminate an organism with a typically blue or violet light source, then look at it through a yellow filter. The yellow color indicates that the filter is blocking the short-wavelength blue light, but letting longer green, yellow, and red wavelengths pass through. (Remember green + red = yellow when working with emitted light.) Photos taken this way can be both enlightening and misleading, since most animals will not be viewing their environment through yellow lenses. This diagram shows the set-up as it was used to generate the picture on page 203 (and others throughout this chapter).

under natural illumination and without filters. This is sometimes true, but usually not. However, because fluorescence changes the color of the light that hits it, even this small amount of light can be obvious in the right habitat. By a combination of factors, the perfect habitat is the ocean.

As we've discussed, and as you've seen in the scuba diving photos here, the ocean is relentlessly blue. Except near the coast, where it can be relentlessly green, the vast bulk of light underwater is as blue as a blue LED. The blue may be paler when you look up, and more violet when you look down, but the main impression you get when diving in the ocean is of being inside an enormous blue jewel. Although beautiful, this purity of color limits the colors that can be seen. A red-pigmented animal will only appear red if there is red in the illumination, otherwise it will appear black. Even an animal that is red via iridescence will look black because both processes can only show colors that are in the illumination to begin with. Bioluminescence can, of course, be many different colors (some of which we have shown here), but it is usually blue in the open ocean and green in the green waters near the coast. This is likely because the eyes of animals in these habitats have evolved to be most sensitive to the colors that are in the illumination. There is rarely much point in using red bioluminescence in a world where nearly every animal can only see blue or green. However, in this monochrome blue world, there are actually a fair number of animals that can tell blue and green apart. Their color vision is seldom as good as that found in colorful reef fish or birds, but they can have good color vision in the blue-green range. Coastal fish are similarly good at determining green from red. Therefore, something green in the blue world of the open ocean (or yellow or red in green coastal waters) can be both detectable and obvious.

It turns out that many fluorescent materials found in animals are excited by blue light and emit green light. In addition, some very common tissues of ocean animals do this by default, such as the exoskeletons of crustaceans. Some coastal animals can also absorb green light and emit red. Fluorescence is quite common in the marine world. Therefore, although

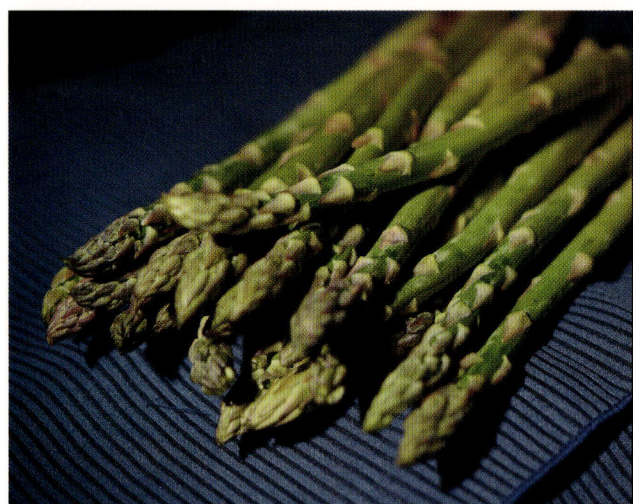

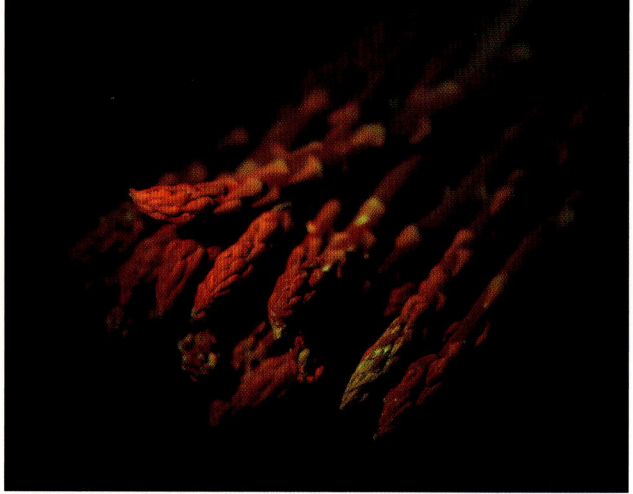

fluorescence may be hard to detect on land, and thus may seldom have a purpose, it is common and potentially useful in the ocean. In addition, it has been known for a number of decades that fluorescence can be used to change the color of the emitted light in bioluminescence. In fact, the most famous molecule connected to bioluminescence, which won its discoverer Osamu Shimomura a Nobel Prize in 2008, is known as green fluorescent protein (GFP) and is responsible for changing the emitted light of many jellyfish and jellyfish-like animals from blue into green. In the deep sea, where red light is extraordinarily rare, some use fluorescent molecules (known as fluorophores) to emit this color.

Understandably, there is often confusion between bioluminescence and fluorescence. A couple of reasons that the confusion is understandable are that bioluminescent energy can be used to excite fluorescence, and bioluminescent luciferins are usually also fluorescent! As you will see in some of the photos that follow, you can sometimes localize luminescent photophores in animals using their fluorescence, even though you are not seeing the actual bioluminescence. For completeness and to round out the confusion, phosphorescence is a delayed-emission variation of fluorescence. Here, electrons are excited by photons, but instead of immediately decaying back down and re-emitting fluorescence, the electrons have to take a circuitous route back to their resting state, like marbles rolling down a series of tilted ramps instead of falling directly to the ground. The result is a dim steady light used in glow-in-the-dark stickers; the emission is spread out over seconds or even hours after excitation rather than a hundred-millionth of a second, as with fluorescence.

The ultimate question is: What is it all for? In many cases, the fluorescence may serve no purpose in the organism. As we have said, many molecules, especially pigments, are fluorescent. Our own eyes, skin, and teeth are all fluorescent, but nobody attributes any function to that. Chlorophyll, possibly the most important molecule for life on earth, fluoresces bright red due to its chemical structure. This fluorescence, while

←
Because chlorophyll fluoresces a deep red, nearly all vegetables will show some level of fluorescence.

→
Brown eggs have a porphyrin pigment in their shells, which is the same family of molecules that gives some fish their bright fluorescence.

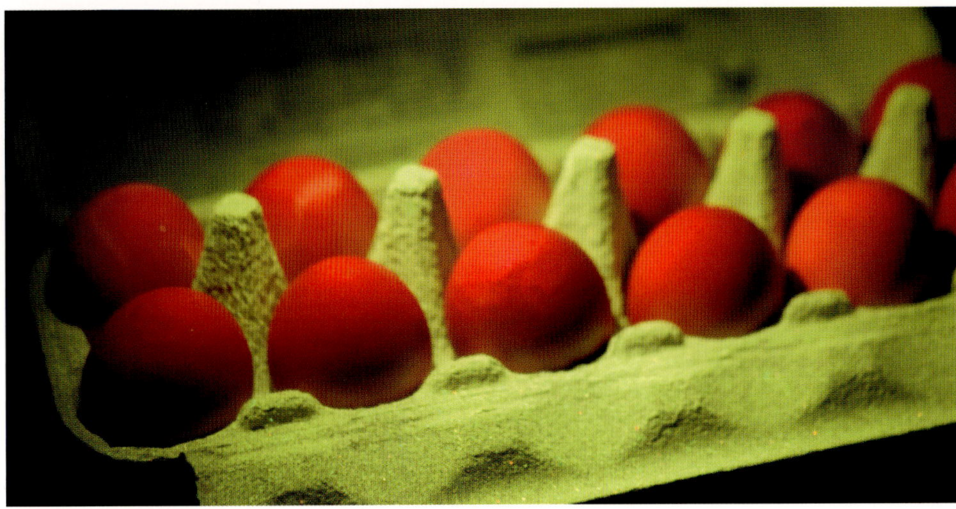

not functional for the plants or algae, has turned out to be very useful to oceanographers, because they can measure it to determine how abundant algae are in different parts and depths of the ocean.

In shallow waters, fluorescence may be a way of getting rid of energy that is dangerous to an organism. As we know from sunburn, ultraviolet light can damage tissues, sometimes severely. Even blue light can be dangerous. This is because absorption of their high energies can break or change important molecules such as proteins and DNA. Using fluorescence, a pigment can absorb only part of the energy of one of these more dangerous photons, and send part of the energy back out of the tissue in a harmless form. Interestingly, this is also thought to be one of the mechanisms that led to the evolution of useful bioluminescence: dangerous energies of oxygen molecules were sent away as harmless photons, which eventually became useful on their own.

There are a few visual uses of fluorescence in the sea that have been experimentally tested. We later demonstrate that the flower-hat jelly (p. 190) uses fluorescence to attract fish as its prey. This application of bright green fluorescence against a blue background seems to be in use by siphonophores, sea anemones, and other shallow predators.

It is also pretty clear that fluorescent proteins like GFP are used to change the color of bioluminescence, and this pigment is more common in coastal and benthic species. In these cases, the bioluminescence is being turned from blue to green to better match the color of light (and green-sensitive eyes) found in these habitats.

In some conditions, the uses of fluorescence may mimic those of bioluminescence, but there are significant differences between the two processes. First, fluorescence requires light, while bioluminescence does not. Therefore fluorescence—except when coupled with bioluminescence—is unlikely to be of use at night or in the deep sea. Second, fluorescence is a passive process. Unlike bioluminescence, it cannot be turned on and off, except by covering and uncovering the fluorescent tissue. Therefore, it

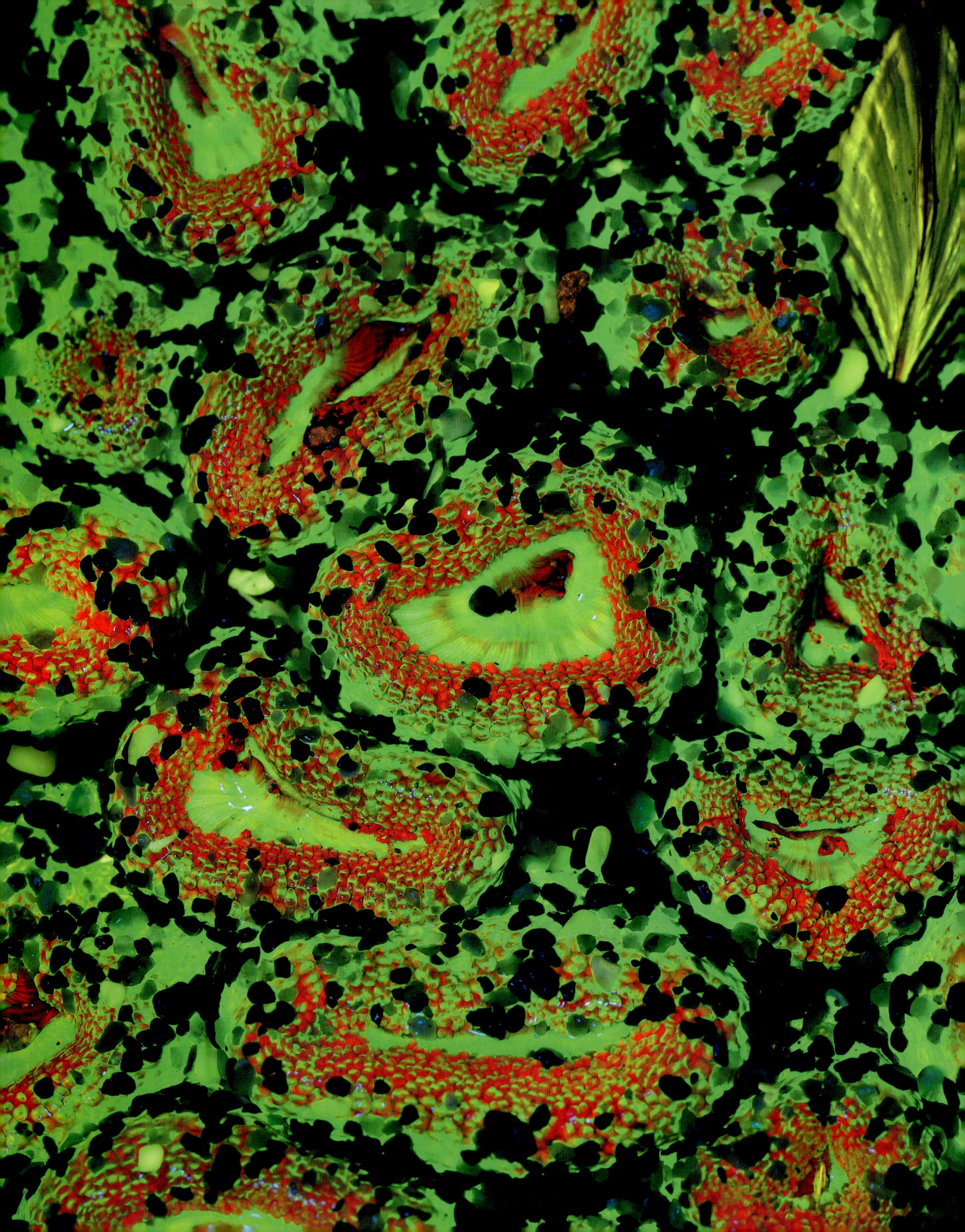

There has been a lot of media coverage of fluorescent vertebrates—notably squirrels, bats, puffins, turtles, and platypuses—stemming from people exploring forests or museum collections with ultraviolet lights. These are often described as "glow in the dark," although, as we have discussed, they really only glow in the light. Many of our own body parts are fluorescent, especially under ultraviolet illumination, but we don't attribute any function to these properties.

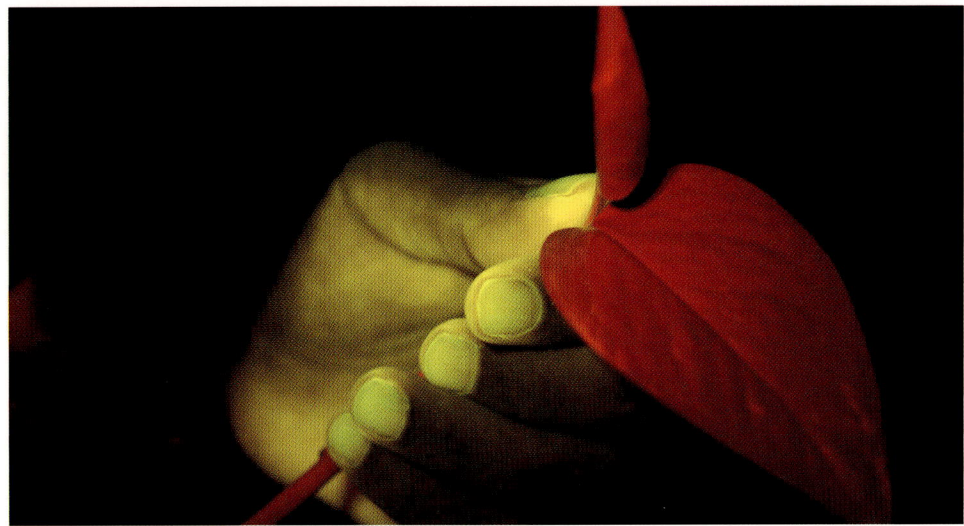

would be less useful for startling a predator or sending a time-coded signal like ostracods do with their blinking light emissions. However, if the excitation light is available, fluorescence typically creates a greater variety of color than bioluminescence does. In addition, fluorescence does not require the animal to expend any energy and will be brighter when the illumination is brighter. Bioluminescence, in contrast, requires both energy and special molecules to produce it, and thus works best in low-light conditions. In many ways, bioluminescence and fluorescence complement each other: the former being more useful in the dark and the latter being more useful in subdued light.

So, in the end, we have a delightful mystery that will keep marine biologists happily busy for decades!

Diphyes dispar

Most of the pictures in this section use blue-blocking filters to make it easier for you to see the fluorescence by itself, without being overwhelmed by the blue excitation light. This photo of the siphonophore *Diphyes dispar* was taken with white light, not blue, and without a filter. The fact that it shows up without any enhancement demonstrates just how bright the fluorescence is in this specimen.

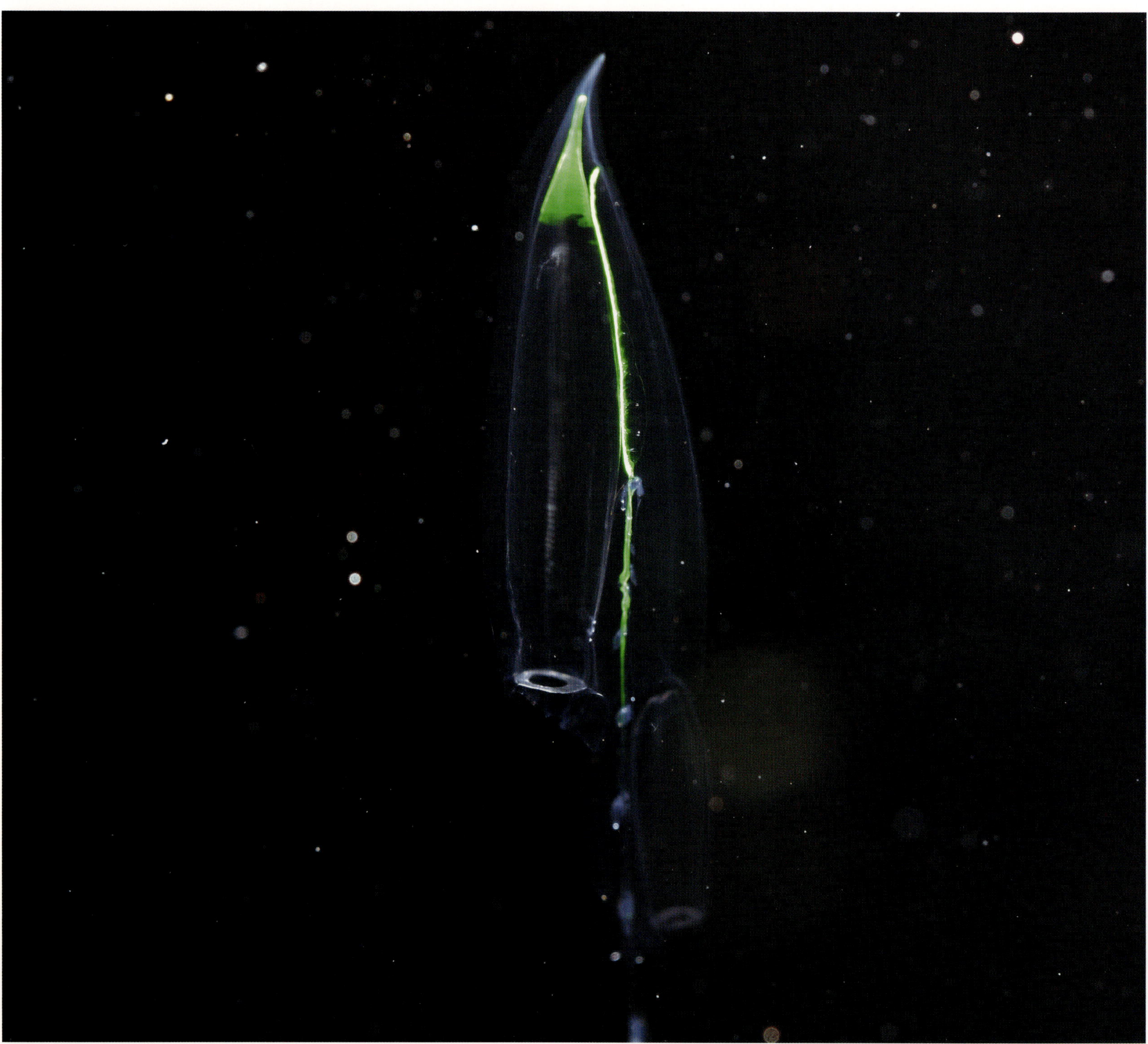

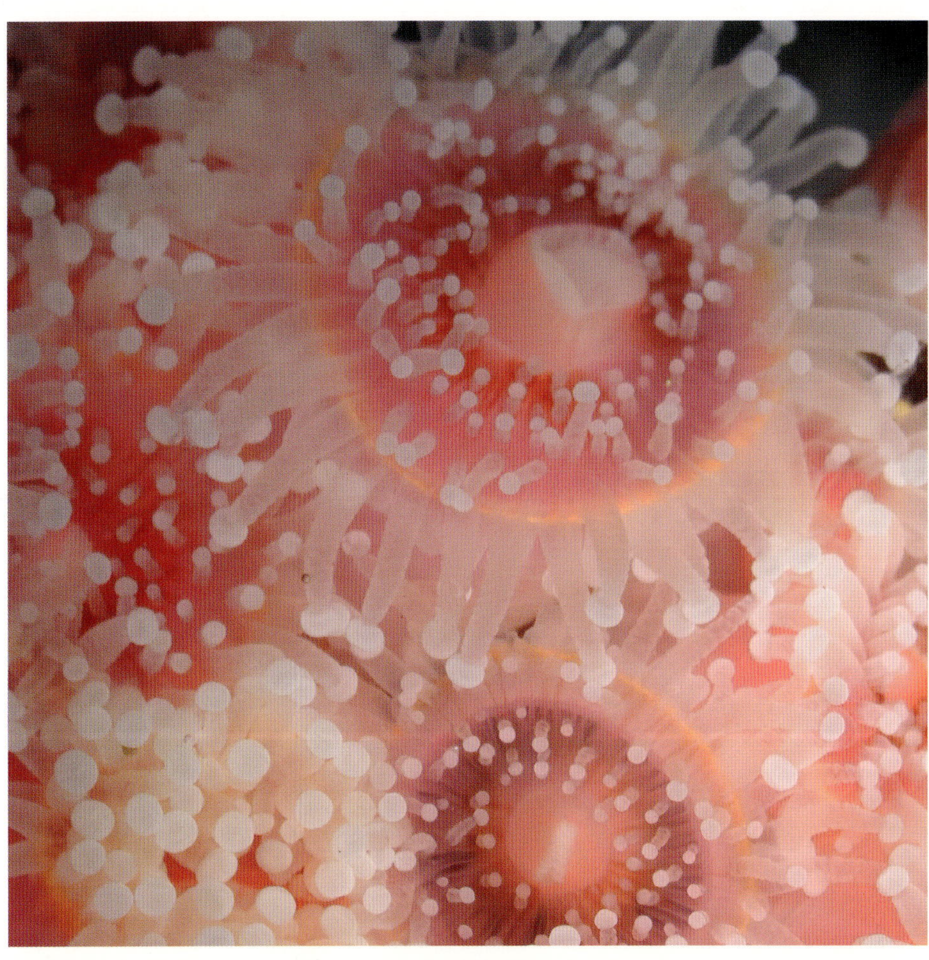

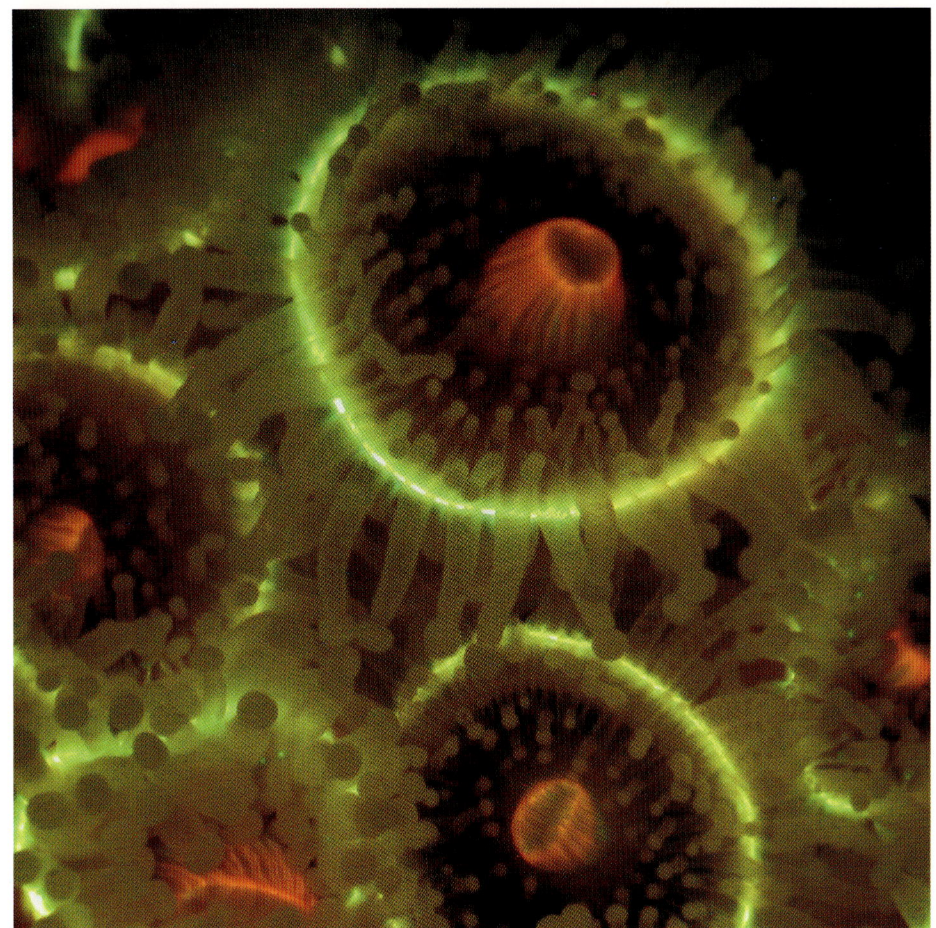

Corynactis californica

Steve: Our open-ocean scuba dives usually take place at depths shallower than about 22 meters underwater, to maximize the time we can spend exploring. On one occasion, when I was diving the California coast at about 30 meters depth, I saw something odd: a patch of strawberry anemones, *Corynactis californica*, that were strawberry colored! Given that I was surrounded by dim blue light, I knew the red color had to be coming from fluorescent proteins (FPs). Working with my then-student, now Dr. Christine Schnitzler, and Dr. Bob Keenan, we cloned the genes responsible for the different colors of fluorescence in *Corynactis.* Most surprisingly, individual anemone polyps had multiple FP genes that could make color patterns spanning green, yellow, orange, and red. In the colony depicted, you can see the yellow and orange colors with flecks of green.

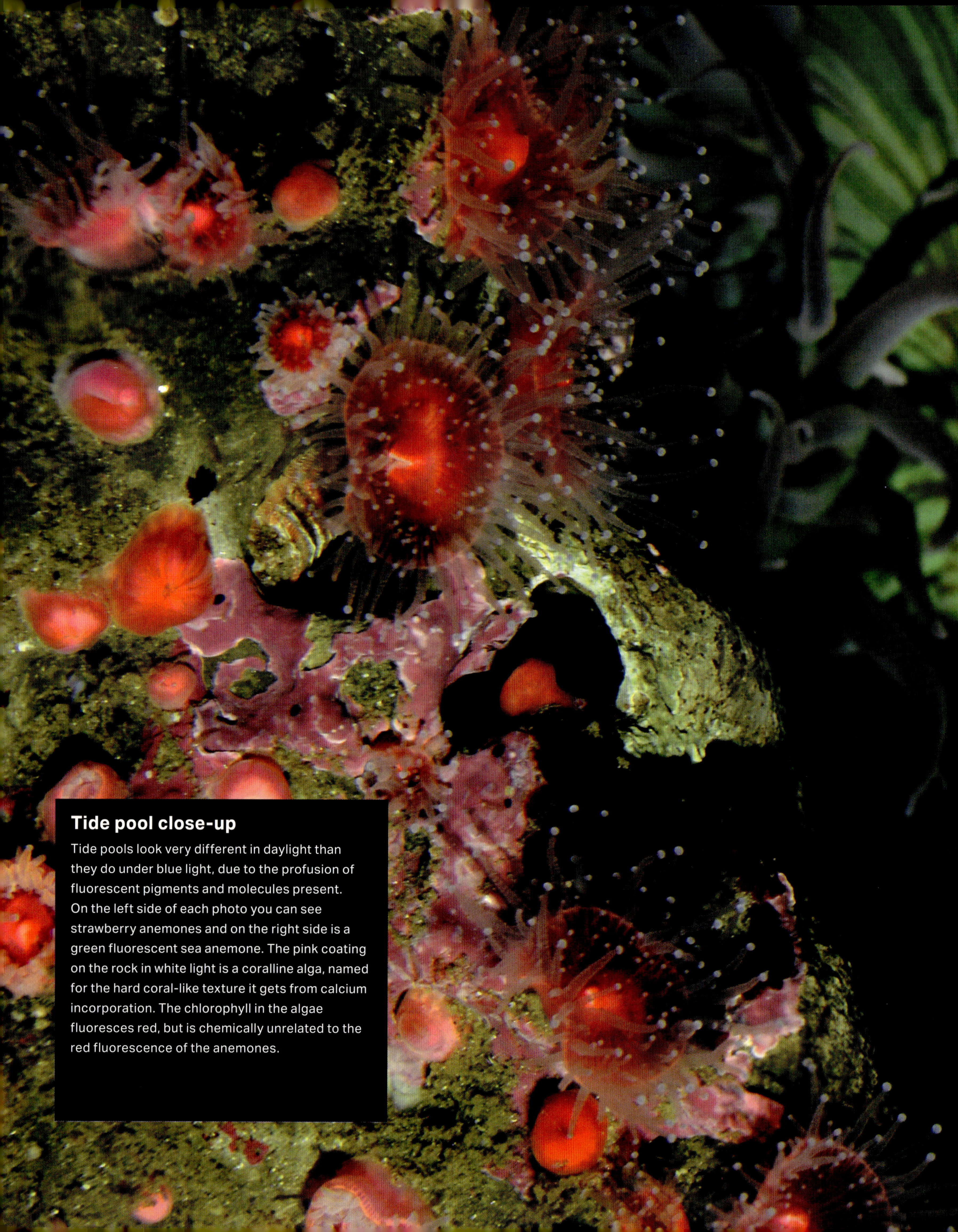

Tide pool close-up

Tide pools look very different in daylight than they do under blue light, due to the profusion of fluorescent pigments and molecules present. On the left side of each photo you can see strawberry anemones and on the right side is a green fluorescent sea anemone. The pink coating on the rock in white light is a coralline alga, named for the hard coral-like texture it gets from calcium incorporation. The chlorophyll in the algae fluoresces red, but is chemically unrelated to the red fluorescence of the anemones.

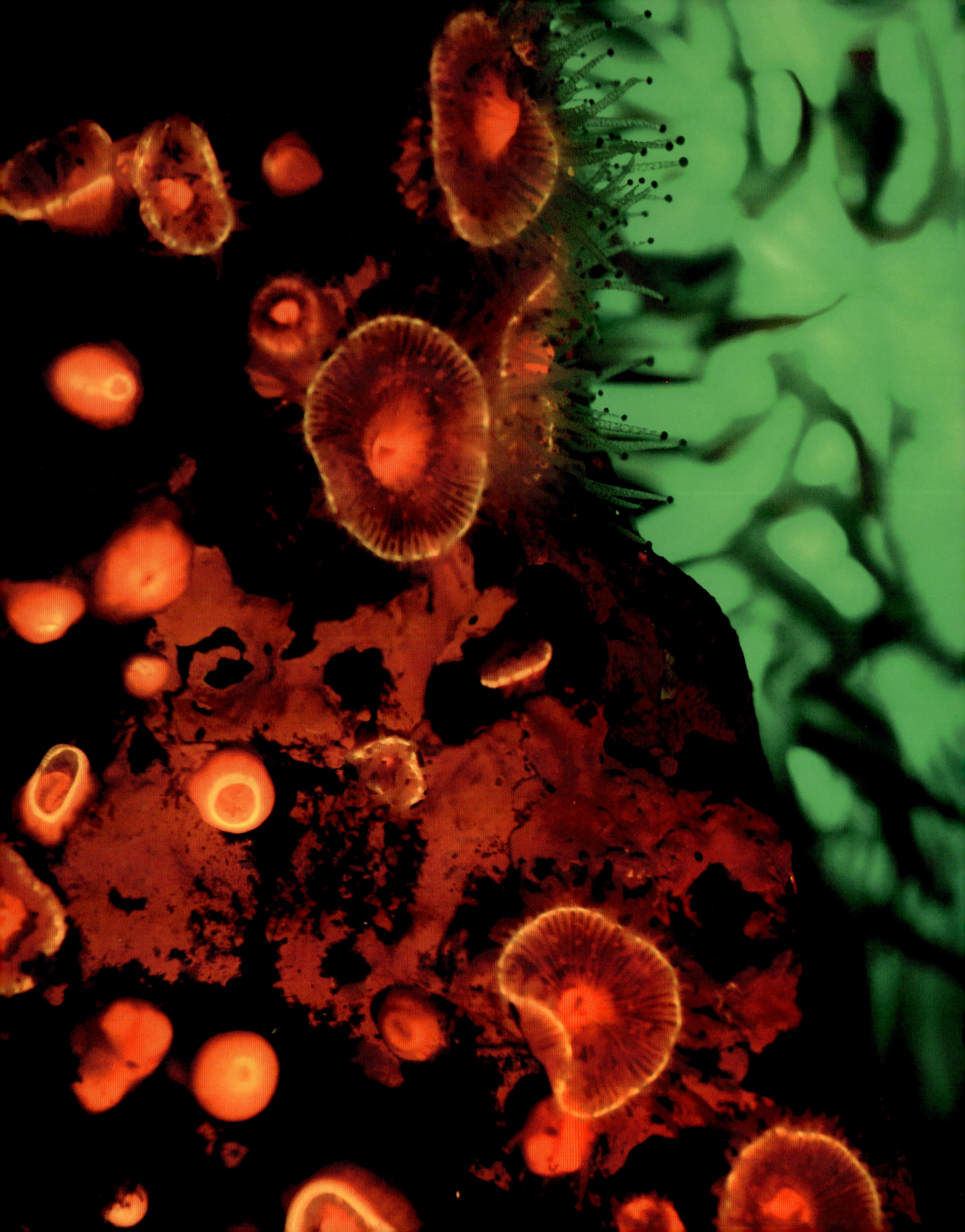

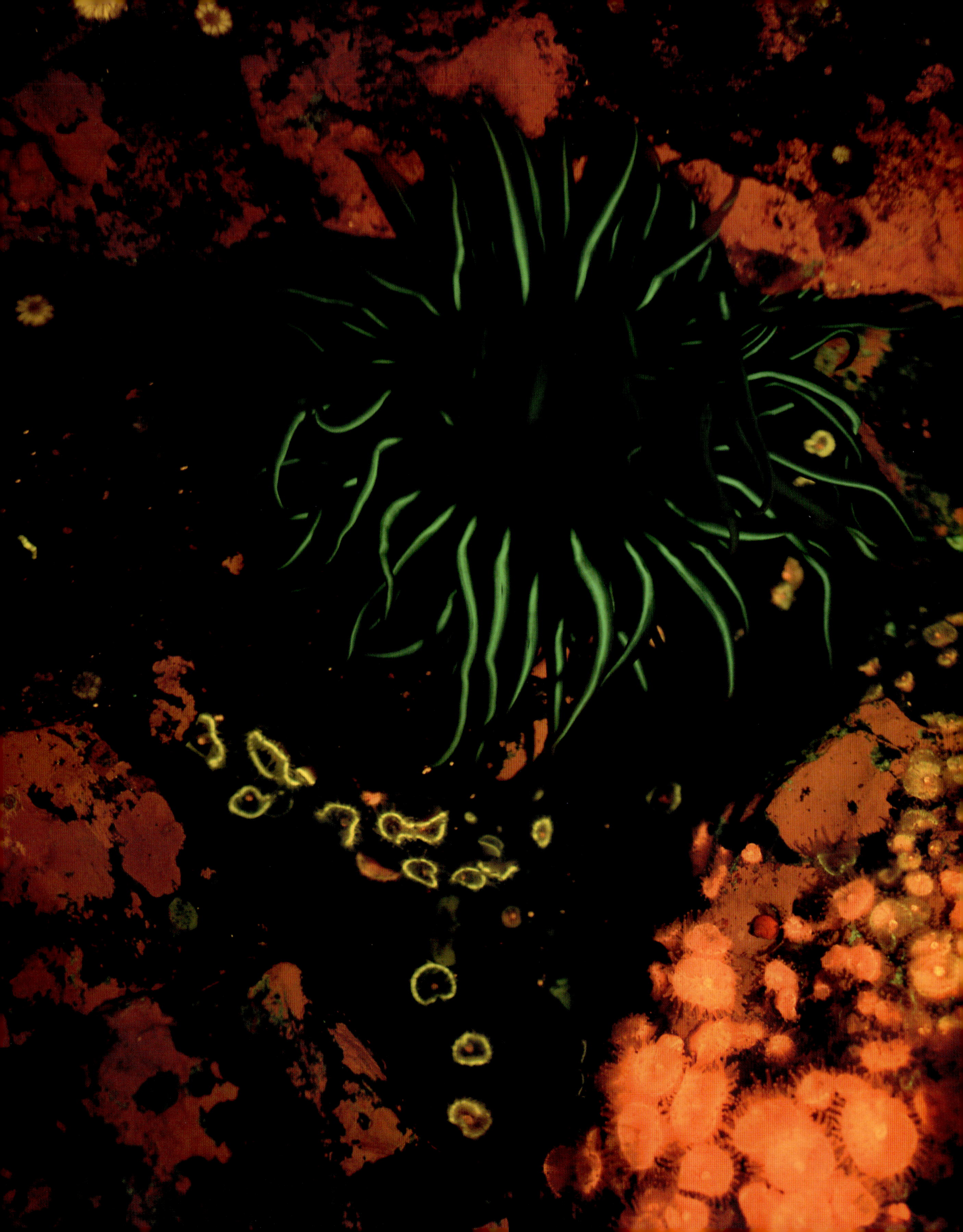

Tide pool overview

Here you can see the diversity of fluorescent anemones and cup corals. Three color variants of the strawberry anemones across the bottom are apparent in white light, and you can see that this pattern carries over to different fluorescence patterns as well.

Corals reproduce by having tiny larvae settle on a rock and then grow and divide to form large colonies. The larvae are cryptic and almost impossible to see by eye, but they are also fluorescent, so researchers can simply scan the reef with a blue light and look for the bright specks on the reef to assess the state of coral settlement.

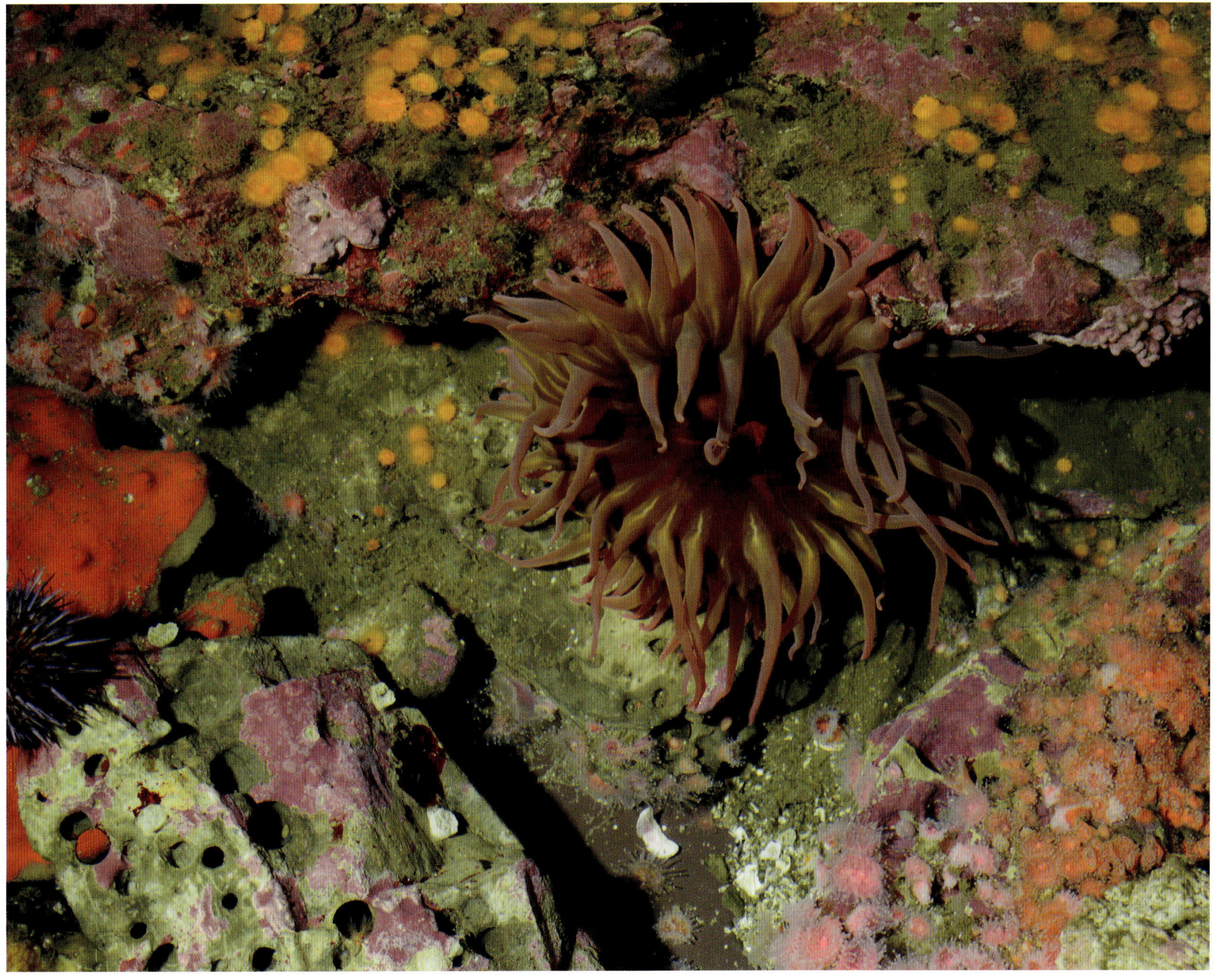

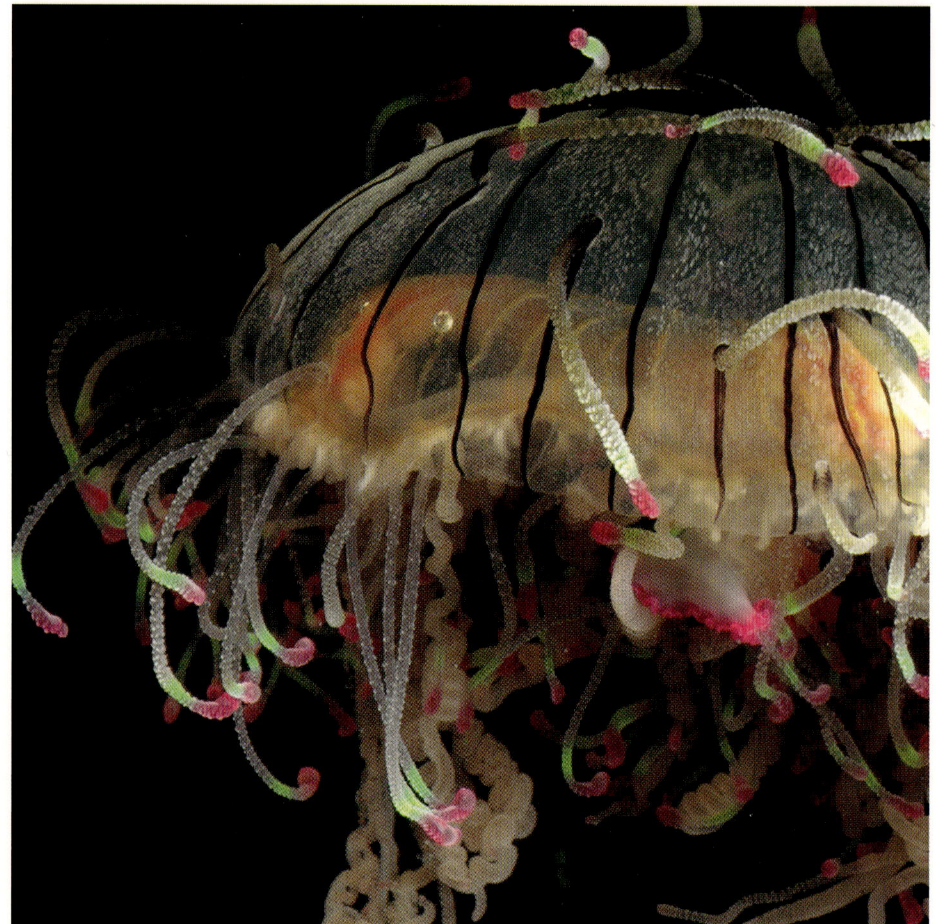

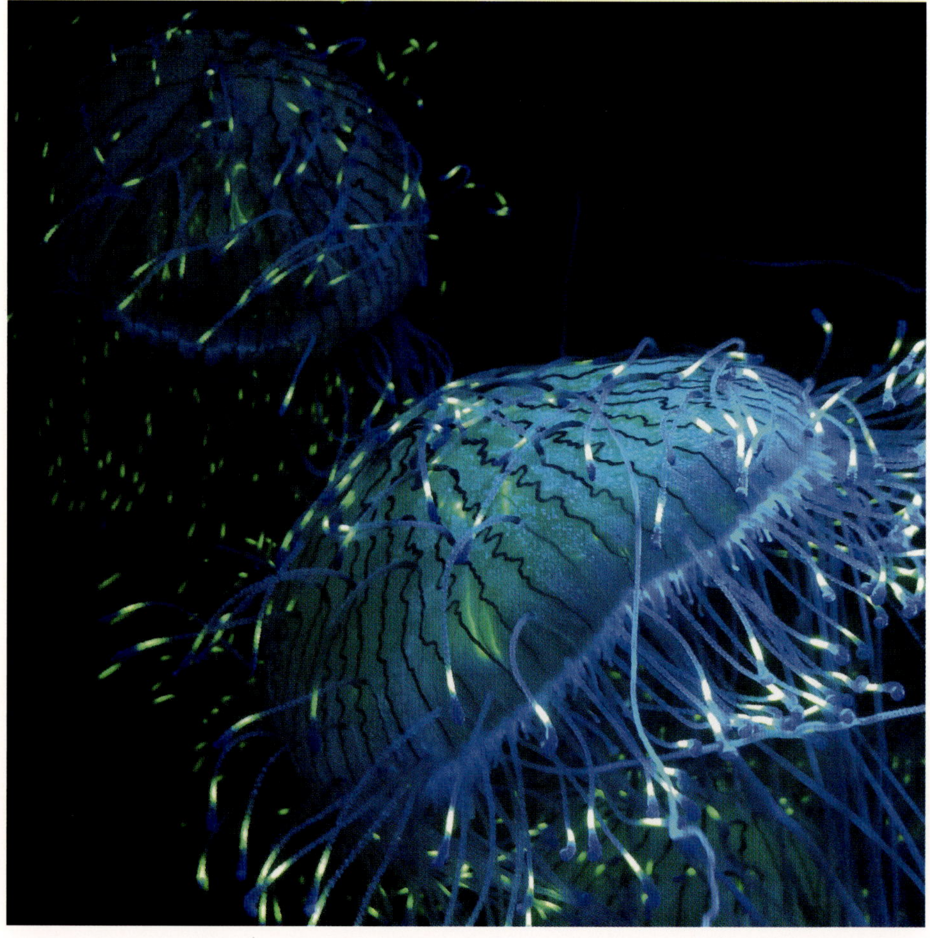

Olindias formosus

Fluorescent proteins (FPs) are fairly widely distributed among animals, especially within anthozoans (the group containing corals and sea anemones). However, not many functions have been reliably demonstrated for FPs. Some are certainly used to modify the color of bioluminescence, and researchers have suggested that they could enhance or protect algal symbionts living inside the anthozoans. What about an animal like the flower-hat jelly, which is not bioluminescent and doesn't have symbionts but is brightly fluorescent?

In *Olindias* there are green FPs near the tips of the tentacles, and a bright pink spot right at the end of the tentacle. (Notice that the pink spot is not lit up in the fluorescence image.) These are non-fluorescent FPs called chromoproteins. They absorb blue light but do not re-emit it. Remember that pink colors in white light indicate that the pigment is absorbing everything *except* pink light. So, chromoproteins under blue light look almost black, and positioned next to the green FPs they enhance the contrast.

Steve: Using the unique properties of *Olindias*, we did some experiments to test the functions of fluorescent proteins. *Olindias* eats fish and is a relatively strong stinger for a hydromedusa, making it an occasional nuisance to swimmers. We set up a tank with jellies on one side of a clear barrier and fish on the other. It was a gladiator's arena, but without any bloodshed. We tested the fishes' behavior under three different lighting conditions: white light, green light (which didn't excite fluorescence), and blue light. Under blue light, when the fluorescence was maximally visible, fish would strike at the barrier, trying to catch the wafting tentacles.

These experiments convinced us that, in some cases, fluorescence was a very effective way to attract prey that can see. Once we saw these results, we started to suspect this mechanism was at play in many other unrelated organisms, some of which are shown in the following pages.

Sea anemone

This beautiful anemone from the California coast has two fluorescent pigments that are localized to different parts of its body. Some FPs will start out green and then "mature" to a red color over time; others can be switched from green to red by exposure to light. When researchers discovered one such protein, they named it "Kaede" for the Japanese word for maple trees, to evoke the image of leaves changing colors in the fall.

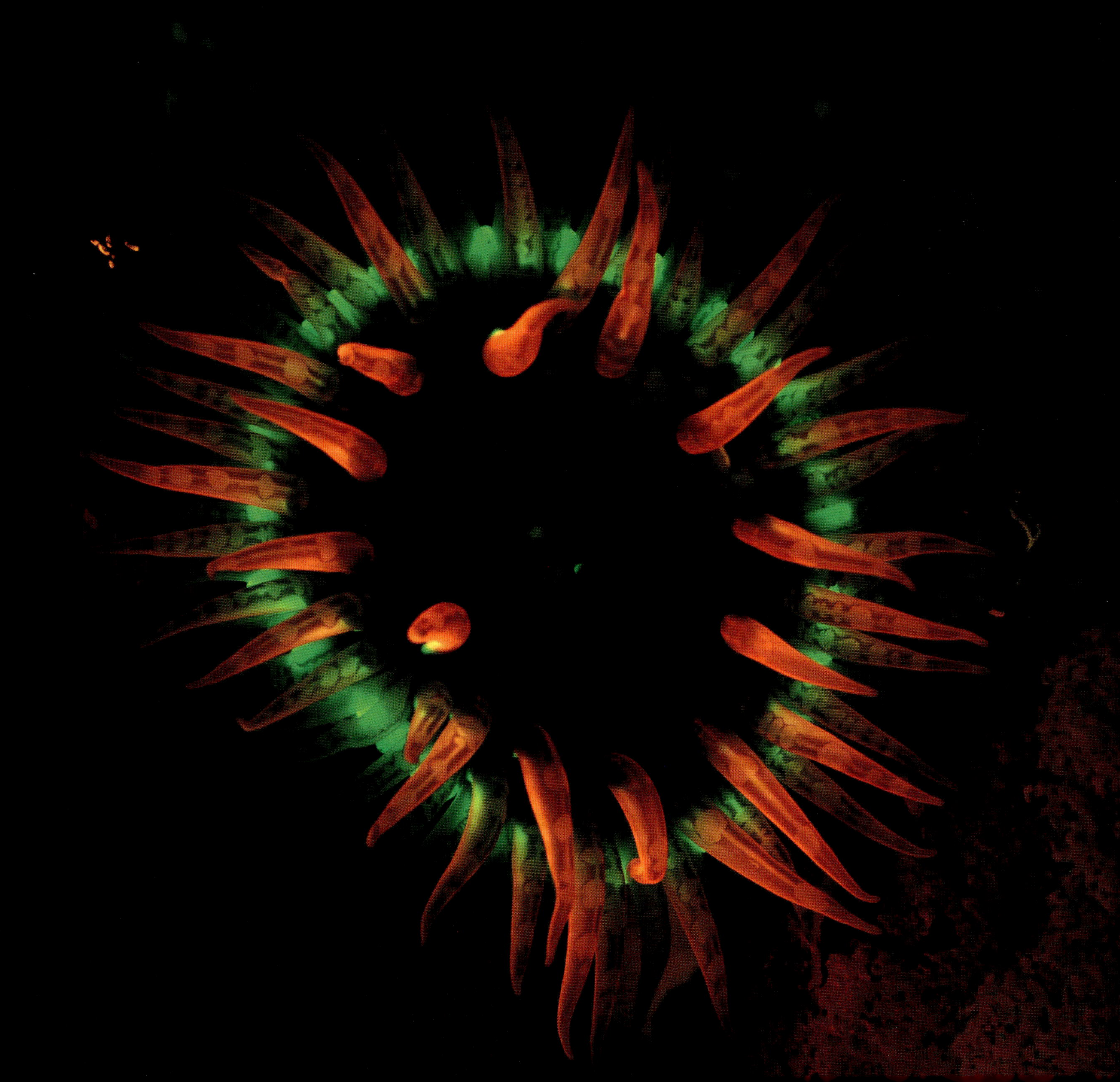

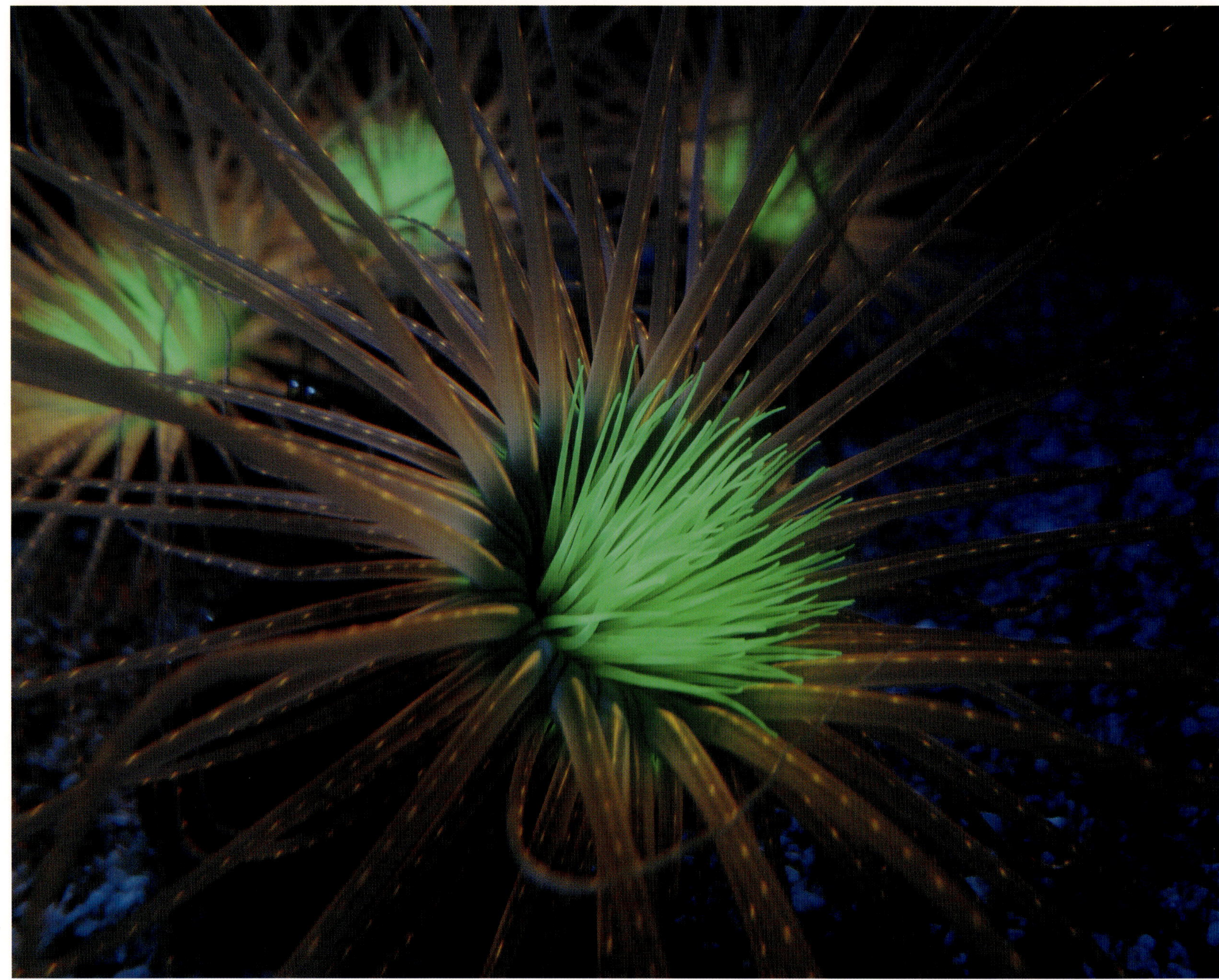

Cerianthus

Unlike their rocky-reef cousins, tube anemones live on sandy, soft seabeds and deeper in the ocean—from just below low tide to beyond 3,000 meters. They have a mane of long tentacles around the outside, surrounding a tuft of shorter tentacles at the center of their tube. When something bumps into the tentacles, the anemone will retract with startling speed. Take a look at where fluorescence is brightest in this species; right where you would expect a potential prey item to investigate. We believe that the fluorescent proteins in these anemones are serving to attract prey in their normally dim blue environment.

Resomia ornicephala

Steve: During our explorations, we often come upon an animal that we haven't seen before, and we give it a placeholder name to use until we can formally describe the species. In this case, we always saw this siphonophore at a depth of 200 meters, give or take 10 meters. We knew it was in the genus *Resomia*, so for years we called it R200. (An R600 and R1200 were also ultimately described in the same paper.)

As with the angler jellyfish *Erenna* (p. 143), one of the easiest ways to distinguish these species was by looking at the side branches of the tentilla. Because the tentilla house the stinging cells, they are the part that most directly interacts with potential prey, and they differentiate, becoming optimized for attracting and subduing different types of animals. In R200, the tentilla were very strange (which is saying a lot for siphonophores, which lean toward strange by default). Each folded band of stinging cells had a thin filament connected to a multicolored flag that floated above it like a kite in the breeze. These flags turned out to contain fluorescent pigments, with a green central pad and eyelike orange pads on either side. As far as we could determine, these appendages are only fluorescent, not bioluminescent, so to emit light they had to be excited by ambient blue light. Aha, 200 meters! We surmised that their constrained vertical distribution was because this depth has the optimal light field during the day for their lures to appeal to their preferred prey of krill.

When it came time to name the species, we decided on *R. ornicephala*, due to the bird's-head appearance of these unique structures. (*R. fluoroflapicus* would have been a bit too undignified, although we did name a siphonophore *Lilyopsis fluoracantha* after its impressive whole-body fluorescence.)

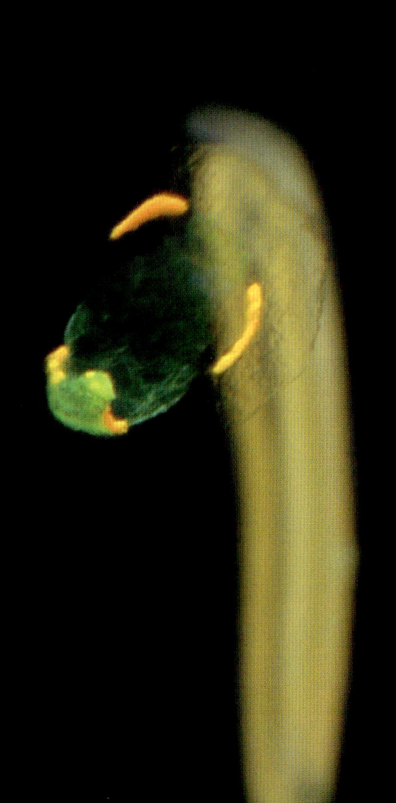

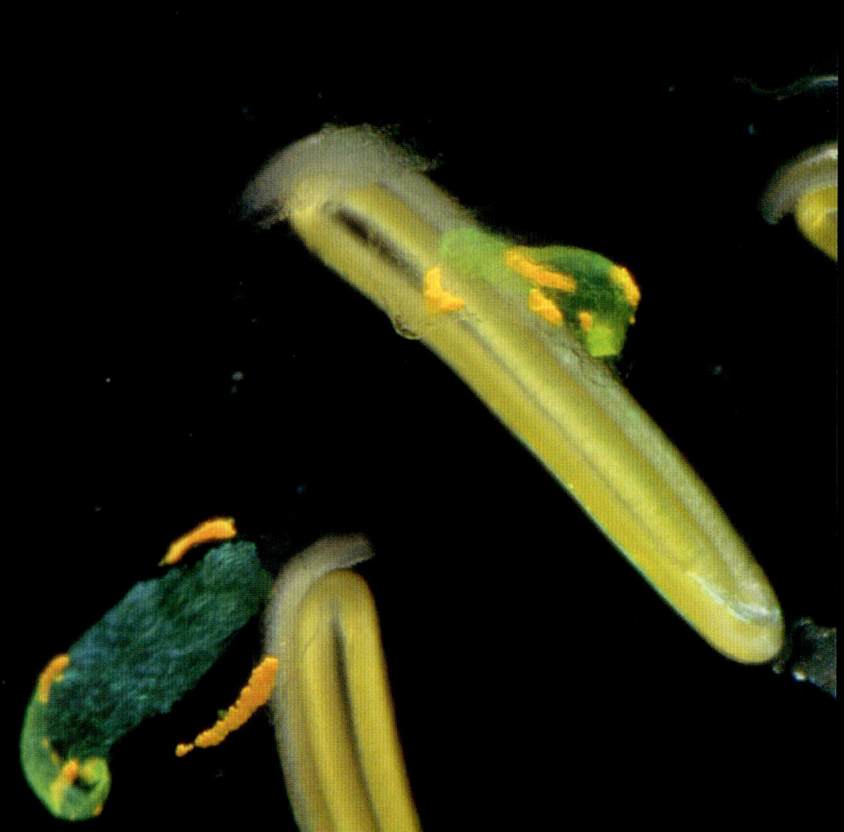

Gonodactylus

From a behavioral and optical perspective, mantis shrimp (stomatopods) are truly superstars. Their attack strikes are among the fastest and most powerful in the sea, and their visual capabilities are unsurpassed. If that weren't enough, they also have complex fluorescent patterns on various parts of their bodies. Among these are the paddle-shaped "antennal scales" that dangle from either side like the flags of a semaphore operator. In addition to being polarized, these paddles are also fluorescent. In a video of a stomatopod taken under blue illumination, a fish can be seen repeatedly darting toward one of these paddles, until on the third pass the mantis lunges and captures its prey.

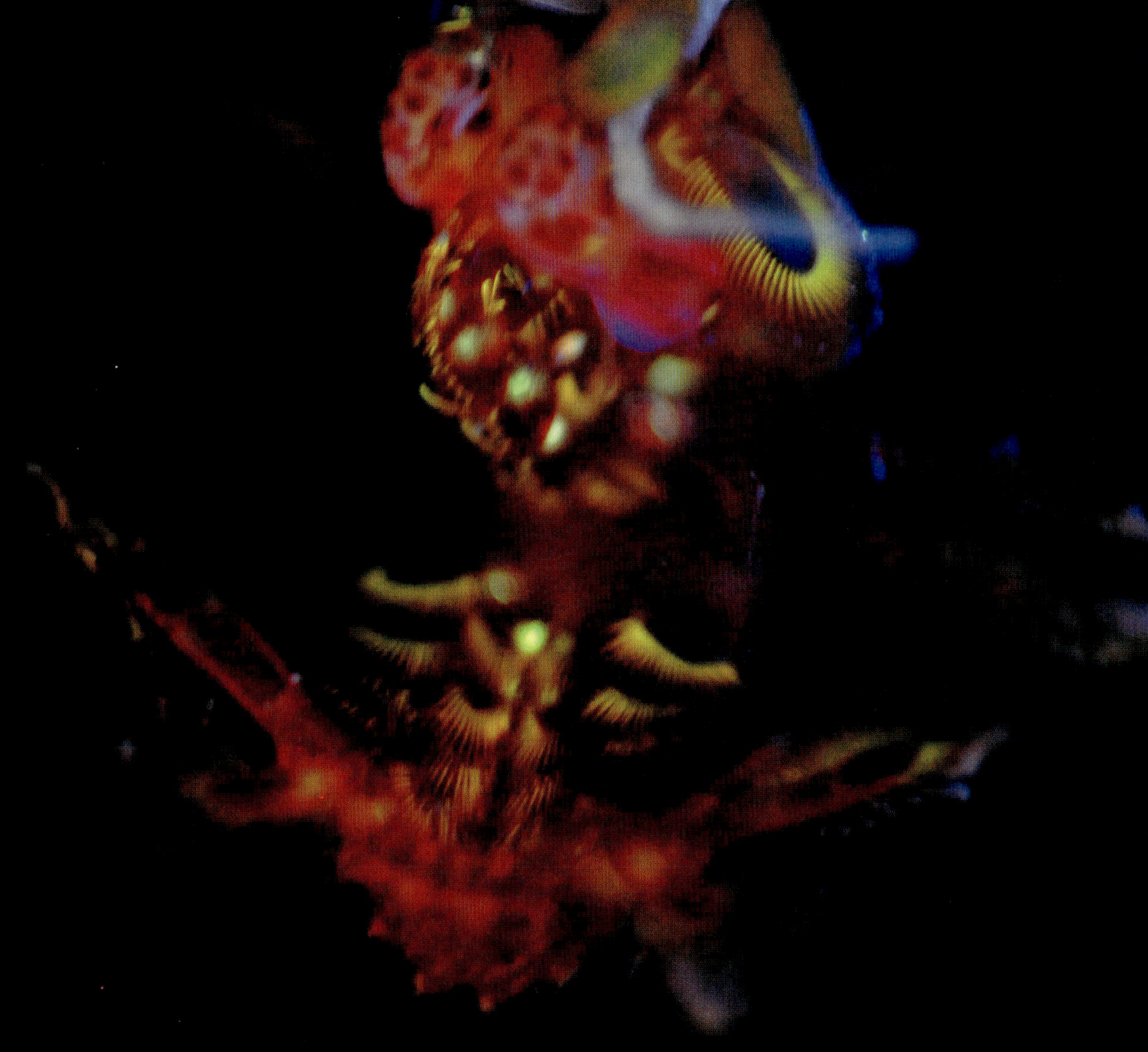

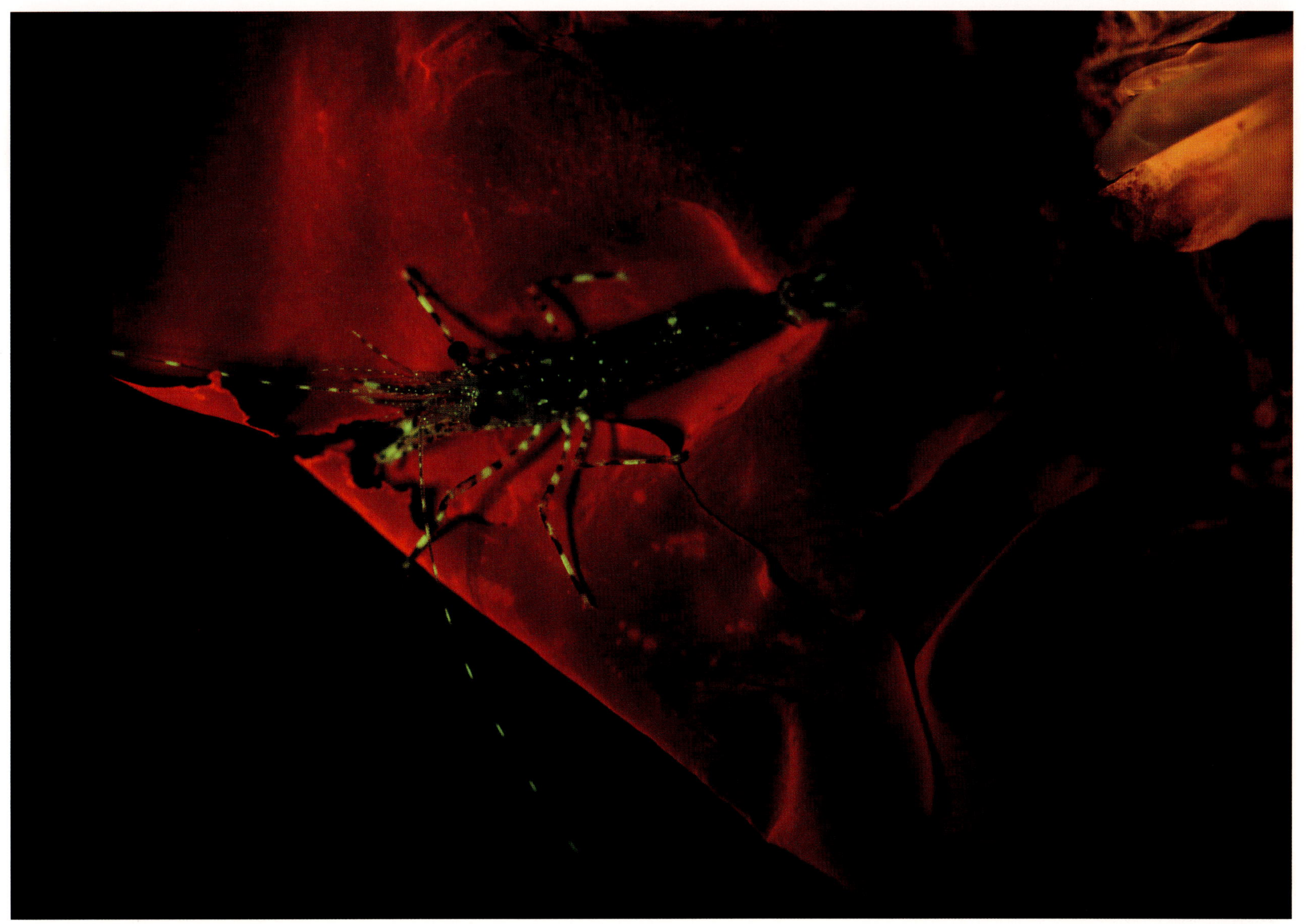

Pandalus danae on algae

As we have emphasized, many biological building blocks are inherently fluorescent, including the exoskeleton of crustaceans. Just because something looks fluorescent when we explore it with our bright blue lights doesn't mean that fluorescence is serving a function. (It also doesn't mean it is not functional—only that you have to demonstrate that it would form a visual signal under natural conditions.) This shrimp perched on a red-fluorescing blade of kelp certainly looks fancy, but we don't have any evidence that its fluorescence serves a role.

Dichotomia

We have seen many examples of algae fluorescing due to their photosynthetic chlorophyll. Algae also form symbioses with many marine animals, providing sugars in exchange for nutrients and a place to call home. The internal canals of this hydromedusa house red-fluorescent algae, which help the jellyfish survive in shallow tropical seas.

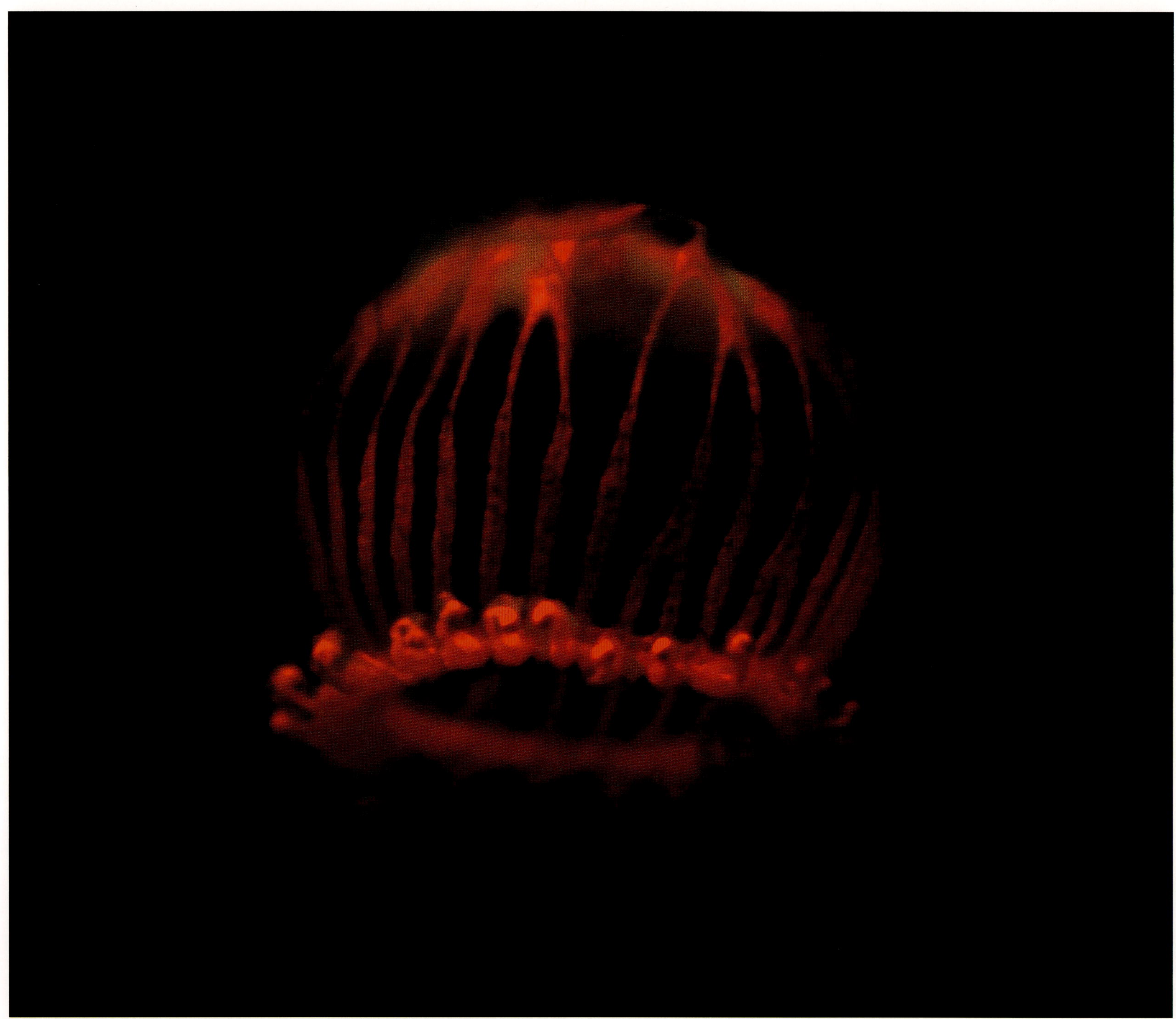

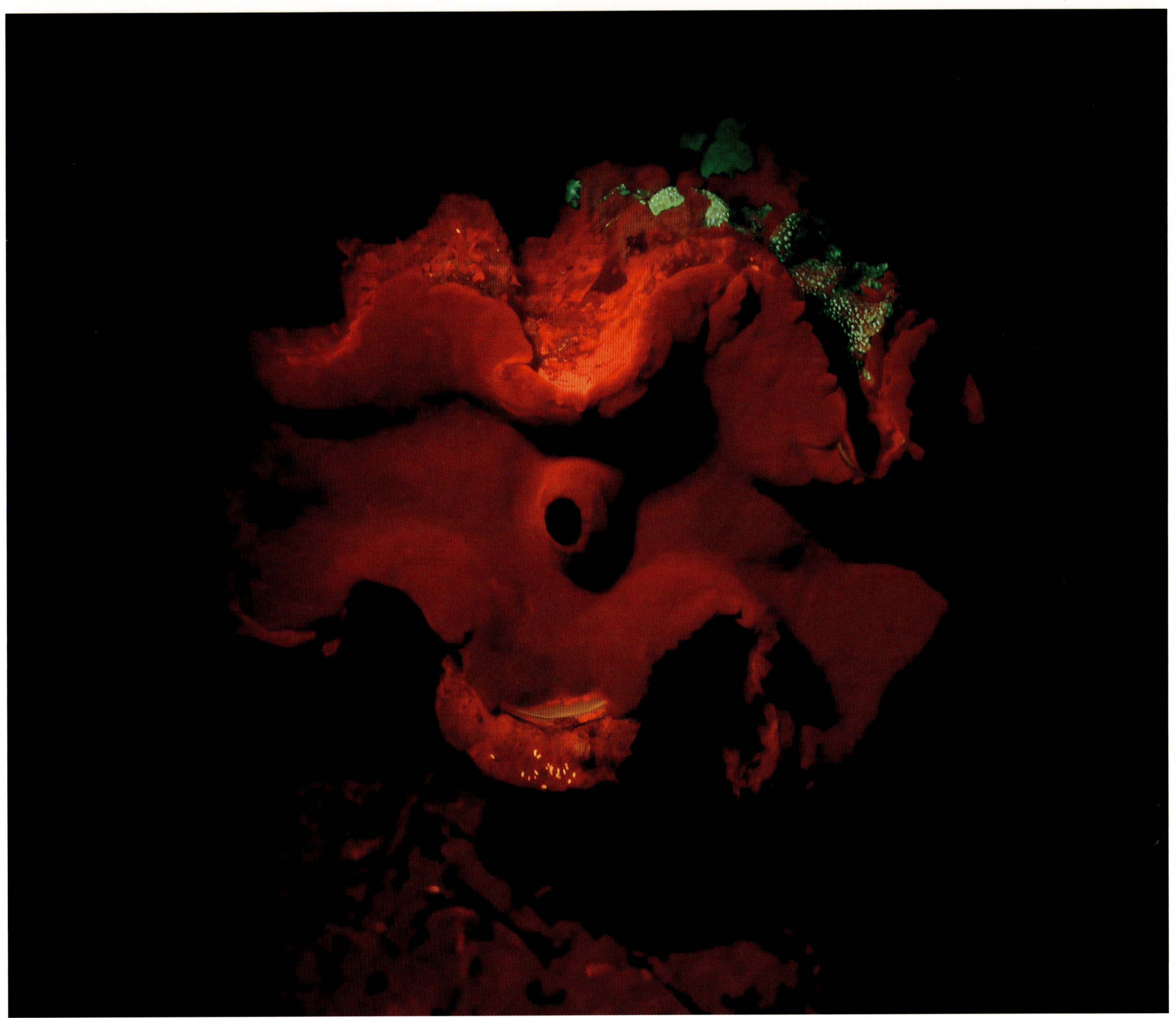

Tridacna

Giant clams may get a bad reputation due to unfounded fears that they will slam closed on a snorkeler's foot and hold them underwater. In fact, they are slow-moving and would rather keep their valves open to expose their algal symbionts to the tropical sunshine. The red sheet of fluorescence is from algae, the green at the top is from a nearby coral, and the orange spots at the bottom are from some curious isopods, which are inexplicably conspicuous orange throughout tropical reefs.

Anthopleura and *Tegula*

Tidepool anemones can also have their own green fluorescence combined with the red fluorescence of their algal symbionts. How much of this is normally visible in their shallow, white-light habitat and the potential functions remain to be explored. It is possible that the fluorescent proteins (FPs) shift the incoming light from unusable wavelengths to colors more suitable for photosynthesis. The turban snail in this scene has built its shell from fluorescent material, but the color is almost certainly related to its chemical composition and not to its function.

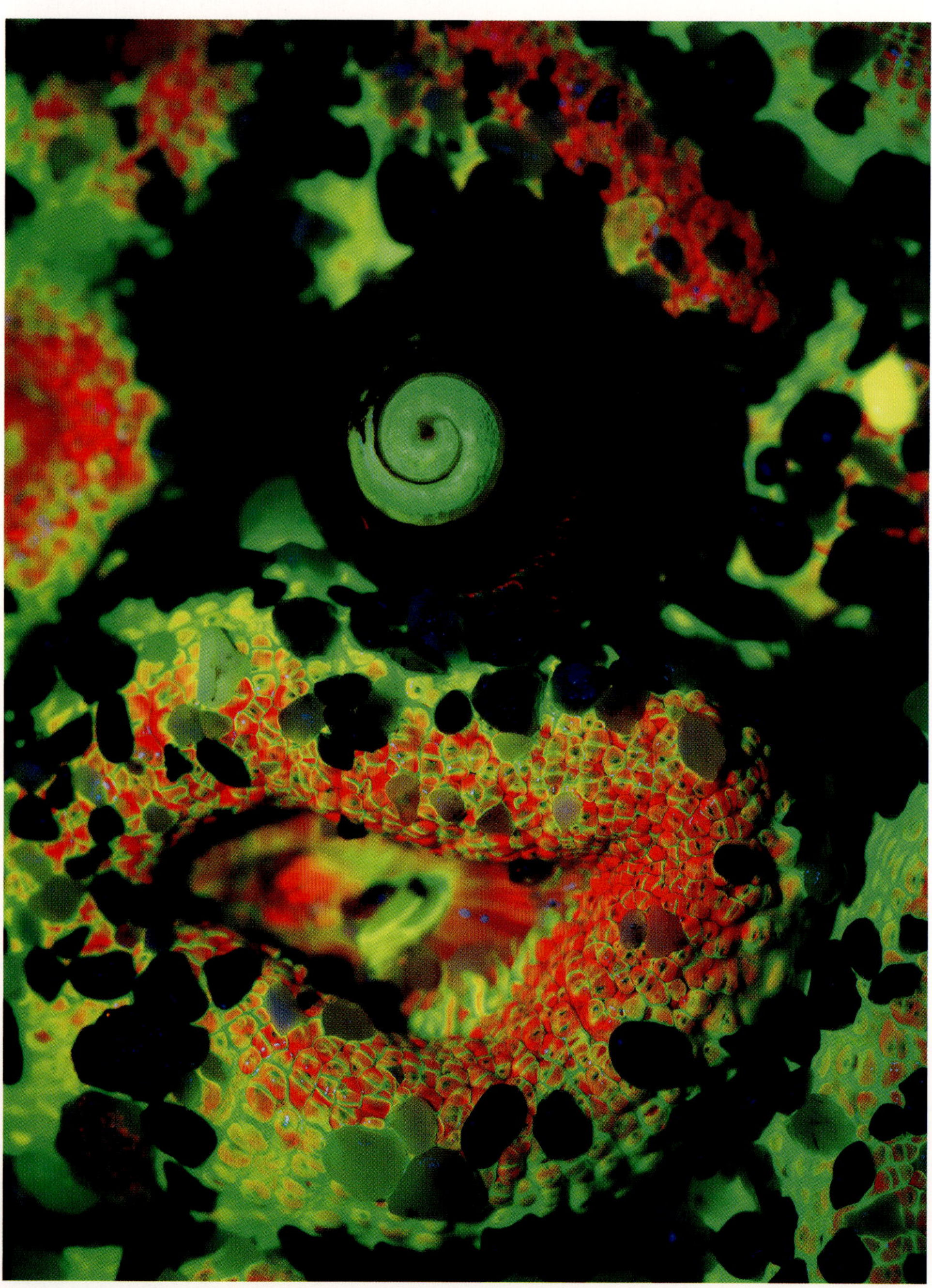

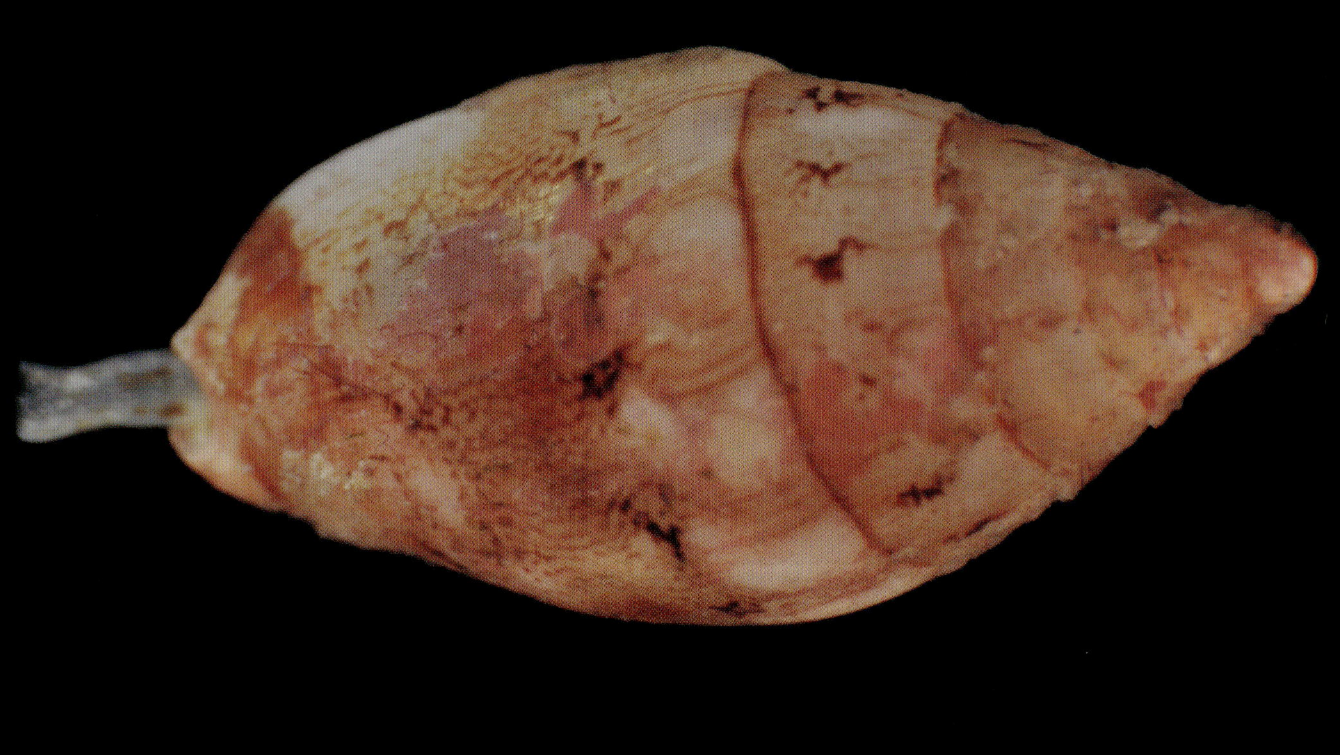

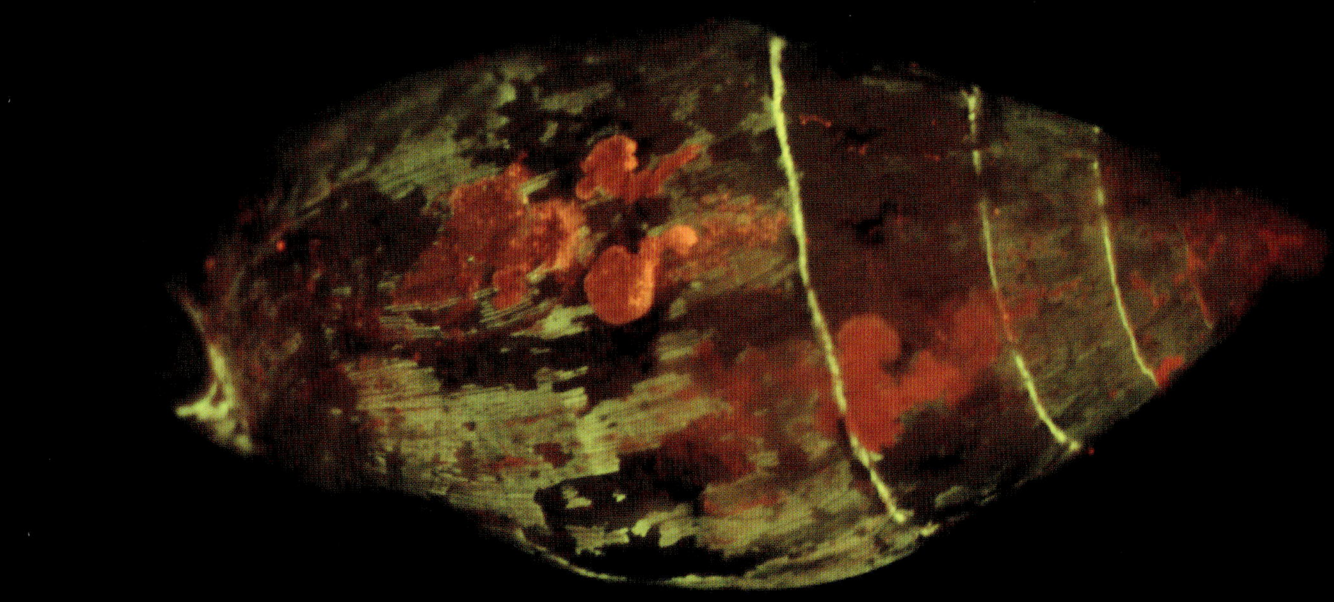

Shallow-living gastropod snail

Crustaceans aren't the only animals that are built from fluorescent materials. Like the tidepool snail, this snail's shell is laced with fluorescence, and it has patches of fluorescent algae growing on its surface.

Pterygioteuthis sucker disks

If you illuminate with even shorter wavelengths, such as violet or ultraviolet light, you can reveal some of the many compounds that fluoresce blue. The suckers of this squid are lined with chitinous rings that exhibit a similar blue color as the fluorescent exoskeletons of terrestrial scorpions.

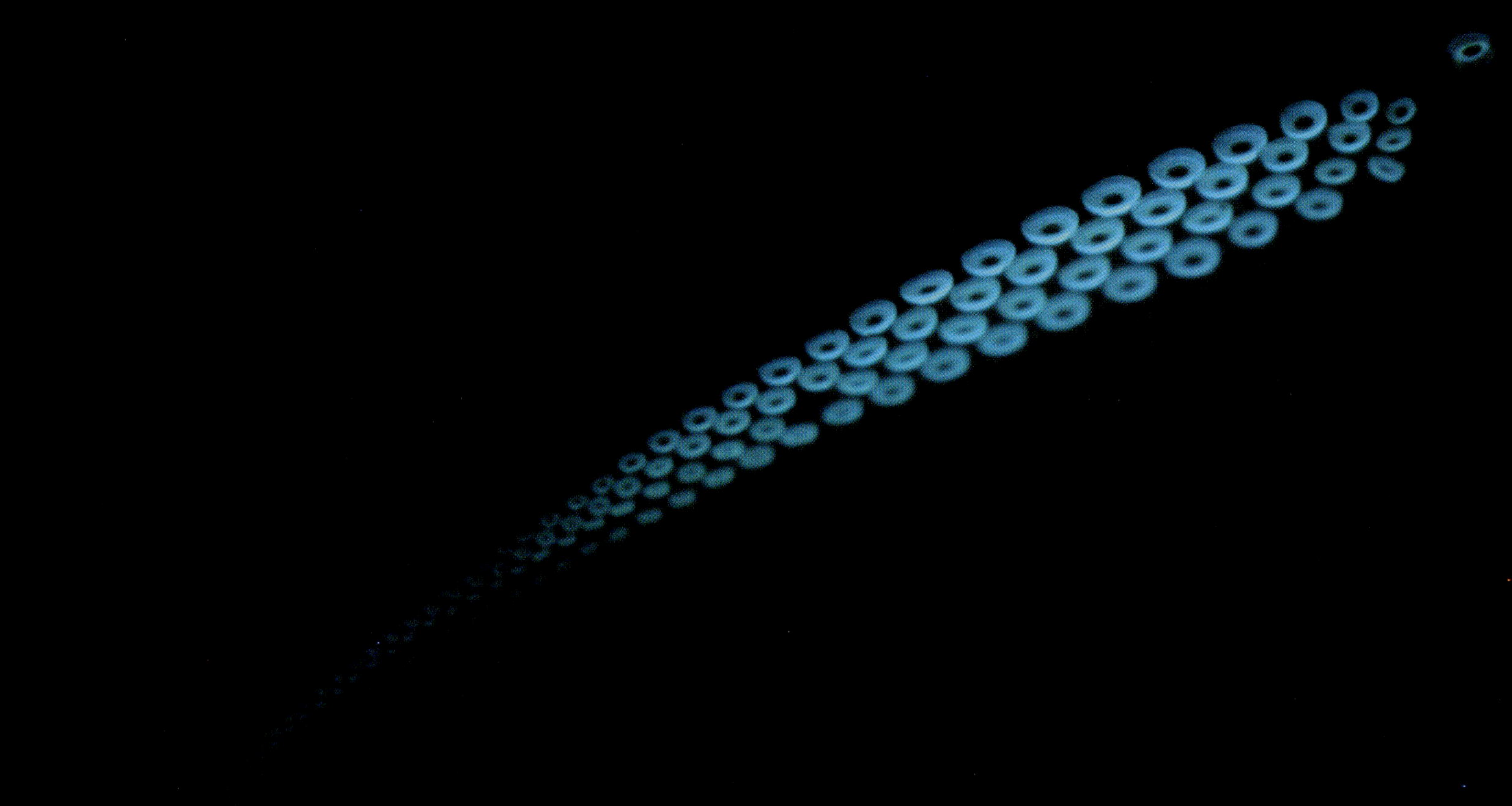

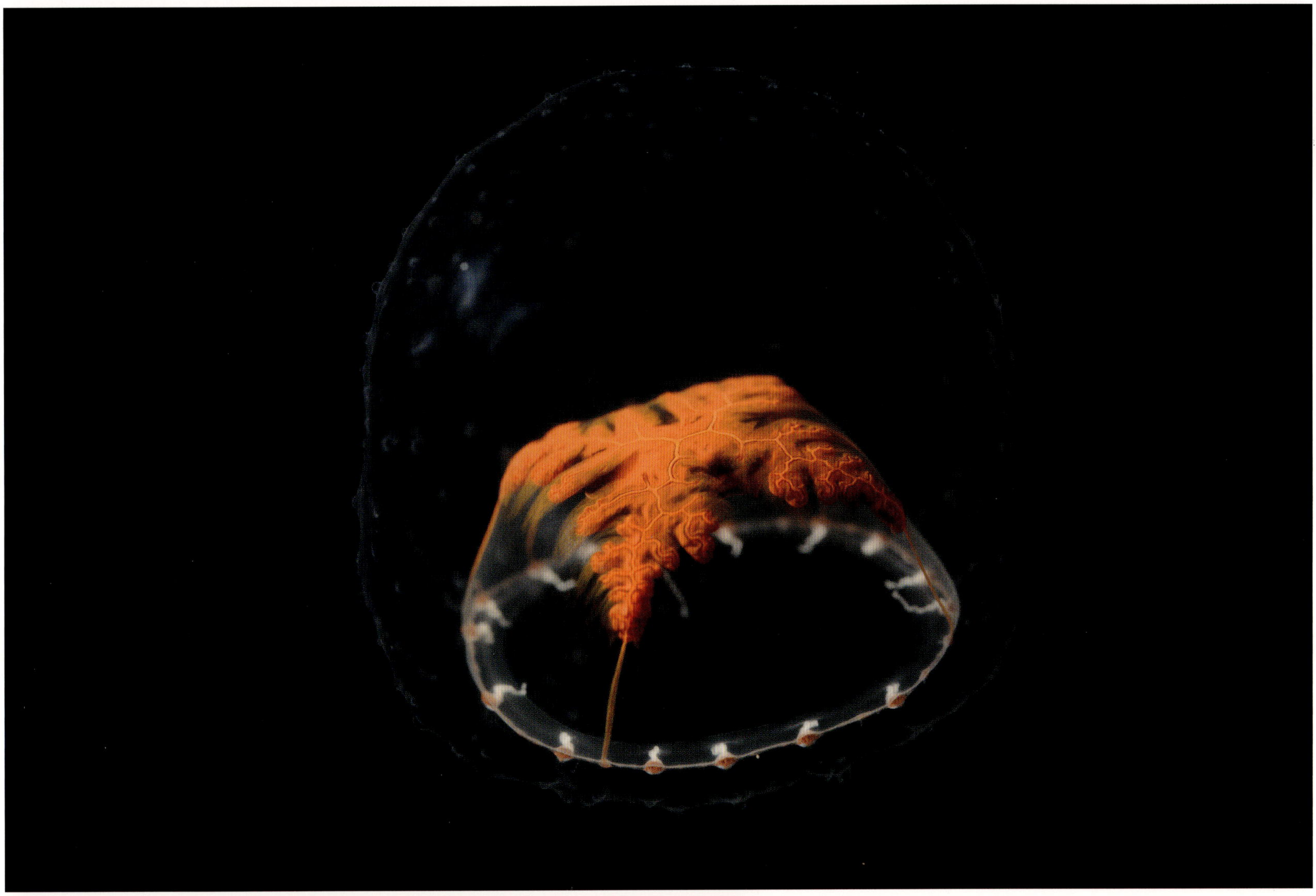

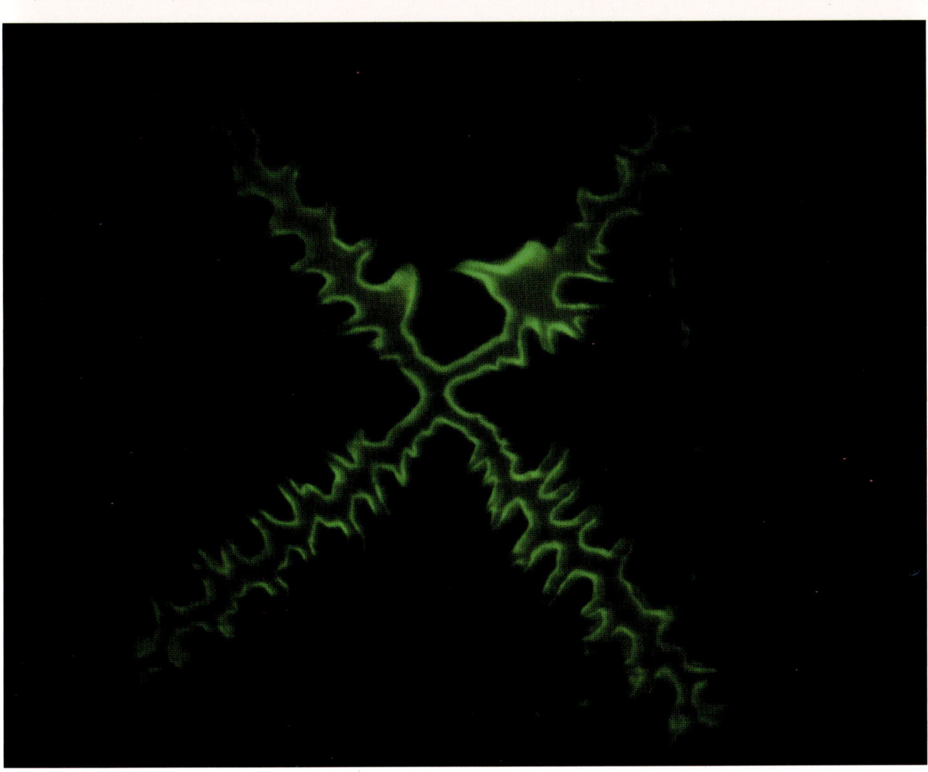

Hydromedusa and gut fluorescence

We have seen some cases of "you are what you eat" in the Pigmentation and Bioluminescence sections, but it can also happen with fluorescence. This jellyfish seems to have eaten a fluorescent prey item. We have even made the embarrassing gaffe of cloning a fluorescent protein from an animal that had actually originated from its prey. In that case, how the DNA survived and how the proteins were incorporated into the predator's tissues remains a mystery.

Tripterygiidae

Many tropical reef fish are surprisingly fluorescent, including this common triplefin blenny. They really stand out, even under natural lighting conditions, which seems like a risky couture for such a diminutive fish. Several functions have been proposed for this fluorescence, but some of these—such as illuminating the visual field with orange light—seem far-fetched, or at least not near-fetched.

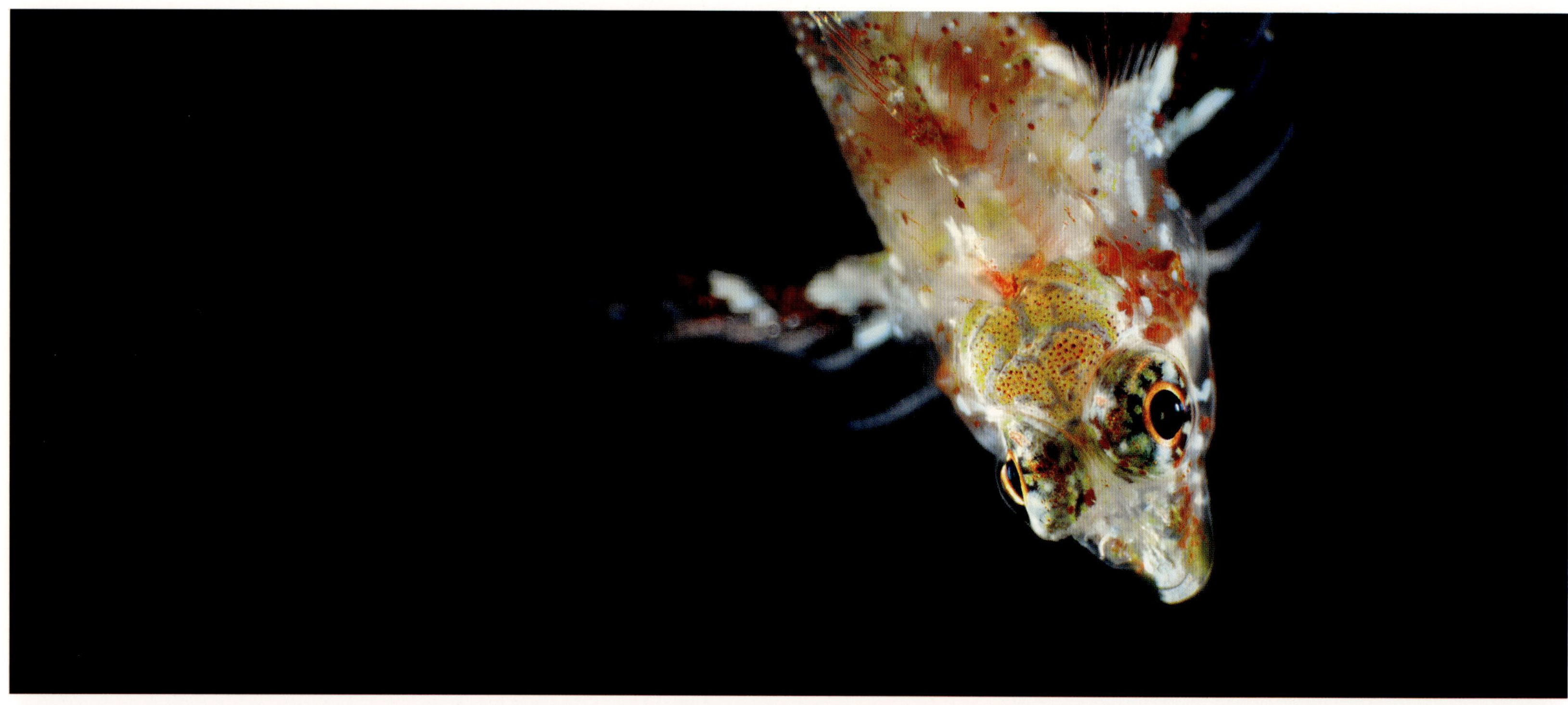

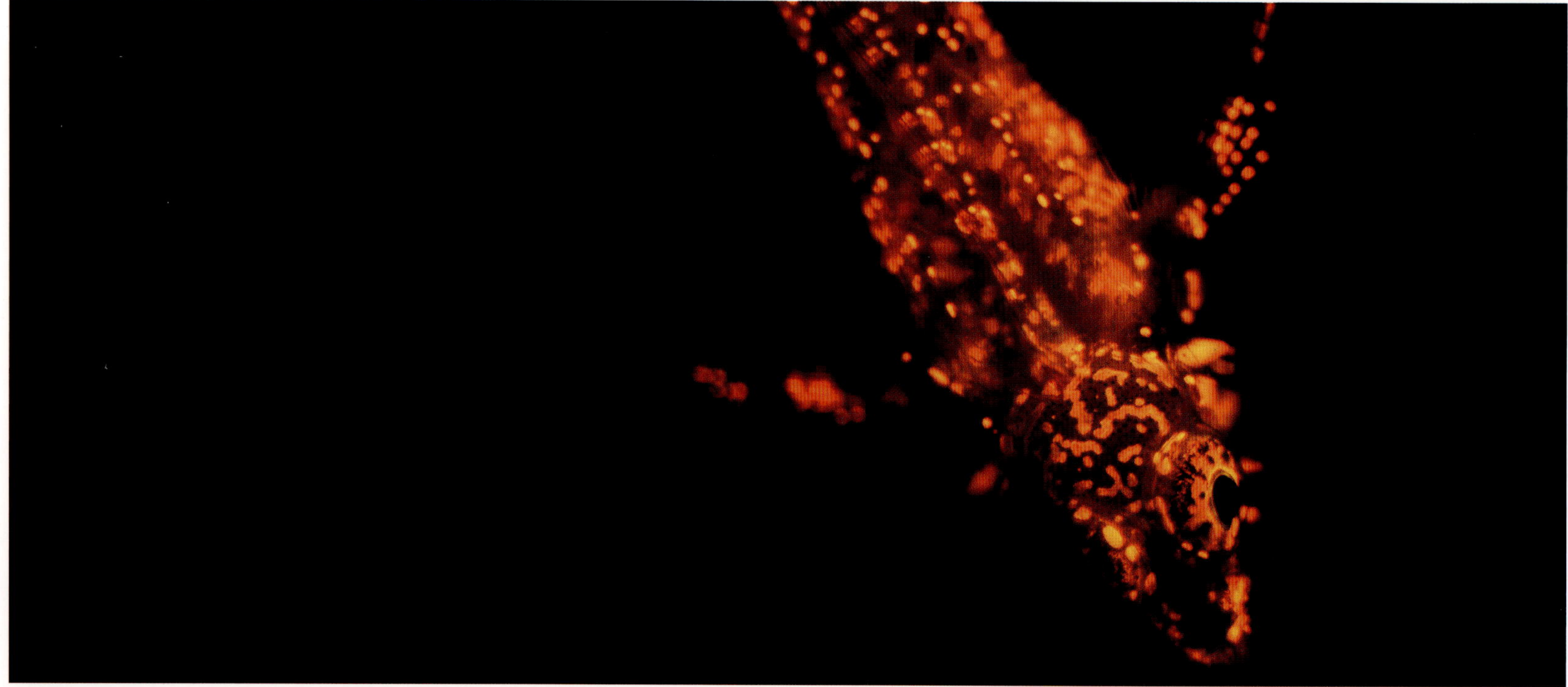

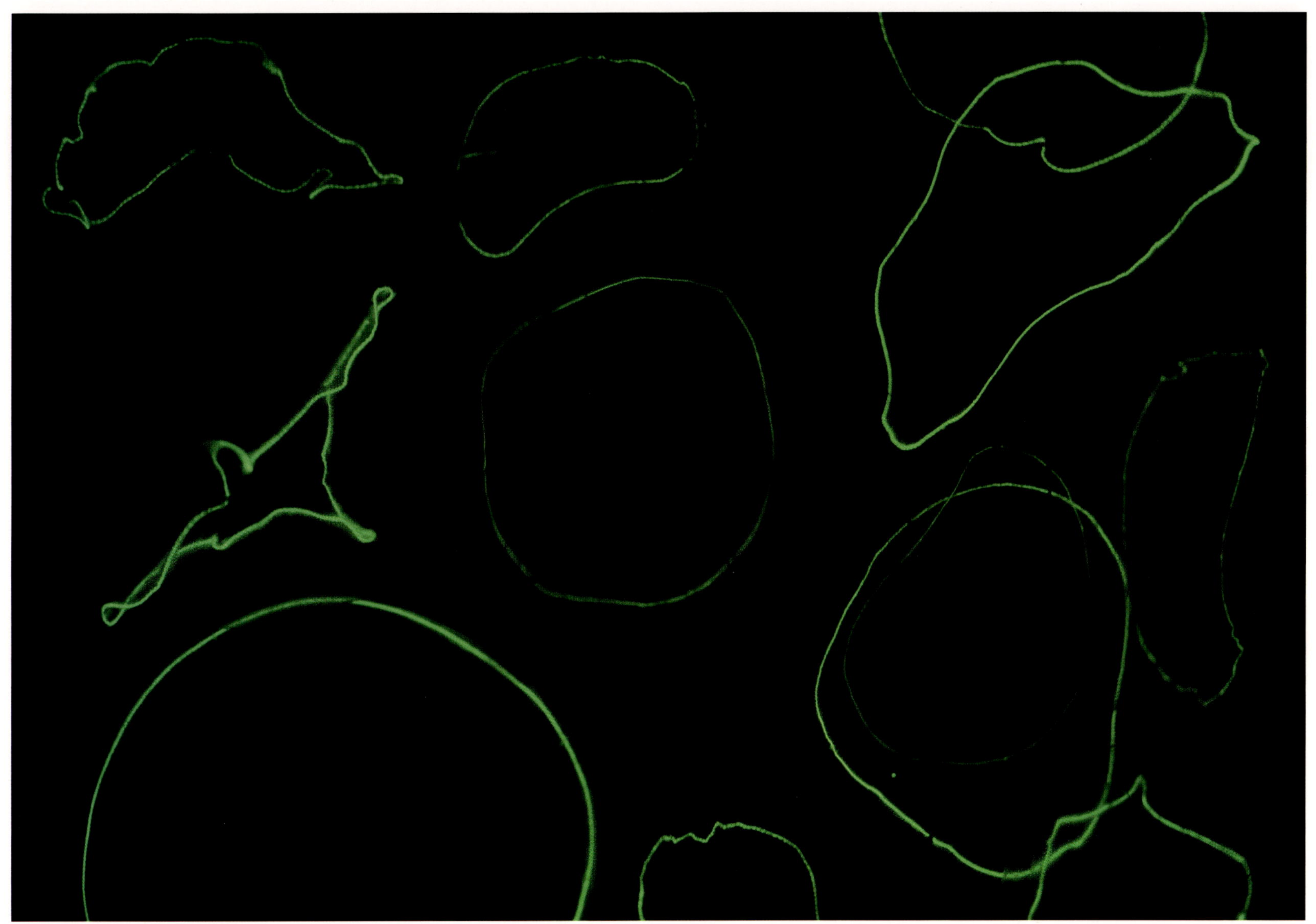

Many *Mitrocoma cellularia* in a tank

One of the best-demonstrated functions of fluorescence is as a partner to the bioluminescence system, converting normally blue luminescence to green emission via GFP. One fascinating aspect of this relationship between the two systems is that the energy transfer happens directly from one molecule to another; they are so closely associated that a blue photon does not need to be emitted and reabsorbed.

Most GFP-related bioluminescence is found in shallow-water or coastal species, like this tank full of cross-jellies. This makes sense considering that these habitats have more ambient green light, as you saw in the Introduction.

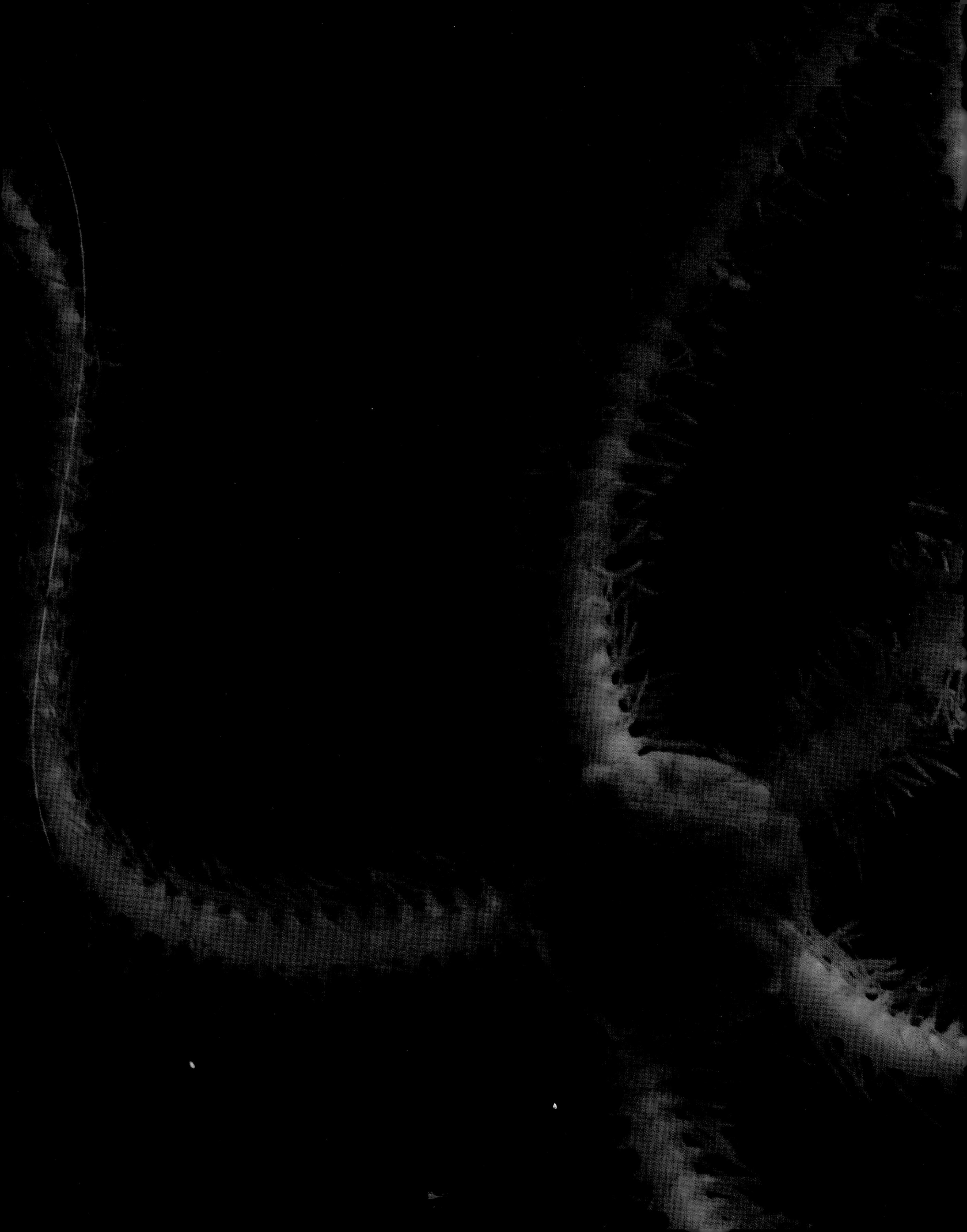

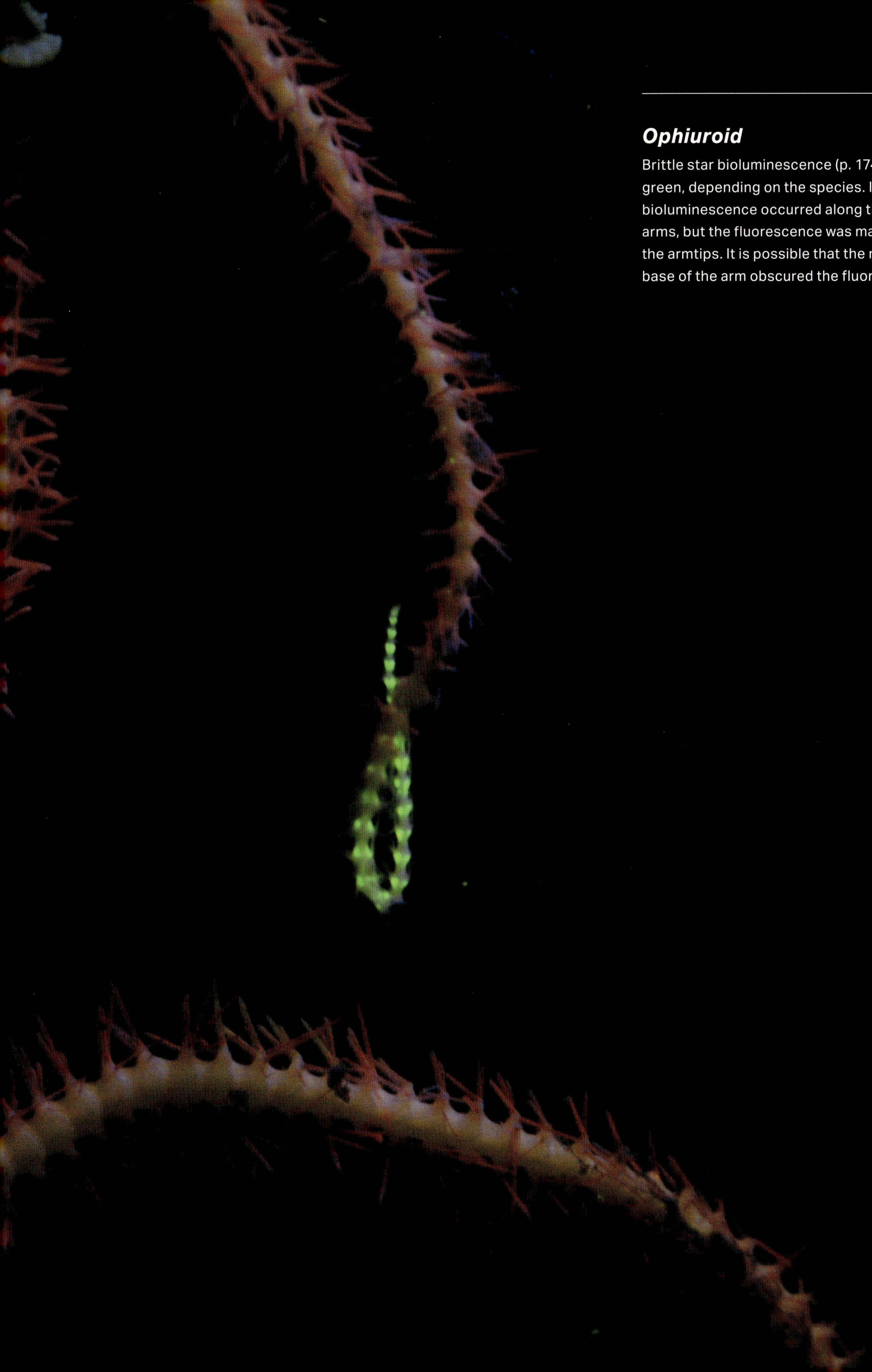

Ophiuroid

Brittle star bioluminescence (p. 174) can be blue or green, depending on the species. In this specimen, bioluminescence occurred along the length of the arms, but the fluorescence was mainly visible only at the armtips. It is possible that the more heavily calcified base of the arm obscured the fluorescence.

Umbellula

These sea pens anchor a bulbous "foot" into the seafloor and then have their polyps suspended atop a long stalk. Like many bioluminescent octocorals, they have green fluorescence associated with their light emission. Some individuals apparently have yellow variants of the fluorescence genes.

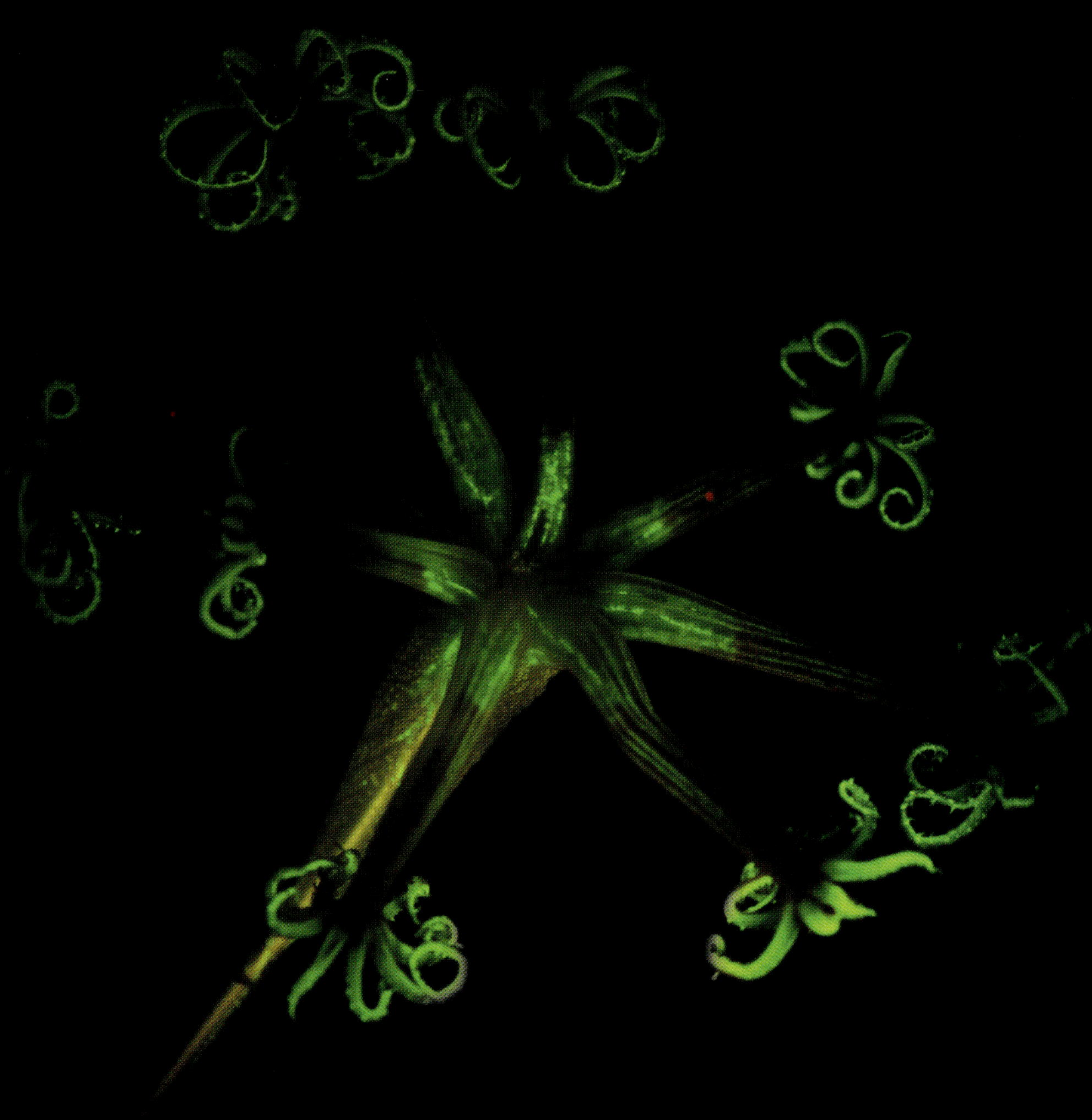

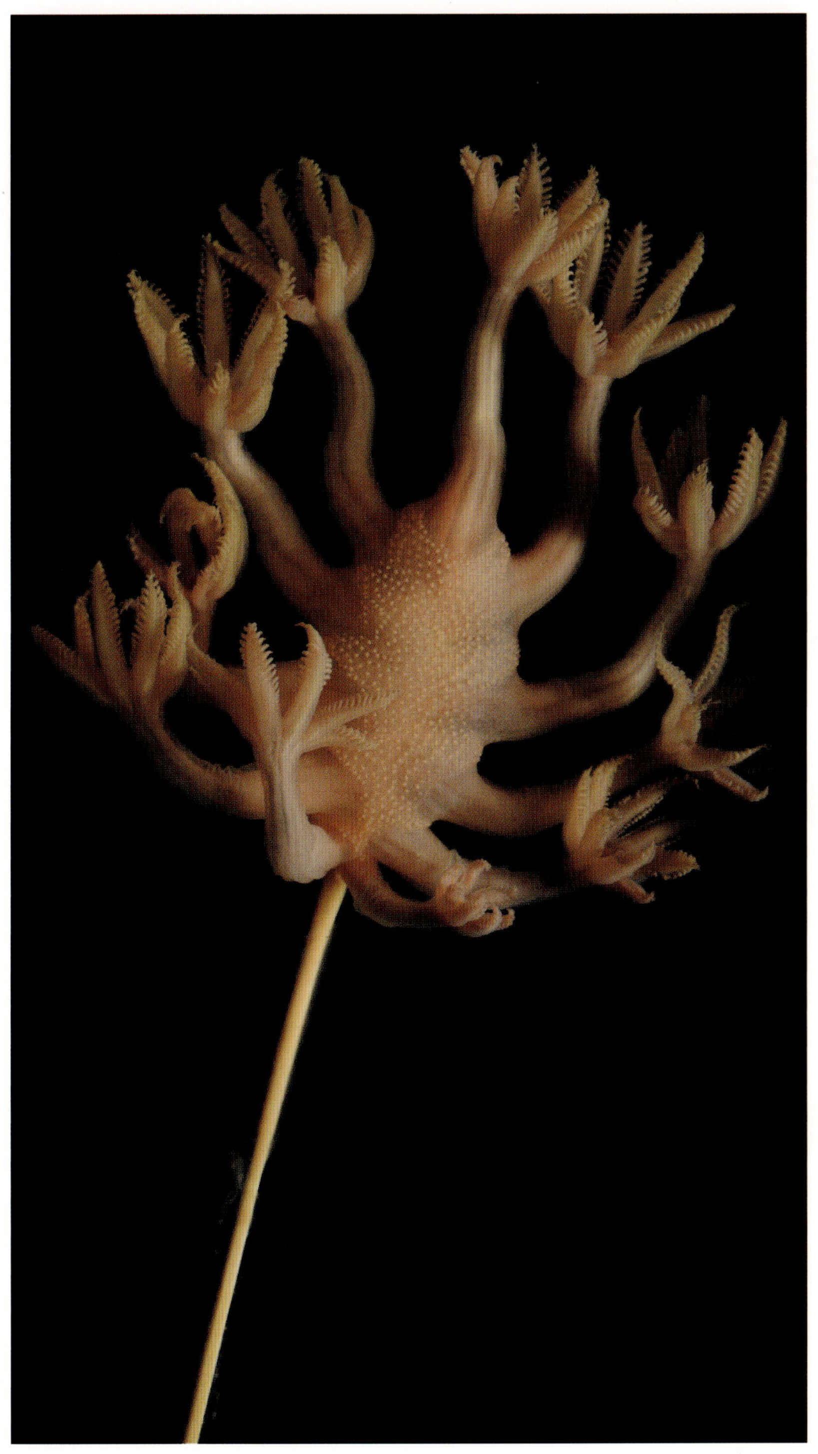

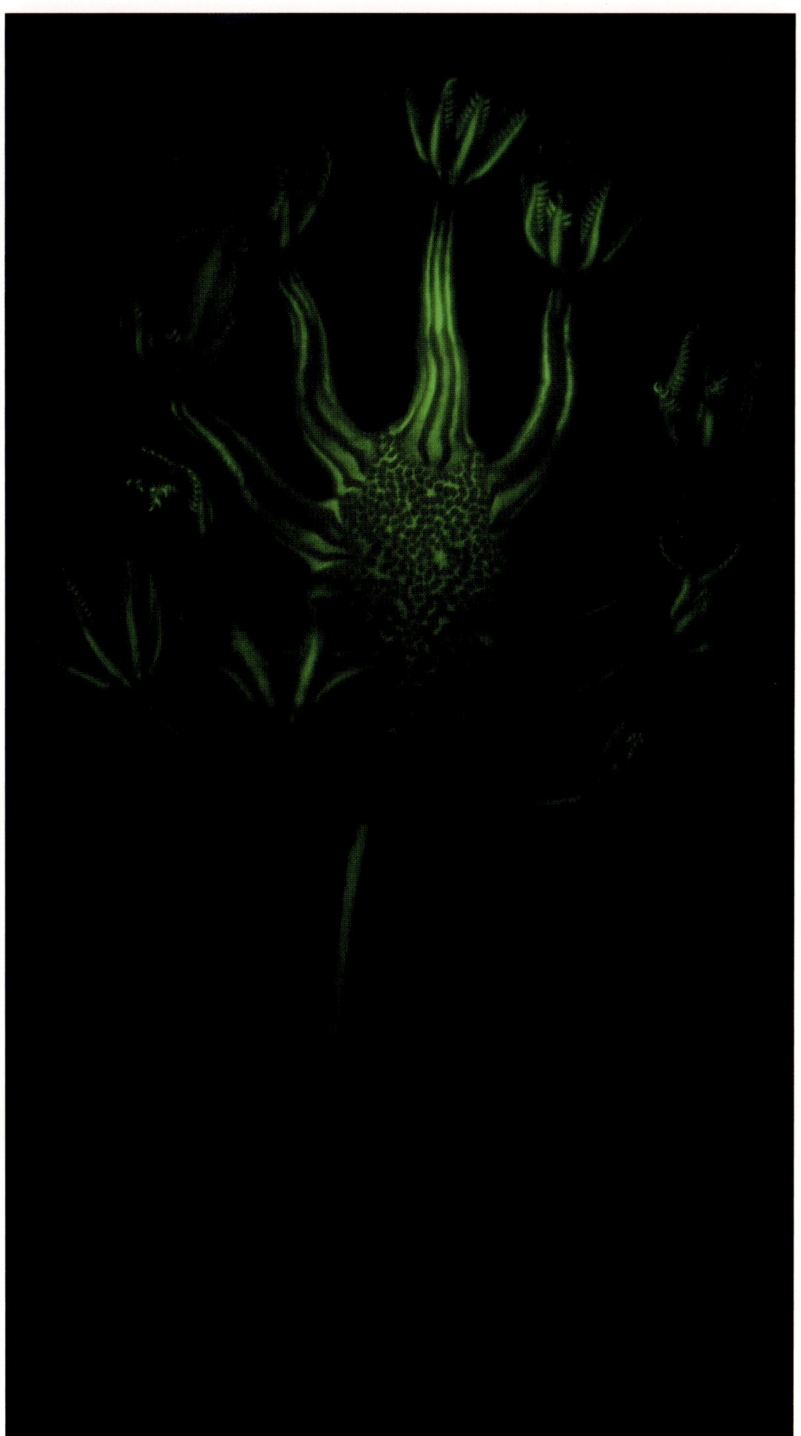

Umbellula species 2

Different anatomical features are highlighted when you look at a sea pen under white light, under blue light (as is shown here), or from the bioluminescence alone. While some *Umbellula* are common at around 800 meters depth, this species was observed 3,100 meters below the surface, where the only light is that generated by the animals themselves.

Sternoptyx

A curious feature of many distantly related deep-sea fishes is that their photophores are often capped with a red-fluorescent filter, as in the belly lights of this hatchetfish. The effect of this red fluorescence on the bioluminescent emission remains unknown, but it gives a convenient way to visualize the light organs when bioluminescence cannot be recorded.

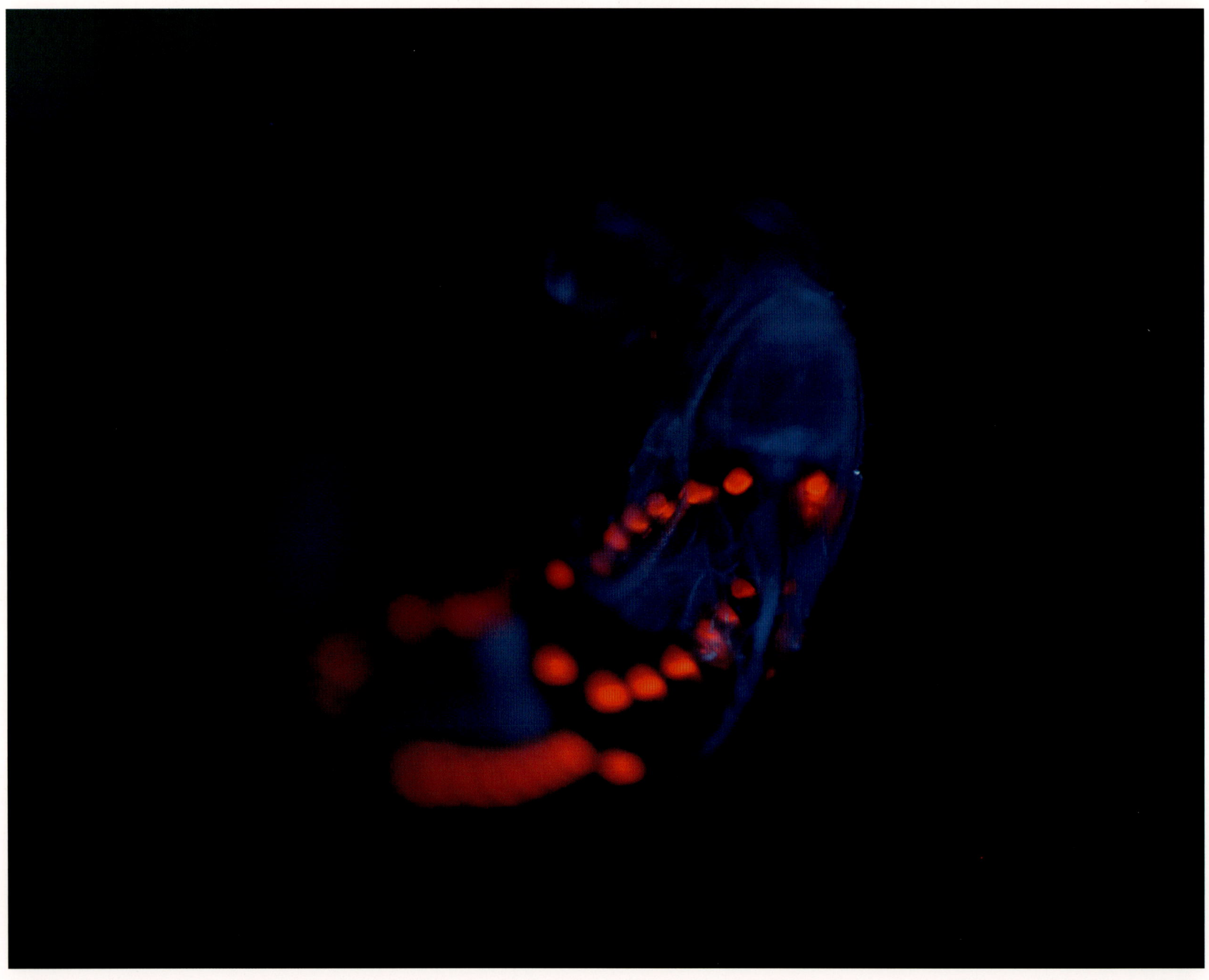

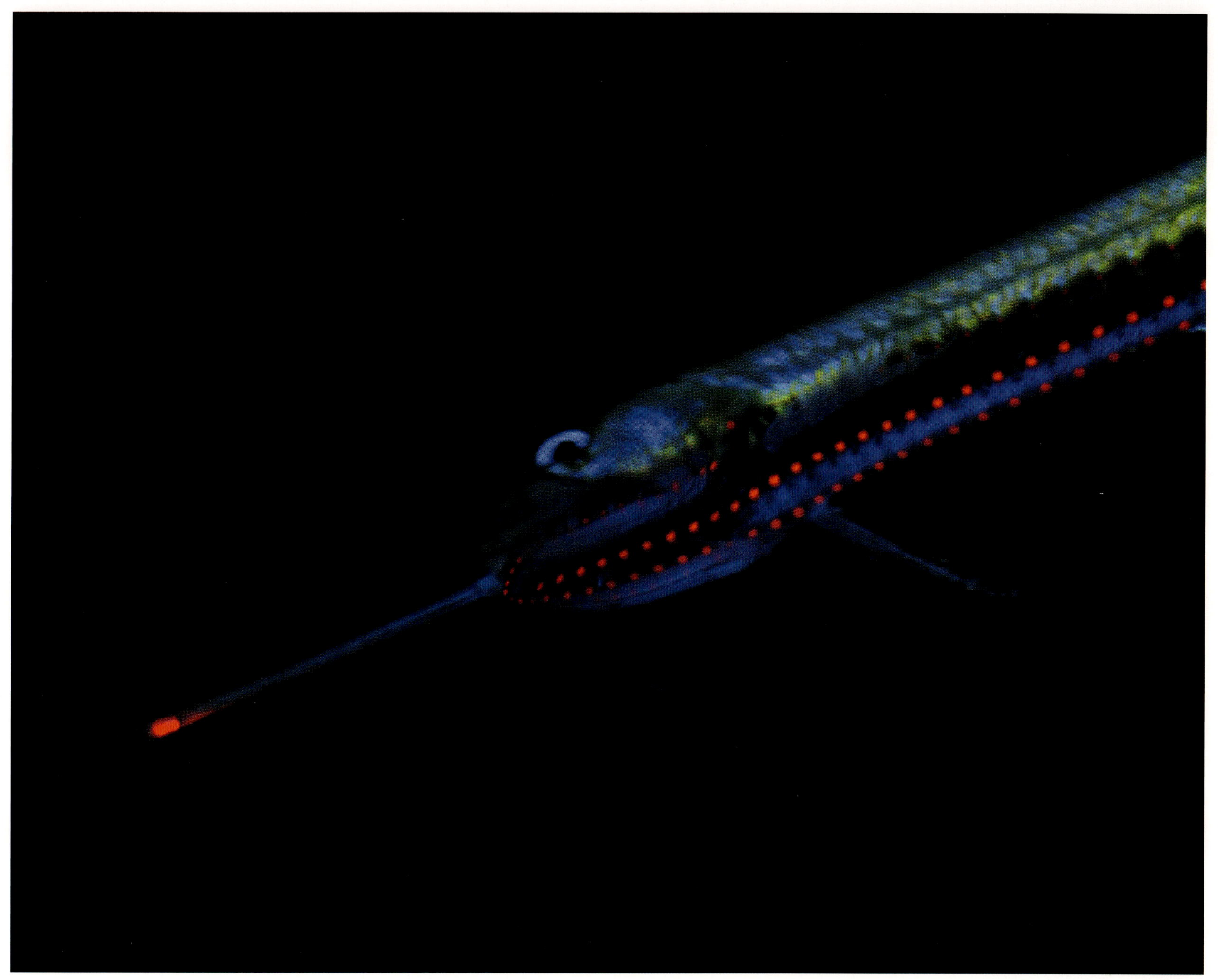

Stomias

Anglerfishes and siphonophores are not the only deep-sea inhabitants to dangle glowing lures. Many dragonfish have luminous barbels, usually at the end of a stiff stalk, held out in front like a magic wand. Again, the photophores have a red fluorescence of unknown function. This species is brightly luminous and certainly uses its light both for defense (the rows of lights along the lower surface) and offense (the prey-luring barbel).

Bathophilus

This formidable fish has been featured in both the Transparency and Iridescence sections, and could have been mentioned in Bioluminescence as well. Here we can admire the fluorescence of its many kinds of photophores. Small diffuse spots cover the body surface, and a green photophore under the eye is used as a headlamp and emergency beacon. (We know this because they flash these lights when wiggling to escape.) What is potentially novel is that this species also has a red light organ under its eye, likely serving the same night-vision function that is found in the dragonfish *Aristostomias* (p. 119).

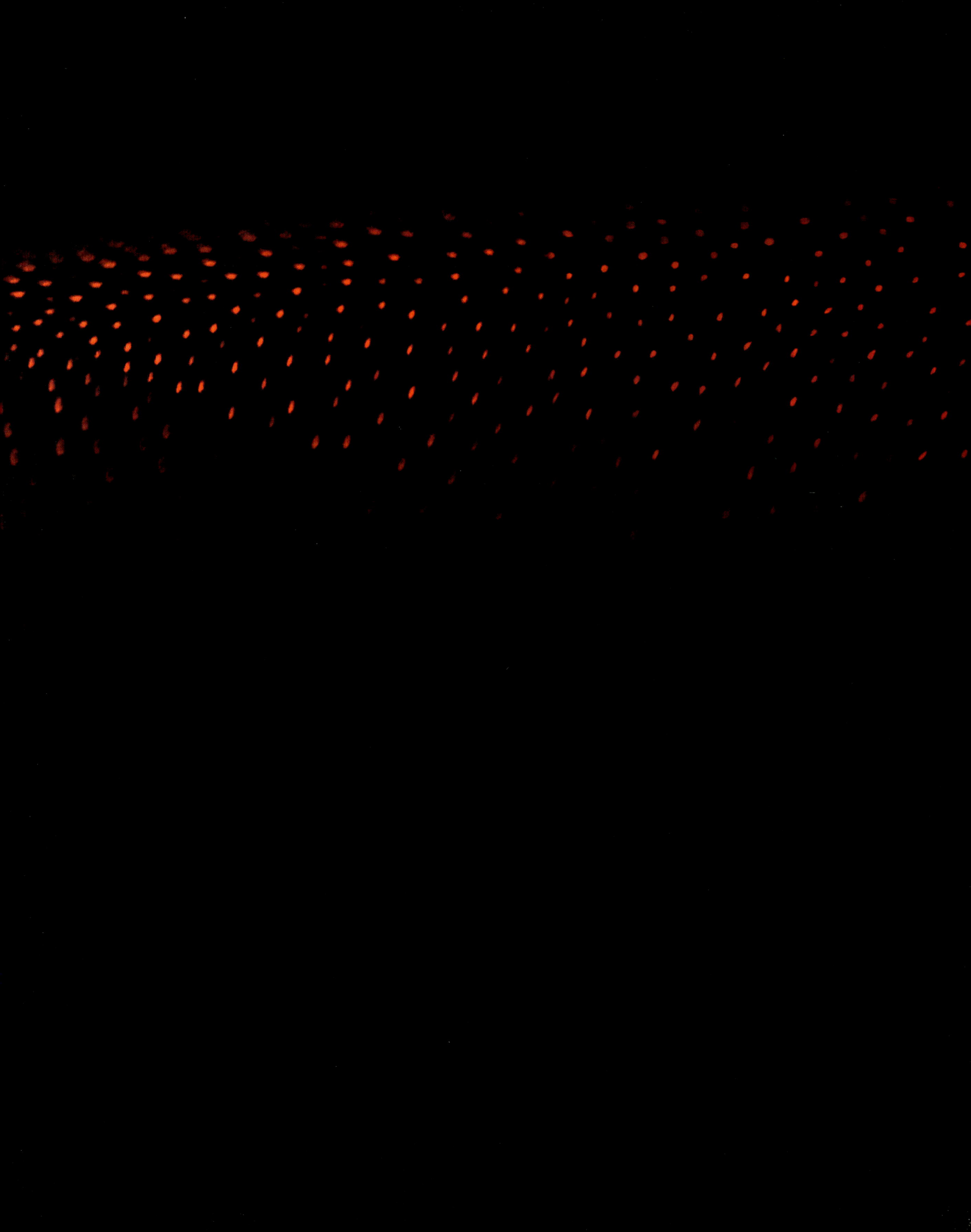

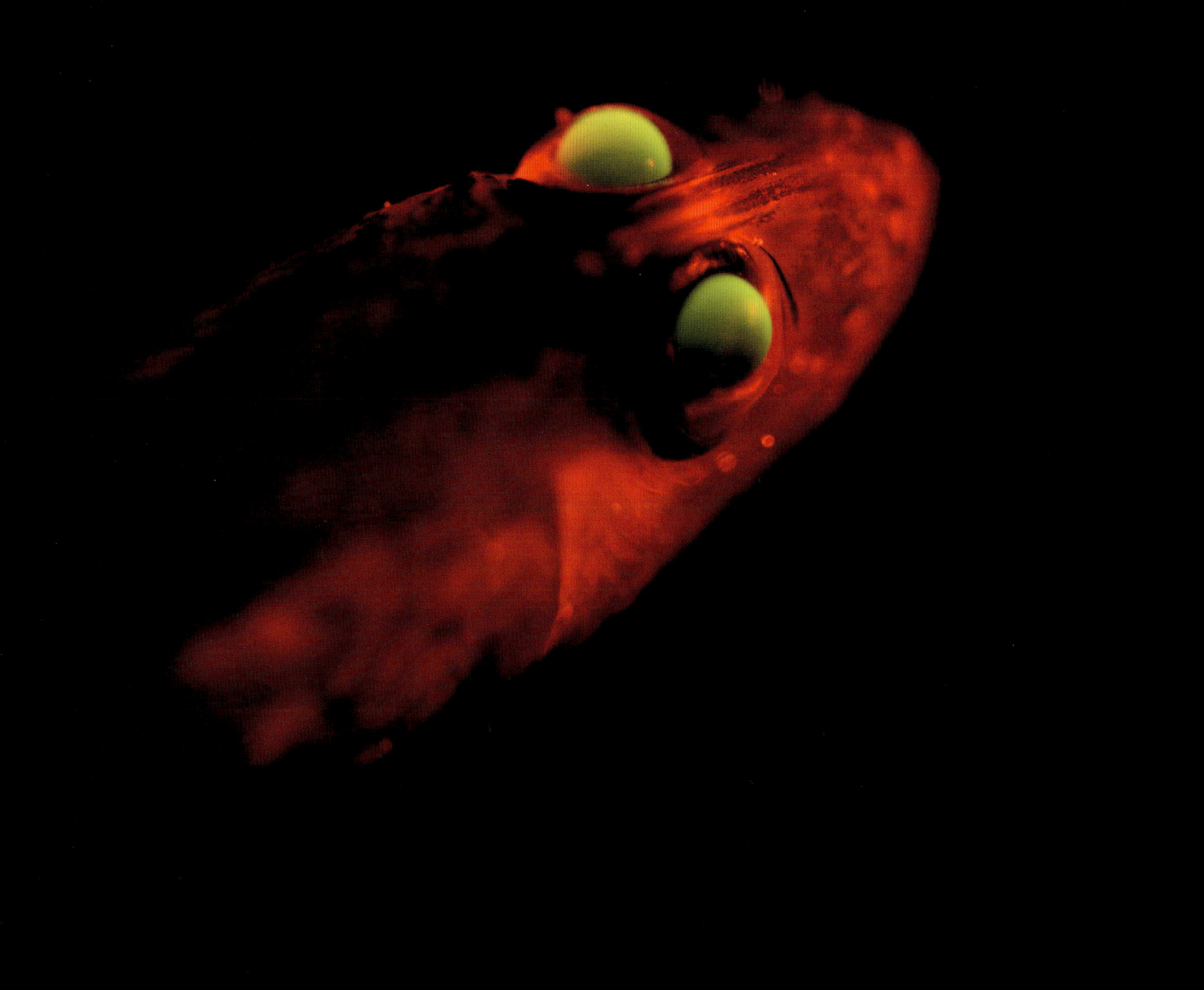

Scopelarchus

The pearleye fish is another upward-looking predator that seems to focus its attention only on the waters above its head. Like the strawberry squid (p. 214), it has fluorescent pigments in its eyes, lending credence to the idea that this is used to break luminescent camouflage. It is possible, or even likely, that the fluorescence is a by-product of the pigmentation and not part of its function. Fluorescent light is typically emitted isotropically, meaning that a photon is equally likely to be emitted in any direction, regardless of the angle that the exciting photon came from—a fluorescent lens will not function based solely on its fluorescence.

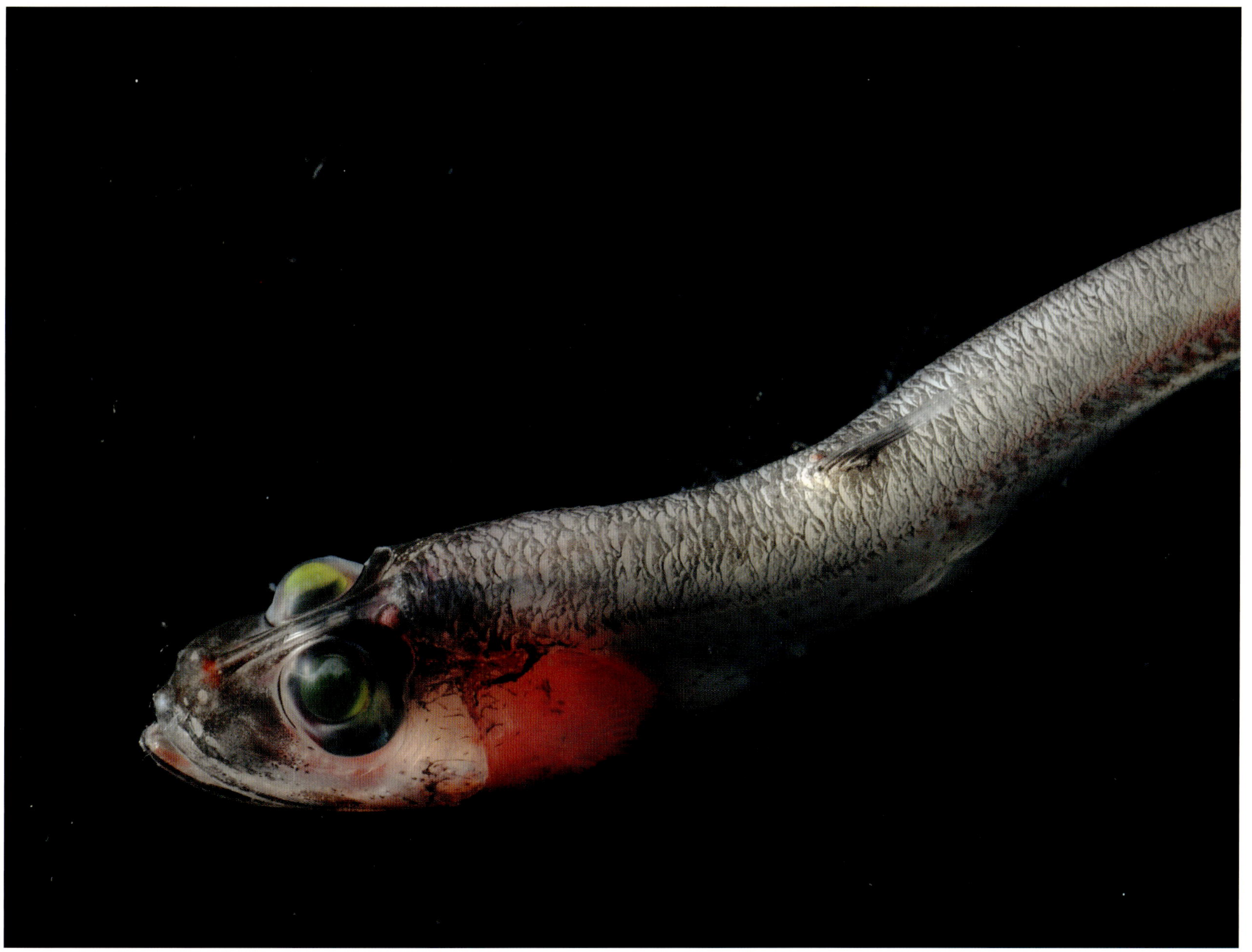

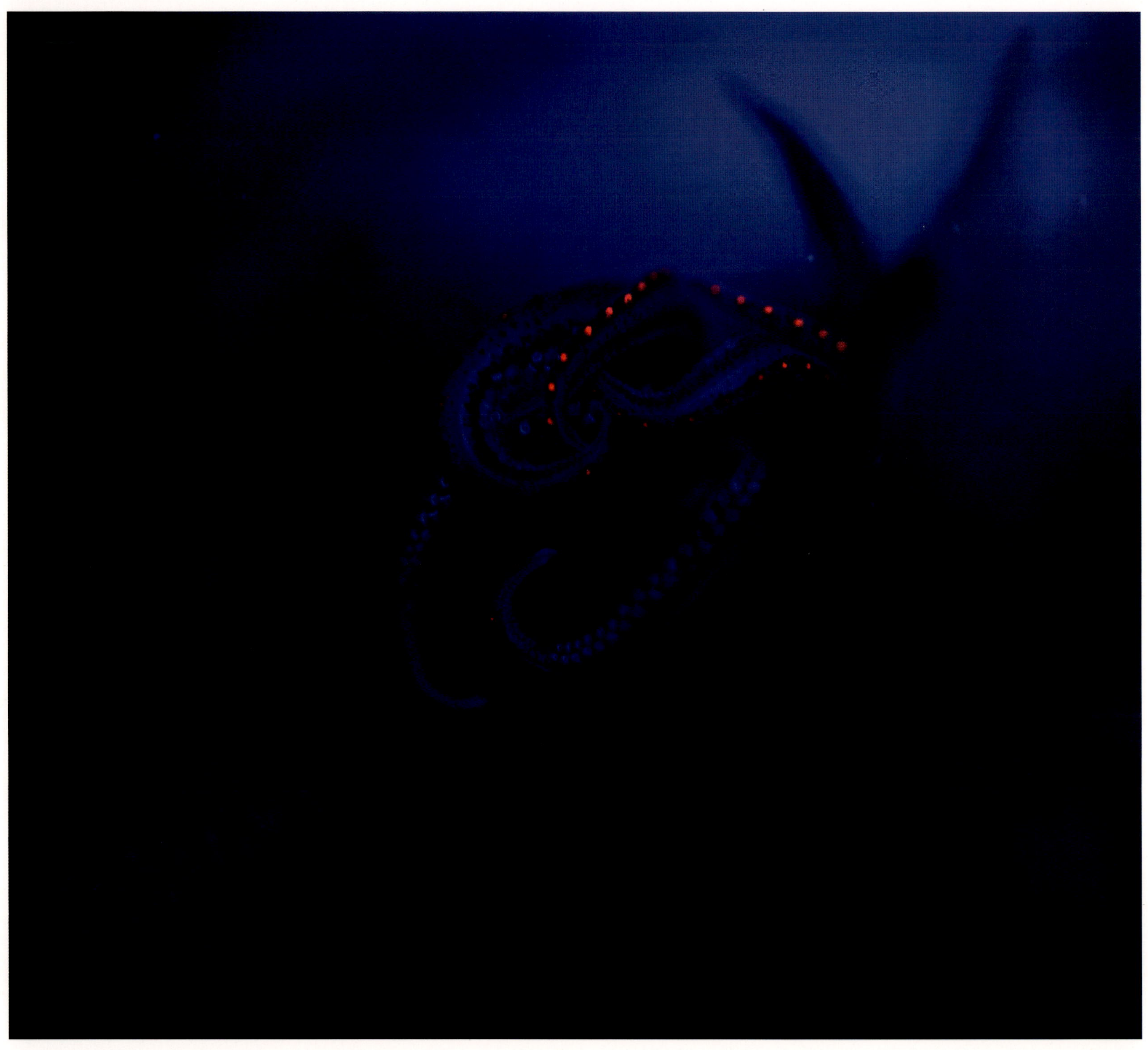

Histioteuthis heterops

Fishes are not the only ones with red-fluorescent filters on their light organs. This strawberry squid, looming toward us out of the darkness, has light organs distributed over its body, which are coated with red-fluorescent pigments. As far as we can tell, this squid primarily emits blue light, but the red pigment may modify the ultimate properties of the transmitted light.

Histioteuthis eye

One of the most fun-to-observe fluorescent phenomena is the enlarged fluorescent eye of *Histioteuthis*. Like many animals, these squid do not typically look forward, but they use one giant eye to look up toward the surface, hoping to spot the silhouettes of potential prey. Since prey can use bioluminescence to counterilluminate their shadows, there is an escalating evolutionary competition to overcome this camouflage. The thought is that the fluorescence of their large lens is indicative of a blue-absorbing pigment. This lens—like the yellow filter that we use for photography—may filter out the ambient light differently from bioluminescence emissions. If there is any mismatch between the color of the prey's photophores and the downwelling light, the squid could see the prey lit up against the darker background. Of course, this hypothesis is easier to demonstrate on the back of an envelope than it is to observe in action.

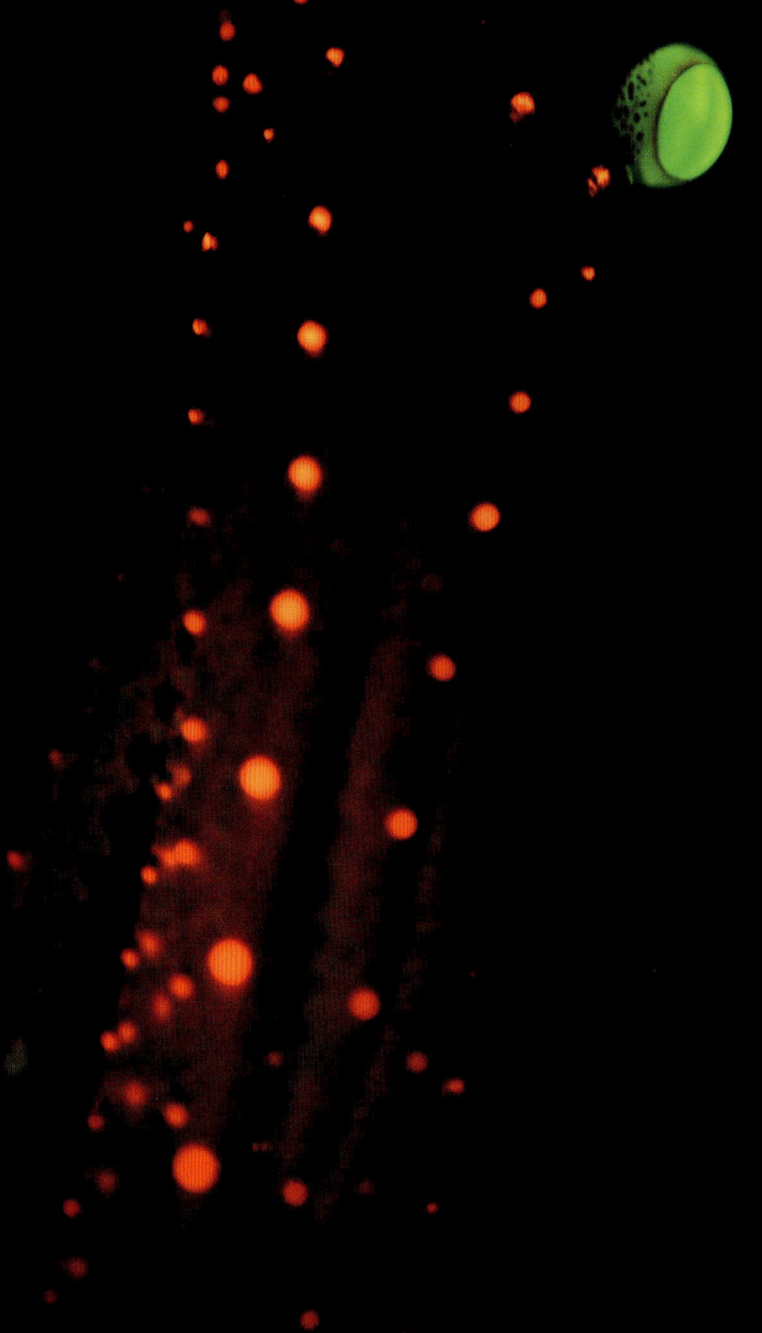

AFTERWORD

In this book, we have focused (literally) on the visual world because we find optical phenomena wondrous, and because together we have spent more than fifty years studying how animals interact with light. However, organisms also employ many other senses, some that humans share and some we can hardly imagine: chemicals, vibration, pressure, magnetic and electrical fields, and sound all provide cues that animals may use and respond to. The sea is not only radiant with colors and patterns, but also with the songs of whales, the smells of shrimp pheromones, and the electric perceptions of sharks and rays. All of these may be used for the same functions that we have seen for visual displays, and some may even be perceived as vision inside the minds of various animals. In addition, the seaweeds, algae, protists, and even the bacteria are communicating with each other in numerous ways, often via chemicals that are either smelled or tasted. Nature is, perhaps more than anything else, a world of communication, and we are aware of only the tiniest part of the conversations that are occurring all around us. Whether scuba diving on a reef, strolling through a park, or even sitting on your couch reading this book, you are surrounded by voices saying —in the languages of sound, light, smell, and so much more:

> I am here
> I want to find you
> I am someone else
> I think I'm in love

So, we hope that by showing you how organisms speak with light, we have also helped open your ears, noses, and minds to the countless stories unfolding every moment in our precious world.

—Steve and Sönke

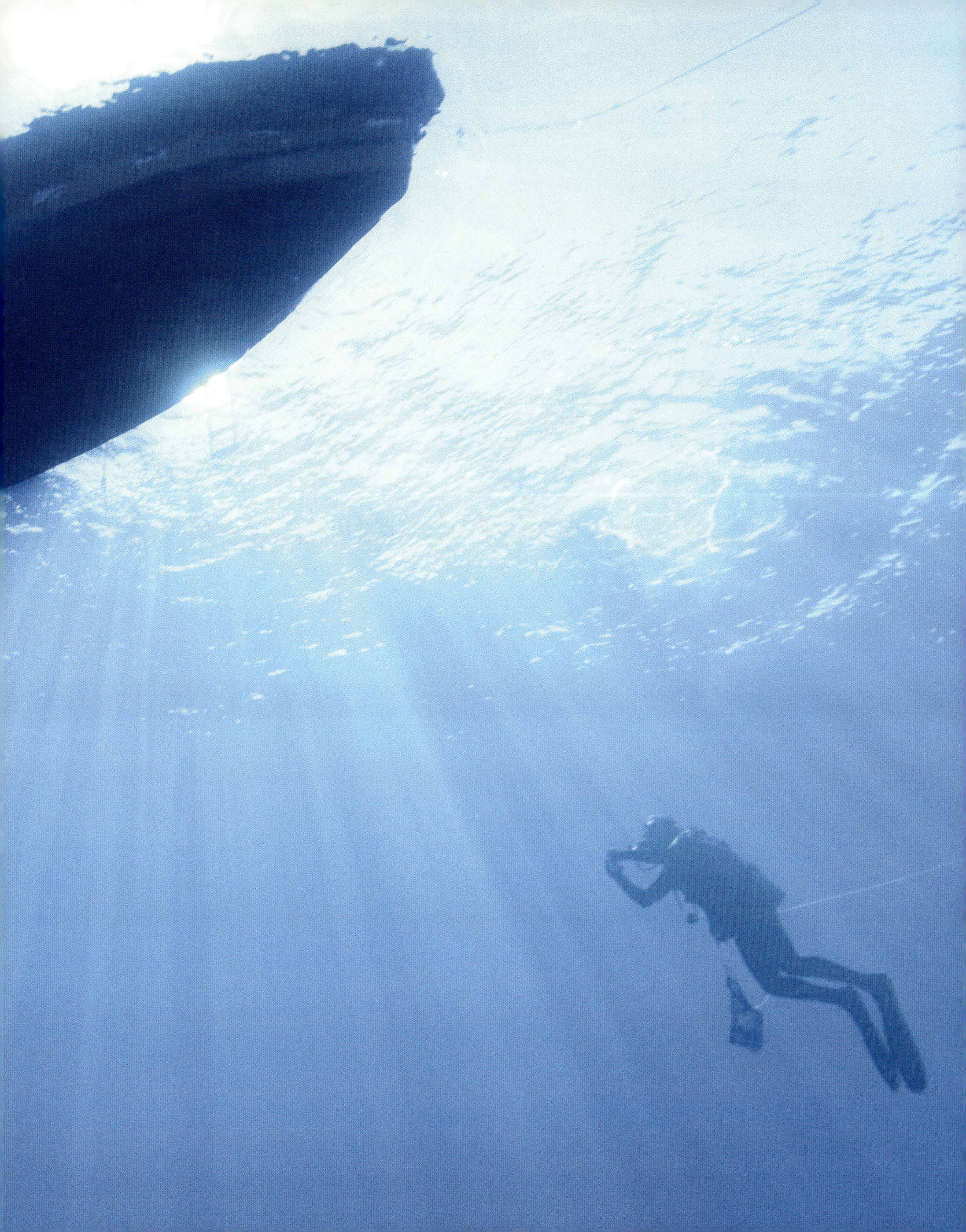

RESOURCES

BOOKS

Fox, Denis L. *Animal biochromes and structural colours*. 2nd edition. University of California Press, 2023.

Herring, Peter J., Anthony K. Campbell, Michael Whitfield, and Linda Maddock. *Light and Life in the Sea*. Cambridge University Press, 1990.

Johnsen, Sönke. *Into the Great Wide Ocean: Life in the Least Known Habitat on Earth*. Princeton University Press, 2024.

Johnsen, Sönke. *The Optics of Life: A Biologist's Guide to Light in Nature*. Princeton University Press, 2012.

Lynch, D. K., and W. C. Livingston. *Color and Light in Nature*. Cambridge University Press, 2001.

Marshall, N. J., A. Hurlbert, J. Boddy, T. W. Cronin, R. Douglas, S. Johnsen, and F. Cortesi. *Color in Nature.* Princeton University Press, 2024.

Scales, Helen. *The Brilliant Abyss: True Tales of Exploring the Deep Sea, Discovering Hidden Life, and Selling the Seabed*. Bloomsbury Sigma, 2022.

Shimomura, Osamu and Ilia Yampolsky. *Bioluminescence: Chemical Principles and Methods*. 3rd edition. World Scientific Publishing Company, 2019.

ARTICLES

Davis, Alexander L., Kate N. Thomas, Freya E. Goetz, Bruce H. Robison, Sönke Johnsen, and Karen J. Osborn. "Ultra-black camouflage in deep-sea fishes." *Current Biology* 30, no. 17 (2020): 3470–3476. https://doi.org/10.1016/j.cub.2020.06.044.

Haddock, Steven H. D., and C. Anela Choy. "Life in the Midwater: The Ecology of Deep Pelagic Animals." *Annual Review of Marine Science* 16 (2024): 383–416. https://doi.org/10.1146/annurev-marine-031623-095435.

Haddock, Steven H. D., and Casey W. Dunn. "Fluorescent proteins function as a prey attractant: experimental evidence from the hydromedusa *Olindias formosus* and other marine organisms." *Biology Open* 4, no. 9 (2015):1094–1104. https://doi.org/10.1242/bio.012138.

Haddock, Steven H. D., Mark A. Moline, and James F. Case. "Bioluminescence in the Sea." *Annual Review of Marine Science* 2 (2010): 443–493. https://doi.org/10.1146/annurev-marine-120308-081028.

Johnsen, Sönke. "Hidden in plain sight: the ecology and physiology of organismal transparency." *Biological Bulletin* 201, no. 3 (2001): 301–318. https://doi.org/ 10.2307/1543609.

Johnsen, Sönke. "The red and the black: Bioluminescence and the color of animals in the deep sea." *Integrative and Comparative Biology* 45, no. 2 (2005): 234–246. https://doi.org/10.1093/icb/45.2.234.

Johnsen, Sönke. "Transparent Animals." *Scientific American*, February 2000, 62–71.

Marshall, Justin, and Sönke Johnsen. "An evaluation of fluorescence as a means of colour signal enhancement." *Philosophical Transactions of the Royal Society of London* 372, no. 1724 (2017). https://doi.org/10.1098/rstb.2016.0335.

Mazel, Charles. "Method for Determining the Contribution of Fluorescence to an Optical Signature, with Implications for Postulating a Visual Function." *Frontiers of Marine Science* 4 (2017): 266. https://doi.org/10.3389/fmars.2017.00266.

WEB RESOURCES

The Bioluminescence Web Page [biolum.eemb.ucsb.edu/]

Video: The Allure of Fluorescence in the Ocean [bit.ly/fluor-lure]

Video: The Case of the Green-eyed Squid [bit.ly/youhist]

INDEX

MAKING THIS BOOK

During this era of biotechnological revolution, it's easy to become fixated on studying drops of liquid in a tube or letters representing genetic sequences on a computer screen. As biologists, we benefit from these breakthroughs, but we also believe that important insights are gained by studying animals alive in their own environment. When your animals live in the open ocean, the way to access them is through scuba and submarine dives—either human-occupied or remotely operated. We've been fortunate to use these methods at locations around the globe, from the North Atlantic to the South Pacific.

The scuba method we use is called blue-water diving. It is similar to a space walk, in that we use a system of tethers to guard against drifting off into the abyss. We've taken many colleagues on their first blue-water dives, and even those who have spent years studying oceanography find that it gives them unexpected perspective and insight into life in the open ocean. They see with their own eyes that the high concentration of algae below the surface, which they've only seen as a blip from one of their instruments, is actually a key feature that structures the otherwise homogeneous environment. They can feel the sudden change in temperature that oceanography textbooks call the thermocline, and realize that organisms and even sinking particles perceive this as a boundary.

Human-occupied submersibles have been largely replaced by remotely operated ones, but there is still no substitute for looking out at the deep ocean with your own eyes. Our eyes are marvelous sensors, able to perceive low light levels, rapidly change focus from broad perspective to close-up scrutiny, and able to accommodate dramatic changes, like bright flashes against a dim background, the green tint of coastal waters, or the gradually dimming surface light in the sapphire-blue open sea. But a submarine dive can be a chilly, isolating, and potentially risky experience. With remotely operated vehicles, we can share the dive with a dozen of our colleagues—each a specialist in a different kind of organism or a different phenomenon. It is easy to take lunch or visit the restroom, watching the dive from various monitors on the ship. This is undoubtedly the most convenient and efficient way to experience the deep sea, and we've been fortunate to lead hundreds of such dives.

Our research benefits from the ability to do comparative studies. This means we are attuned to examples of convergence—when two organisms arrive at the same adaptation by different evolutionary paths—and diversification—when two closely related animals have diverged to exhibit different behaviors and have different body plans. For convergence, think of birds and bats and bees all figuring out how to fly. For diversification, think about jumping spiders versus orb weaving spiders, which use different strategies to catch different types of prey. To find cases where this has happened in the ocean, particularly with bioluminescent or fluorescent organisms, we collect across the world and from the surface to the deep sea. We can compare closely related species from the frigid Arctic to the warm tropics, or animals collected while blue-water diving near the surface to those caught at the maximum operating depth of our submarines: 4,000 meters, or 2.5 miles.

When we collect the animals we're interested in, it is our responsibility to extract (sometimes literally) as much information as we can from the precious specimens. This means taking videos while they're underwater and photographs from all angles while they're still in good shape. In fact, most of our colleagues who study deep-sea jellies have, by necessity, become adept photographers. This is one of the required steps in documenting the diversity of what we find. Most of the photos in this book were motivated by this process of scientific documentation. From there we will often try to get genetic or biochemical information to provide insights about where these creatures fit into the animal tree of life, or how they are able to produce bioluminescence or exhibit fluorescence.

Unfortunately, pictures of live deep-sea animals, even in nature books and magazines, commonly show damaged or preserved museum specimens, with vacuous milky eyes and faded brown skin. These images perpetuate the idea that the deep ocean is full of grotesque and scary creatures. We are privileged to have access to healthy live creatures, which gives opportunities for unique science, but also for sharing the true wonders of open ocean diversity.

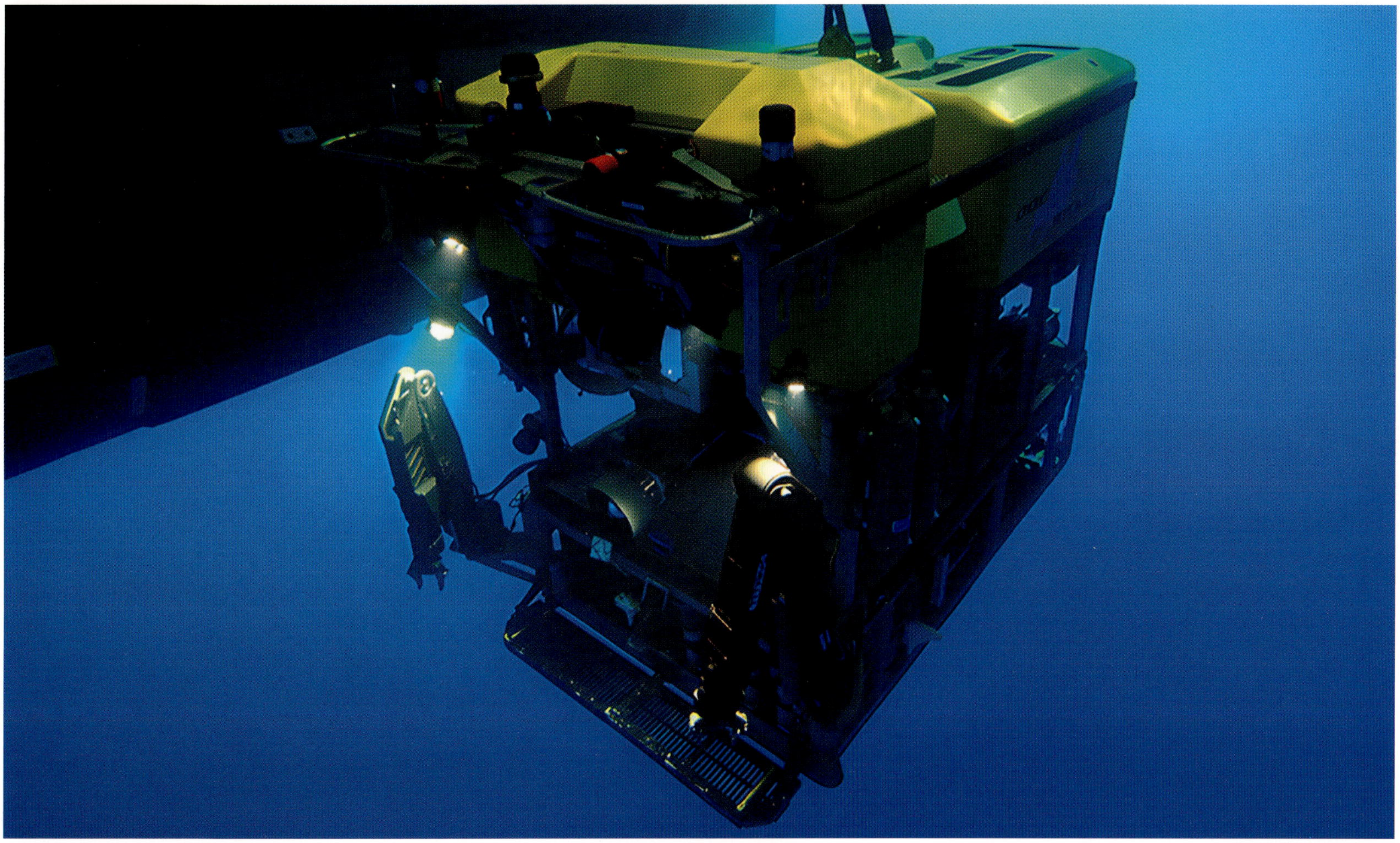

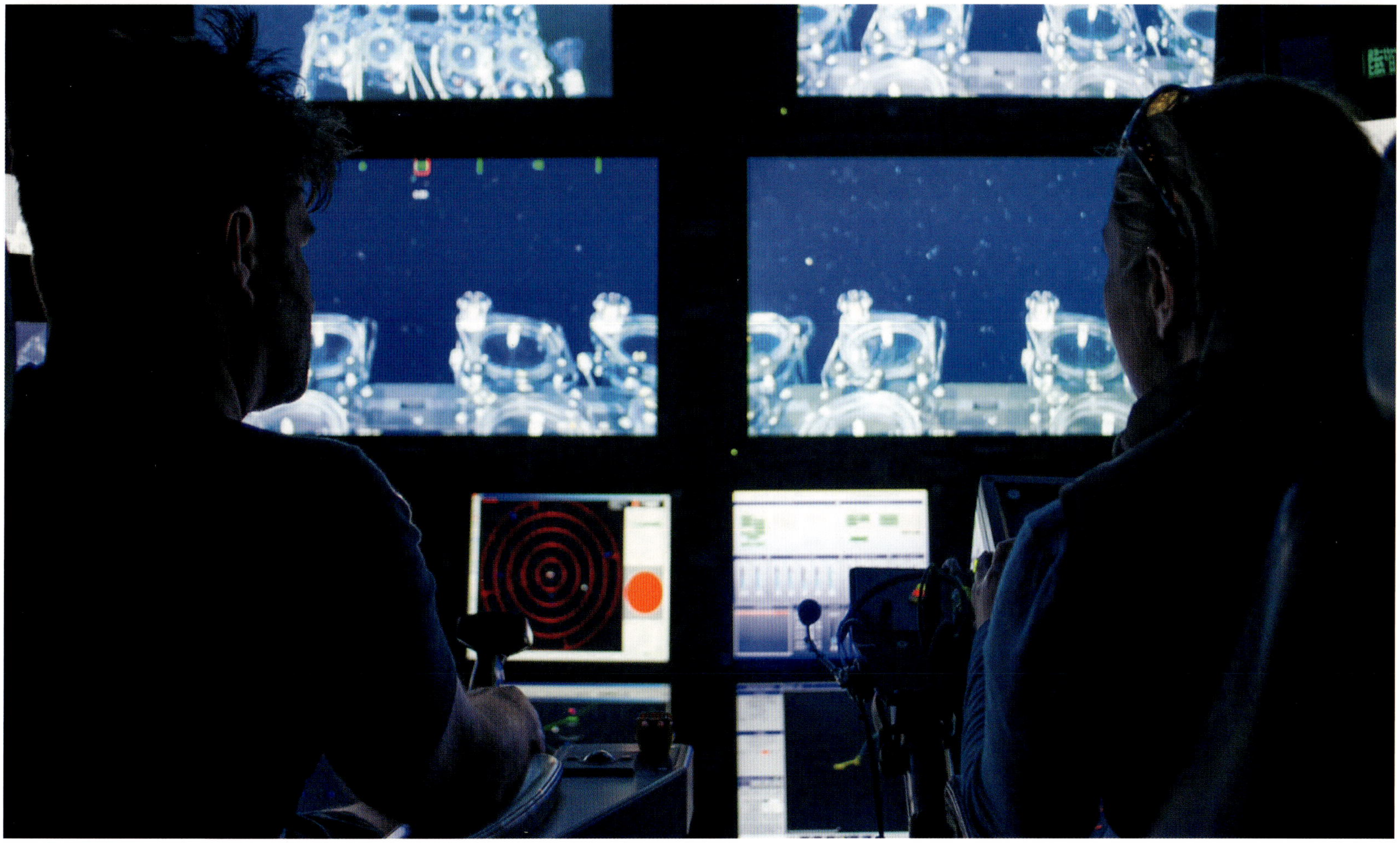

ACKNOWLEDGMENTS

We thank our families, Danielle and Griffin, Lynn and Zoe, for their loving support. Our parents nurtured our curiosity and love of science, and gave us the sense that anything was possible. We benefitted from many opportunities to go to sea through the years, with the support of Edie Widder, Bruce Robison, Jim Childress, Larry Madin, Alice Alldredge, Peter Herring, Tammy Frank, Craig Young, Jose Torres, James Case, Justin Marshall, and many others. The crews of research vessels, submersibles, and ROVs conducted us safely and expertly far from shore and deep below the surface, including the ships *Western Flyer*, *Edwin Link*, *Seward Johnson*, *New Horizon*, *Point Sur*, *Kilo Moana*, *Endeavor*, and the *Rachel Carson*. Our students, postdocs, and collaborators brought vitality, joy, humor, and fresh ideas to our scientific and personal lives. We appreciate Jenny Manstead and the wizards at UniPress Books and Abrams Books for their willingness to bring our long-standing idea of a book into existence.

PICTURE CREDITS

The publisher would like to thank the following for permission to reproduce copyright material and for use of reference material for illustrations.

Alamy: Biosphoto 122; /David Fleetham 110; /Stocktrek Images, Inc. 123 | **Blue Planet Archive: Doug Perrine** 162 (bottom) | **Danté Fenolio:** 162 | **Getty Images:** Humberto Ramirez 114; /Tdub_Video 127, 148; /Torsten Velden 104; /Zhen Li 134 | **Shane Gross Photography:** 115 | **Steven Haddock Ÿ MBARI:** 4, 8, 12, 13, 14, 19, 24, 25, 27, 28, 29, 30, 32-33, 34, 35, 36, 38, 39, 42, 43, 44, 45, 46, 47, 48, 49, 50, 52, 55, 56, 58, 59, 60, 61, 62, 64, 65, 66, 67, 68, 69, 70, 72, 73, 74, 75, 76, 77, 78, 79, 80, 81, 82, 84, 86, 88, 89, 90, 91, 92, 93, 94, 97, 102, 103, 105, 106, 107, 108, 109, 112, 113, 116, 117, 118, 119, 120, 121, 129, 141, 132, 135, 136, 137, 138, 139, 140, 141, 142, 143, 144-145, 146, 151, 152-153, 154, 155, 156, 159, 160, 161, 163, 164, 166, 168, 176, 178, 179, 180, 181, 183, 184, 185, 186, 187, 188, 189, 190, 192, 193, 194, 195, 196, 197, 199, 200, 201, 202, 203, 204, 206, 207, 209, 209, 210, 212, 213, 214, 215, 216 | **Jeff Hester:** 133 | **Sönke Johnsen:** 50, 169, 170, 171 | **Elliot Lowndes:** 147 | **MBARI:** 85, 223, **Jeffrey Milisen:** 16, 20, 26, 40, 41 | **Nature Picture Library:** David Fleetham 111; /Russell Laman 37; / Solvin Zankl 83 | **Science Photo Library:** Danté Fenolio: 165 | **Stefan Siebert:** 101, 182, 191, 198 | **Image from the documentary *David Attenborough's Light on Earth*** produced in 2016 by Terra Mater Studios GmbH and Ammonite Films: 151 (bottom), 158 | **Sid Tamm:** 99 | **David Wrobel:** 31

All reasonable efforts have been made to contact copyright holders and to obtain their permission for the use of copyright material. The publisher apologizes for any errors or omissions and will gladly incorporate any corrections in future reprints if notified.

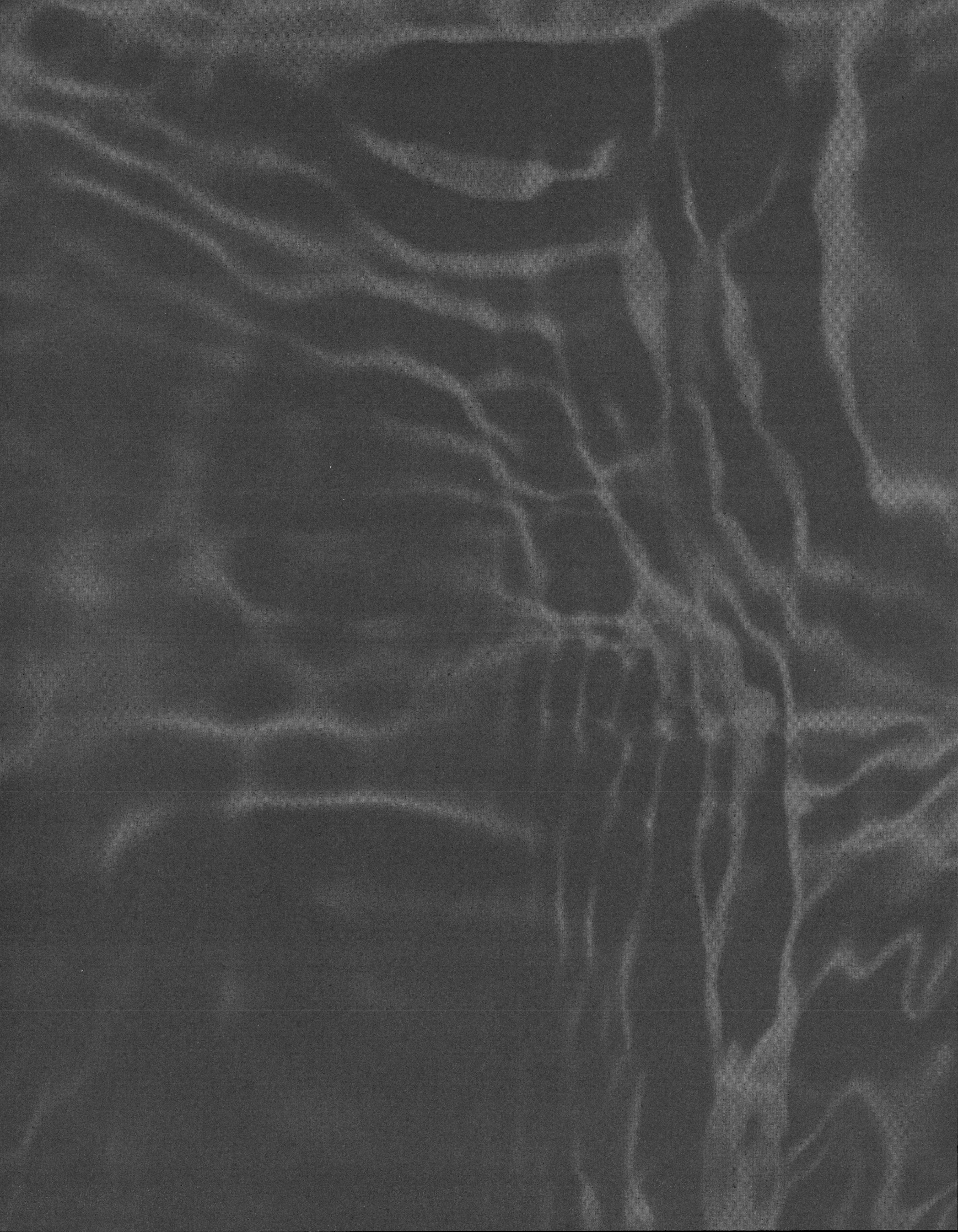

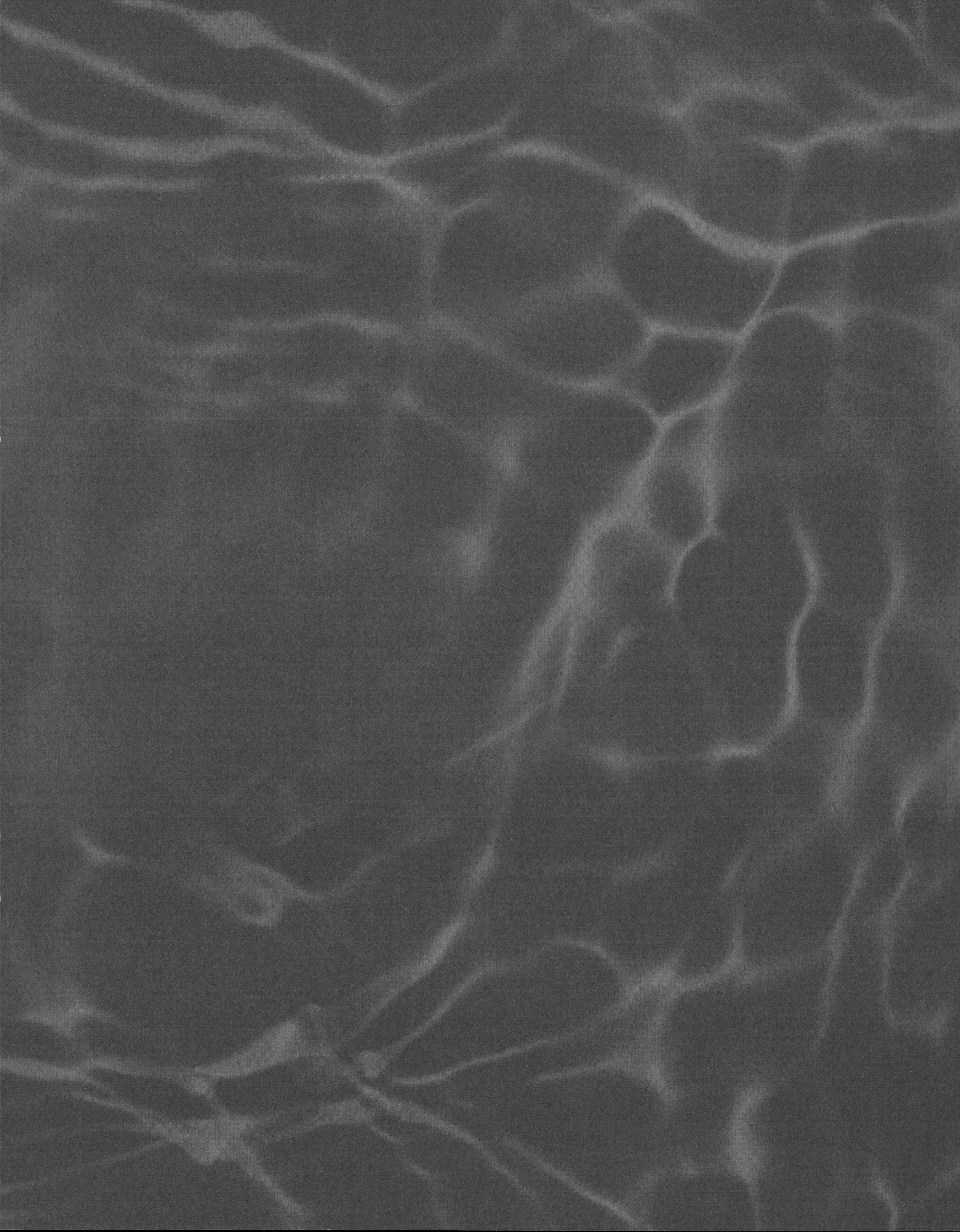